City in a Garden

City in a Garden

Environmental Transformations and Racial Justice in Twentieth-Century Austin, Texas

ANDREW M. BUSCH

The University of North Carolina Press Chapel Hill

This book was published with the assistance of the Wells Fargo Fund for Excellence of the University of North Carolina Press.

Set in Charis Regular by Westchester Publishing Services
Manufactured in the United States of America

The University of North Carolina Press has been a member of the Green Press Initiative since 2003.

Library of Congress Cataloging-in-Publication Data
Names: Busch, Andrew M., author.
Title: City in a garden : environmental transformations and racial justice in twentieth-century Austin, Texas / Andrew M. Busch.
Description: Chapel Hill : University of North Carolina Press, [2017] | Includes bibliographical references and index.
Identifiers: LCCN 2016047656 | ISBN 9781469632636 (cloth : alk. paper) | ISBN 9781469632643 (pbk : alk. paper) | ISBN 9781469632650 (ebook)
Subjects: LCSH: Austin (Tex.)—History. | Austin (Tex.)—Race relations. | Sustainable urban development—Social aspects—Texas—Austin. | City Planning—Environmental aspects—Texas—Austin. | City planning—Social Aspects—Texas—Austin.
Classification: LCC F394.A957 B87 2017 | DDC 976.4/31—dc23
LC record available at https://lccn.loc.gov/2016047656

Cover illustration: Skyline of Austin, Texas, at dusk © Adobe Stock/ Brocreative.

Chapter 6 was originally published in a different form as "Building 'A City of Upper-Middle-Class Citizens': Labor Markets, Segregation, and Growth in Austin, Texas, 1950–1973," *Journal of Urban History* 39, no. 5 (September 2013): 975–996. Chapter 7 was originally published in a different form as "The Perils of Participatory Planning: Space, Race, Environmentalism, and History in 'Austin Tomorrow,'" *Journal of Planning History* 15, no. 2 (May 2016): 87–107. Both are used here with permission.

To Renee and Frank, my people

Contents

Illustrations and Maps

Illustrations

Maps

Acknowledgments

This book began as a hazy idea conceived in a graduate seminar at the University of Texas in 2005. Since then a wide array of people have helped me turn that idea into a book. Without their generous assistance and support I never could have finished this project, and I wish to thank them.

At the University of Texas, my most immediate thanks go to Jeff Meikle, who guided me with positivity, good humor, and wisdom. Other members of the Department of American Studies faculty, especially Steve Hoelscher, Elizabeth Engelhardt, and Janet Davis, mentored me. Ella Schwartz was indispensable in the process as well. Among my graduate student cohort, I wish to especially thank Tony Fassi, Jason Mellard, Gavin Benke, Andi Gustavson, Jeremy Dean, Robin O'Sullivan, John Cline, Ben Lisle, and Danny Gerling, all of whom either read part of the manuscript or discussed it in depth with me. Eliot Tretter has been a generous guide and mentor and has shared a wealth of knowledge about Austin. I would also like to thank Randy Lewis, Stan Friedman, Josh Long, Elizabeth Mueller, Sarah Dooling, Patrick Vitale, Scott Swearingen, and Bob Fairbanks for great conversations and input on this project.

Other scholars have helped me on my intellectual journey. At Illinois Wesleyan, Brian Hatcher and Mike Weis both stoked my interests in history and introduced me to a world larger than I had previously grasped. At Purdue, Elliott Gorn, Nancy Gabin, Mike Morrison, Jon Teaford, and Susan Curtis introduced me to American studies and challenged me to become a scholar. At Miami University, Kimberly Hamlin, Damon Scott, and Peggy Shaffer mentored me as a teacher and scholar and were generous in reading and discussing my work. I just arrived at the University of Texas at Dallas, and my colleagues here have been wonderful. Chris Sellers and Shana Bernstein provided valuable feedback as anonymous readers and improved the book immeasurably with their thoughtful comments.

The University of North Carolina Press has been wonderful and supportive through the entire process. My editor, Brandon Proia, as well as Jad Adkins have been a joy to work with. The wonderful people at the Austin History Center helped me locate countless collections, photos, and maps.

Thanks to all of them, and especially to Mike Miller, Molly Hults, Susan Rittereiser, and Nicole Davis. I would also like to thank all the staff at the Dolph Briscoe Center for American History, the Lyndon Baines Johnson Presidential Library, and at the Lower Colorado River Authority.

I would like to thank my family for their unconditional support. My grandparents, Dick and Connie Houck and Martha Busch, taught me to be curious and were always encouraging. My siblings, Eric, Betsy, and Maggie, put up with me and often challenged me to be a better person. My parents, Ellen and Mike Busch, never lost faith in me despite my many shortcomings and have been consistently selfless and supportive. Finally, I want to thank my wife, Renee Searfoss, for her unending support personally, professionally, and intellectually, and my son, Frank, for being a pretty cool three-year-old. I dedicate this book to you two, my people.

Introduction

The Trouble with Green

In the "new economy" we have to make sure we do not create unintended results.

—Austin mayor Kirk Watson, quoted in Rosenblum, "Mayor Offers Strategy"

In the twenty-first century, Austin, Texas, has become a model of dynamic, sustainable urban development. While most American cities declined under the weight of the Great Recession, Austin flourished. A litany of sources, such as *Forbes*, *Time*, and CBS, called the city a boomtown and named it the top metropolitan region for economic growth and small businesses in 2011.[1] Demographically, Austin grew by over 30 percent from 2000 to 2013; in 2013 it was the fastest growing of the fifty largest U.S. cities. *Forbes* replicated the distinction in 2016.[2] While some of this growth is due to the continued ascension of Texas and the Sunbelt more broadly, Austin's core strategy reflects a set of principles that are much different from those of its regional neighbors. In Austin, environmental sustainability and a "green" urban planning philosophy are linked to quality of life and economic growth, part of the "new economy" that assumes that people and businesses factor general well-being into locational choices.

Along with its economic resiliency, Austin has recently been praised for its creativity, forward-thinking urban planning, and sustainability.[3] The city council plans to be carbon neutral by 2020, and city-run Austin Energy is the nation's leading seller of renewable energy.[4] Austin Energy also offers generous rebates for rainwater collection, green insulation, low-flow toilets, solar panels, and many other environmentally friendly, though often costly, upgrades. Austin is among the nation's leaders in park space and urban trails, one of the top twenty-five Solar American Cities according to the U.S. Department of Energy, and renowned as one of the fifteen greenest cities in the world.[5] The American Planning Association gave *Imagine Austin*, the city's newest comprehensive plan that emphasizes New Urbanism concepts and green development, its Sustainable Plan Award for 2014.[6] National publications laud Austin as a "clean tech hub," and academics have held the city up as an example of community-driven sustainability in recent years.[7]

Yet urban plaudits are rarely without tensions, contradictions, and externalities that expose the limits of sustainable development. Amid robust economic growth, there is abject poverty and flight among minorities in Austin. African Americans have lost population share in Austin every decade since 1920 and experienced a real decline from 2000 to 2010, even as the city as a whole grew by 20 percent during this same period. Austin is one of the few metropolitan regions in the United States with a higher percentage of African Americans in suburbs than in the central city—and poverty in Austin's suburbs rose by 143 percent between 2000 and 2011, the second-fastest increase in the United States. African Americans who live in the city continue to lag, averaging half the household income of whites. The city's population as a whole had a poverty rate of around 23 percent in 2011, among which a majority are people of color. In some transitional neighborhoods, poverty is 2,000 percent greater among African Americans than among whites.[8] And a recent study demonstrates that the Austin metropolitan area has the highest level of economic segregation among all 350 U.S. metropolitan areas.[9] These statistics indicate that, despite economic growth, sustainability awards, and an overall strong quality of life, Austin is not a particularly sustainable place for its historically disadvantaged residents. It makes sense, then, to ask why the concept of urban sustainability has developed in a way that seems to exclude so many people from its benefits and why Austin is a city so tied to its natural environment yet so bifurcated by race.

Answering this question demands a perspective that traces how both the natural environment and the social environment have changed through time and examines the relationship between these two categories that are often analyzed separately but in reality are extremely interdependent. Although Austin has emerged as a global hot spot in the twenty-first century, policy decisions stretching back to the early twentieth century illuminate how such a city came to be. The process of improving on nature has engendered social inequities that reflect the city's, as well as the nation's, changing attitudes about the relationship between nature and urbanity and who gets to enjoy, transform, and profit from urban nature and who does not. In this book, my foremost concern is to explain the connections and tensions between the landscapes and practices that have epitomized sustainability in Austin and the less visible practices and policies that have generated and perpetuated racial segregation and discrimination. I argue that urban environmental improvements have been inseparable from racial discrimination and have resulted in geographic restructuring that has tied minorities to underfunded, less visible spaces and Anglos to cleaner, picturesque, and regulated spaces with more access to nature and fewer urban problems. To create and maintain a desirable environmental city (and later a sustainable one)—what I call the garden—necessitated augmenting the natural world in ways familiar to most urban environmental historians: putting in roads and sewerage systems; building single-family homes, parks, and schools; using technology to control water systems; and siting employment facilities. Yet Austin's leaders also created and perpetuated its shadow, the place that contained the city's surplus—the industrial facilities, trash dumps, wastewater facilities, power plants, and other unsightly and hazardous necessities—as well as most of the city's minority residents.[10]

The story of Austin's growth exists at the intersection of twentieth-century urban planning and city building, the changing nature of racial discrimination and segregation, and the transformation of human-environment relationships. Like all cities, Austin is a man-made agglomeration of resources where space is structured and differentiated by fixed capital investments on the landscape, from industrial facilities and office parks to shopping malls, houses, and roads.[11] Power relationships are inscribed on these landscapes, and often the landscape itself becomes a source of domination and difference. In cities in the early twentieth-century, infrastructural improvements, siting decisions, policy decisions, bonds, and zoning privileged certain groups and disadvantaged others. Infrastructure increased the value of real estate and, conversely, lacking infrastructure decreased value. Restrictive covenants, zoning laws, state laws, and federal mortgage and tax policy ensured for whites access to home ownership, stable real estate values, socioeconomic and racial homogeneity, and single-family homes on uniform properties. They made getting loans easier for whites and more difficult for minorities. Any white person who chose to bridge the racial divide not only gave up a privileged place in the social landscape but also suffered a severe economic penalty.

During the 1960s, and despite the political victories of the civil rights movement, existing racial disparities intensified and new inequities emerged. Federally sponsored urban renewal projects and interstate highway construction disproportionately dispossessed minorities, often forcing them into residences even less desirable than those that were destroyed and not building enough affordable housing. White flight to suburbs decimated urban tax bases, and infrastructure and social programs deteriorated. Deindustrialization and mechanization had harsher effects on minorities, who were often the first to be laid off during downsizing or decentralization; by the 1970s and 1980s a "new urban poverty" was created, where poor minorities were increasingly cut off from any chance at economic improvement.[12] In these and other ways, city building in the twentieth century reflects what George Lipsitz called a "possessive investment in whiteness," where conforming to racial norms and excluding minorities socially as well as spatially conferred economic and structural benefits to whites, who were able to generate wealth from federal subsidies, count on opportunities from jobs and education, and gain advantages from racial politics in ways not afforded to other groups. Racial ideology also structures how land itself is valued. For Lipsitz, the liberal emphasis on individualism obscured the ways that minorities were systematically excluded from the positive aspects of growth. "Collective exercises of group power relentlessly channeling rewards, resources, and opportunities," he wrote, "from one group to another will not appear to be 'racist' from [the perspective of liberal individualism] because they rarely announce their intention to discriminate against individuals. But they work to construct racial identities by giving people of different races vastly different life chances." This process is well documented, and its deleterious effects have been demonstrated in numerous American cities.[13]

Many of these discrepancies existed in Austin, and they are an important aspect of the city's tumultuous racial history. Yet in Austin the investment in whiteness also manifested itself in access to natural resources and subjection to environmental risk,[14] in environmentally focused planning and improvements,[15] and eventually in environmental consciousness itself.[16] In a city like Austin, with its large middle class, exceptionally small industrial base, and lack of independent suburbs, racialization emerged as part of an ideology that spatially differentiated urban from nonurban, environmentally improved from not environmentally improved. The investment in whiteness, though, meant different things at different times; it was reconstituted to fit different socioeconomic imperatives and different conceptions of human-environment relationships. Lipsitz reminds us that "racism changes over time, taking on different forms and serving different social purposes in different eras."[17] Thus the process of racialization is best understood by investigating how the city building process emerged and proceeded through the twentieth century.

Scope and Structure

City in a Garden is structured in a way that links changing conceptions of nature with changing conceptions of race through the twentieth century.

The book's chapters are arranged to roughly correspond to three eras: the 1890s through the New Deal, World War II through the early 1970s, and the 1960s through the 1990s. Yet integrating historical change with geographic change is essential to understanding how race and environment were continually reimagined and reconstituted. Social changes can often be seen most clearly in the production of space and in geographic change; in Austin, discrimination was shaped by environmental beliefs and desired environmental improvements during the early decades of the twentieth century. During those years, Austin leaders imagined nature as something to be overcome and "put to work" by human planning, rationality, and technological expertise. Urban environmental planning freed white Austinites from the burden of proximity to industry while also allowing them access to the city's cherished outdoor spaces: naturally fed swimming pools, bluffs overlooking water and hills, improved creeks. White homeowners enjoyed protective zoning, racially restrictive covenants, and access to better parks, schools, and real estate. Keeping white residential neighborhoods free of urban surplus—industry, garbage dumps, dilapidated housing—as well as minorities imparted significant economic value to these areas. City leaders imagined large infrastructural projects, such as the dam system on the Colorado River, as a regional benefit, yet increases in socioeconomic value and environmental stability favored whites exclusively and further differentiated whites from minorities. Conversely, African Americans, and to a lesser degree Latinos, were systematically excluded from these advantageous situations and nicer landscapes through Jim Crow's institutional discrimination. Improvements were designed to attract them to areas marked as urban rather than to make their lives better. Thousands were uprooted. Separate but equal laws incentivized their movement into parts of the city that were more crowded, less improved, less picturesque, and zoned to include industry and other hazards. Minorities were removed from refurbished areas. As in other cities, federal mortgage discrimination inhibited their mobility and made real estate renewal more difficult by the 1930s.

After World War II, possessive investment in whiteness encouraged Austin's businesspeople and politicians to begin treating nature as a fixed capital investment that could be commodified to enhance accumulation. From the 1930s through the 1960s a variety of actors transformed the landscape into a series of envirotechnical systems, where the environment was augmented by technology.[18] Using federal capital, local elites dammed the Colorado River in seven places, creating a vast system of reservoirs, dams, and rivers that altered the region's ecology dramatically but also produced cheap power and water, recreational space, and flood control for humans. By the 1940s, owing to this newly found environmental stability, being progressive meant using natural amenities to grow without industrializing and using government and the University of Texas to the city's advantage. A variety of businesspeople, politicians, and university administrators began to link the commodification of the environment with a lifestyle that catered to knowledge laborers and research and development firms connected to the university. Vast tracts of previously undeveloped or agricultural land in areas no longer susceptible to flooding were annexed and transformed by the construction of suburban-style homes, shopping centers, schools, and office parks that catered to Anglo white-collar workers, many of whom were employed in the growing research and development field or at the University of Texas. Park land was designated and usually improved with ball fields, paths, swimming pools, and playgrounds.

Yet this transformation was attended by a brutal retrenchment of the color line, as selling Austin meant making its less desirable aspects less visible and more physically contained. City and university leaders expanded and refurbished the campus and the central business district using federally sponsored urban renewal programs, which dispossessed and displaced thousands of African Americans and made other areas more dense, congested, and filled with dirty industries. A new central highway increased the barriers between Anglo and minority Austin, and other peripheral expressways encouraged whites to move away from the central city. Decentralization of the population also decentralized resources; tax money went to provide infrastructure and enhance or preserve environmental characteristics in new peripheral areas, as streets, sewers, and other infrastructure went unimproved in minority neighborhoods. Minorities, still segregated in Austin schools, often lacked educational opportunities that would make them competitive in the city's white-collar job market. The parks they had access to were fewer in number, smaller in size, and often in less valuable and picturesque places. Urban growth exacerbated the environmental hazards and pollution that disproportionately affected minorities, because the city produced more garbage, needed more power, and contained more automobiles. By the early 1970s, despite robust economic, demographic growth and the end of de jure segregation, Austin was more segregated and socioeconomically bifurcated than at any time before.

Austin's growth accelerated markedly in the 1970s and especially in the 1980s as it became a national center of technological development and a secondary circuit of real estate investment. When mainstream environmentalism emerged in Austin in the 1960s and 1970s, environmentalists encountered a city already heavily augmented by technology and the logic of nonindustrial capitalism, a hybrid landscape that smoothed nature's rougher edges and hid the city's undesirable surplus functions and people. In this sense, their possessive investment in whiteness allowed them to define the environment as a relationship between themselves and the aspects of the natural environment they valued most. By the 1980s, defending the environment against unscrupulous politicians, greedy developers, and those who benefited from urban growth and environmental degradation became the hallmark of the environmental movement. William Scott Swearingen, who chronicled the movement, wrote that "to be an 'Austinite' in that sense means to be a person who is part of that connection [between the social and natural landscape], to have one's life in part defined by that connection between social life and physical environment."[19] To Swearingen, the core environmental values and sense of place were defined in large part by this relationship.

Yet this logic of environmental meaning failed to consistently resonate with minority groups precisely because they were unable to define their values and their sense of place vis-à-vis the natural environment. Historically, they lacked access to the most prominent civic outdoor spaces and the most picturesque sites and therefore had less emotional and social connection with them. The spaces and places that white environmentalists fought for were symbols of the possessive investment in whiteness for Austin's minorities, places that they were largely excluded from first legally and then geographically. Their environmental movements and sense of place imagined the environment as something that contained minorities and their communities and as something that was part of their oppression. The history of race, geography, and social exclusion contextualized their sense of environmental injustice. They focused on the built environment because that was what undermined their quality of life, created health hazards, and provided their environmental meaning. The wide racial and cultural chasm produced by the possessive investment in whiteness precluded sustained coalition building and further alienated environmental justice groups. The two movements occasionally found common ground but were unable to build a lasting bond. By 2000, minorities were directly contesting sustainable development policies encouraged by environmentalists and the city's Green Council because they drove real estate prices up in minority neighborhoods and engendered widespread displacement.

Framing Race and Environment: Liberalism, Labor, Landscape, and Sustainability

The relationship between environmentalism, liberalism, and race in Austin illuminates how the possessive investment in whiteness is not contingent on overt, individual racial prejudice or mitigated by liberal political agendas that advocate for a commitment to ideas such as individual rights, tolerance, diversity, or collective grassroots action. From a racial perspective, even as a growing number of people accepted the general tenets of liberalism's racial tolerance, little changed on the ground. The concept of intentionality is important here. Before the 1930s, political and business leaders in Central Texas subscribed to Jim Crow and separate but equal philosophies that consciously imagined racial difference as a central feature in the social and geographic landscape. From the 1930s to the 1960s, liberals found common cause in the possibilities for modernization presented by the New Deal order, with its emphasis on government-sponsored improvement and opportunity, and often leaders such as Lyndon Johnson spoke passionately about the plight of minorities. Yet separate but equal was still the norm, and the possessive investment in whiteness was still intentional and overt, supported by law as well as social custom. Whites exclusively benefited from federally sponsored modernization programs as well as from private developments. Racial liberalism found few devotees in Texas or in Austin.

Legally and socially these barriers eroded after the civil rights movement of the 1960s, an era that witnessed the dismantling of a fundamentally racist political and legal order. Public articulations of racial superiority and prejudice declined precipitously as expression of white nationalism was pushed to the margins of acceptable discourse.[20] Since then, Austin has become one of the most liberal cities in the South, if not in the nation.[21] Its political culture has been self-consciously progressive and its economy robust, and its popular social movements have often gained widespread support. Many of Austin's liberal elites, long aware of the city's racial and class tensions, argue that diversity and affordability need to be part of a truly sustainable agenda.[22] Yet the possessive investment in whiteness has continued to operate forcefully despite this lack of intentionality, as demonstrated by the enduring economic, environmental, and social chasm separating whites from African Americans and Latinos in Austin. As elsewhere, housing discrimination, employment discrimination, resistance to school integration, and municipal land use all expressed a strong belief in racial difference.

This continuing gap reflects the fundamental racial inequity that pervades postwar liberalism and sets its ideological limits despite the end of legalized discrimination. In industrial cities the white working class, while strongly Democratic and formerly the core of the New Deal coalition, did not accept racial equality as part of the postwar liberal regime.[23] Southern whites, who often subscribed to a "color blind" discourse, likewise forcefully rejected racial equality as part of the broad liberal platform.[24] The liberal Keynesian state itself provided a framework that facilitated socioeconomic discrimination and racial segregation. Federal and local governments worked with private businesses to create a new suburban-focused geography that neglected central cities and their residents, and industrial decentralization pulled jobs away from black urban areas. Federal initiatives such as the GI Bill, Federal Housing Administration (FHA) mortgage guarantees, and the interstate highway system facilitated white flight to the suburbs. Redlining, discriminatory lending and real estate practices, and racial covenants kept burgeoning suburbs racially homogeneous as capital migrated to metropolitan peripheries.[25] Minorities, frustrated with liberal policies that did little to improve their lives or communities, often jettisoned liberalism in favor of community empowerment.[26] Institutional problems were complemented by ideological inconsistencies. Recently, Daniel Martinez HoSang has questioned whether the conception of liberalism as an ideology that moves people away from racism is valid or whether "white supremacy as an ideological formation has been nourished, rather than attenuated, by notions of progress?"[27] By inverting the traditional assumption that liberalism conveys universal rights to all regardless of race or class, thereby making the world more egalitarian, HoSang opens up new lines of inquiry into the ways racial difference operates within a politically and socially liberal context.[28]

Mainstream environmentalism is an important component of the postwar liberal regime that also reflects assumptions about values that are imbued with the possessive investment in whiteness.[29] In Austin, environmentalists' beliefs about nature mirrored conceptions of class, labor, and leisure that were fraught with racial meaning because of the city's fractured racial and class history. Environmental historian Richard White's work on the tension between nature and labor is valuable in this regard. For White, "environmentalists so often seem self-righteous, privileged, and arrogant because they so readily consent to identifying nature with play and making it by definition a place where leisured humans come only to visit and not to work, stay, or live."[30] Nature is thus juxtaposed with labor, which is seen as damaging to pristine environments and anathema to recreation. Yet the ability to separate nature from work is a privilege that carries with it notions of class and social status. Because of its leading sectors of employment—the university, the state, and the knowledge economy—Austin long had a much more white-collar, middle-class population than most other cities. For Swearingen, two themes define Austin's environmental sense of place. The first is that white-collar occupations, high levels of education, and middle-class status correlate strongly with environmental awareness and organizing. The second is that environmentalists are in geographic proximity to Austin's treasured natural spaces—Barton Springs, Town Lake, and the creeks and greenbelts central to Austinites' leisure pursuits.[31]

Both of these avenues to environmental meaning were largely unavailable to Austin's Latinos and African Americans. Because of institutional discrimination in the labor market and unequal educational opportunities, Latino and African American Austinites rarely had access to the stable white-collar jobs that correlated with Swearingen's type of environmental meaning. In fact, minorities also bore the brunt of environmentally focused policies that sought to limit blue-collar employment by limiting growth more broadly.[32] By fighting against new housing, manufacturing, and construction, environmentalists were fighting against the labor many minorities depended on to survive. Ironically, preserving places of environmental significance also made adjacent neighborhoods more desirable and economically valuable, thereby decreasing their availability to less wealthy residents.

Because the environment is often portrayed as something outside the realm of humanity, as something to be saved from humans, mainstream environmentalism is also susceptible to an ideology of color blindness that flattens historical, cultural, and economic differences. If we do not protect the Earth, after all, we are destined for the same fate, and so differences of black, brown, and white are often subsumed under the priority of green. By favoring the purported universal definition of the environment that minimizes race, mainstream environmentalism tacitly downplayed the difference in access to nature and in subjection to environmental risk. Yet for minority Austinites, race was an inescapable category that conditioned their relationship to nature and circumscribed their social and economic possibilities. For them, overt racism and strict segregation meant that they did not live near Austin's cherished spaces and could not even legally access them until the 1960s. Even afterward, cultural history and urban geography made access to these spaces difficult for minorities. Most importantly, these spaces were symbols of history and the society that actively sought to oppress them. As such, Austin's environmental movement reflects the less than egalitarian history of conservation and environmental politics in the United States, in which all too often the poor and minorities were either relocated or contained, often near undesirable industry, to make way for leisure or economic opportunities for middle-class and wealthy whites.[33]

Labor itself plays a significant role in *City in a Garden*, and Austin's history contributes to our understanding of the knowledge economy, its relationship to the environment, and the way that natural and built environments are increasingly difficult to separate.[34] On the one hand, as this book demonstrates, an incredible amount of labor and socioeconomic capital undergirded the transformation of nature into a safer, more useable form. Austin's natural landscape was not the untouched virgin land that it was cast as by boosters and environmentalists alike; instead, technology and political capital turned it into "second nature," that is, a space that appears pristine but is really a hybrid landscape constructed by humans and reflective of power relations. Almost all of the city's environmental amenities are the product of human activity and are therefore marked by politics and power relations. Like cities, gardens must be built. On the other hand, Austin's economic and political leaders have long imagined the city as a place suited to white-collar work rather than to industrial labor. They devised strategies to lure high-tech firms, to develop knowledge labor at the University of Texas, and to generate spin-off companies. Environmental amenities proved important in this equation; the creative, autonomous, and lucrative culture associated with knowledge labor was linked with an outdoor, suburban lifestyle in close proximity to natural spaces and away from both unseemly urban surplus and poor, minority residents.

During and after World War II, as capital was more mobile, nonindustrial modes of production more lucrative, and consumption more central to the economy, Austin's nonindustrial approach to planning and growth began to pay dividends. The University of Texas emerged as a regional leader in research and generated numerous spin-off companies. The state of Texas and the university made rapid developments. A series of federally funded dams harnessed the Colorado River, making it safer and allowing it and its watershed to be managed and commodified. The chamber of commerce, the city council, and university administrators devised an entrepreneurial strategy for growth that linked knowledge labor with natural amenities and outdoor recreation in an effort to attract and generate knowledge-, university-, and research-related business.[35] Travis County, of which Austin is the county seat, consistently had the lowest percentage of people employed in manufacturing among the one hundred most populated counties in the United States. Heavy industry had no place in this thriving postwar economy. By the 1990s, Austin had become a prototypical creative city and an important node in the emergent cognitive-cultural economy.[36]

From World War II onward, and in many instances from the late nineteenth century, the arguments that Austin leaders used to attract new workers and businesses echoed this implicit opposition to heavy industry. Boosters and other concerned citizens worried incessantly about the physical problems that industrial production would produce, by undermining the city's health and attractiveness, blighting the landscape with smoke and pollution, and using resources in amounts disproportionate to the benefits. Yet they also continually worried about the social impact of industry. Instead of attracting a "laboring class of cultured intellectuals," as the chamber of commerce cheerily described the goals of this growth paradigm in the 1960s, heavy industry meant unskilled and often uneducated laborers unfit for a center of education and culture. The solution was simple: avoid industries that used large numbers of unskilled and semiskilled laborers. City leaders rarely saw the need to invest in or attract low-skilled workers of any color, especially after the agricultural labor force severely contracted in the 1930s and 1940s. Unskilled whites as well as most African Americans and Latinos, who lacked access to good education and the skilled labor market, found few jobs in this type of economy.

This book also addresses contemporary policies regarding urban and environmental sustainability and green urban practices. *Sustainability* has recently become a buzzword loaded with positive implications, a term used to describe an endless array of plans, practices, ideas, and techniques geared toward lasting solutions to environmental and other large-scale problems. Put another way, sustainable development seeks to satisfy the needs of society today without sacrificing its future by conserving and renewing resources. The very ambiguity of this definition is perhaps what makes the idea so attractive; there is an empty quality to it that can be filled in any number of ways. It invites collaboration, indicates forward thinking, and connotes a liberal faith in planning and regulation. It puts the collective ahead of the individual and views environmental degradation as the core of unsustainable practices. When applied to urban development, sustainability has many nodes. The most prominent nodes relate to land use, conservation of resources, creation of more livable spaces and decreased dependence on the automobile, and protection of environmentally sensitive areas.[37] And, according to Kent E. Portney, sustainability has also given cities greater agency to pursue economic development than in previous eras.[38] The idea thus operates as something of a panacea, and Austin's planners have adopted the rhetoric of sustainability as the core of their planning philosophy.[39]

Yet we must be cognizant of externalities generated by sustainable planning and green urbanism that negatively affect vulnerable communities. Historical analysis makes clear that, before the 1970s, efforts to improve the city and make it more livable often overtly discriminated against minorities. But even policy undergirded by sustainable planning ideas, which often consciously sought to include minorities, often undermined the plans' conception of sustainability. New urban design, promoted as "green" because it discouraged automobile use and was more energy efficient, increased real estate values in minority neighborhoods and caused displacement of minority residents and businesses. Sustainable zoning changes made some existing living arrangements illegal. Tax abatements for green infrastructure, green building, and more sustainable improvements were not complemented by rent controls or other assistance for economically disadvantaged citizens. Put simply, as sustainable practices fix some problems, they generate others or repeat and sometimes exacerbate existing social and economic discrepancies. As such, we must learn to think fluidly and develop fluid policies designed to mitigate the inevitable issues that attend sustainable planning initiatives.

As our cities become less industrial, it is also imperative to understand how urban environmental injustice emerges in complicated and multifaceted ways.[40] While pollution from high technology production, power plants, and other industrial facilities drew the ire of marginalized communities, lacking access to natural resources, environmental improvements, and discriminatory planning proved deleterious as well. Relatedly, Austin demonstrates that environmental racism remained largely invisible to even the most fervent mainstream environmentalists because of a history and geography that essentially created separate societies. Planning by its very nature is something that looks forward in time and imagines better communities. Yet creating holistically sustainable communities means understanding and appreciating how uneven development undermines even the most egalitarian planning efforts that fail to address historical discrimination. If we think of the city and planning as a commons, a set of resources available for public use and used via public discretion, we must also acknowledge that the commons have not been available to all equally.

Finally, this book locates Austin as a city that is unique in its specific environmental-social landscape but also representative of the way that many other American cities are growing. As a Sunbelt city located at the crossroads between the South and the West and straddling two distinct ecoregions, Austin provides a compelling case study that illuminates both its local historical particularities and the broad, global concerns of how urban and natural worlds interact. Austin is at once part of a region and a state that have been growing and urbanizing intensely for seventy years, yet it is something of an ecological and social anomaly in that region. Ecologically, the city sits at a place that has marked the transition from one world to the next. Frederick Law Olmsted remarked in his 1850s travelogue that on leaving Austin "we crossed the Colorado into, distinctively, Western Texas" and tellingly ended the section he titled "Route through Eastern Texas" and began "Route though Western Texas" at Austin, noting the difference.[41] For him, Texas's landscape was literally divided at Austin.

Geologists, environmental scientists, and other professionals interested in biology and geography have since validated Olmsted's view. Austin straddles two quite distinct ecoregions. The eastern part of the city lies in the blackland prairie, with its rolling hills that make for fine farmland and get enough rain to support intense agriculture. To the west is the Edwards Plateau, a much drier savannah located on an uplift much less conducive to farming and better for ranching. So Austin is a geological and ecological hybrid, containing a surprisingly wide array of animals and plants, and also the place where, according to the Environmental Protection Agency (EPA), southeastern plains give way to southwestern semiarid prairie. Put more simply, the city serves as a physical link between farmland and arid grazing land.[42] As for all human settlements, the natural world has provided sustenance and beauty, but it has also been an impediment to growth. Austin, like many settlements in the American West, faced its biggest natural challenge from weather and water.[43] While the region receives an abundance of rain compared to most of the areas lying to its west, it is very susceptible to drought. And, owing to its limestone soil and happenstance meteorology, it is one of the most flood-prone areas in the United States.[44] Thus the particular history of human-environment interaction in Austin has revolved around managing water as well as enhancing access to and preserving unique environmental characteristics that have high use and exchange value. Managing water has been the region's greatest challenge, but it also became an issue and a symbol that tied otherwise disparate entities together and generated a more regional environmental and political perspective.[45]

As an economic unit Austin is likewise a hybrid city, at once subject to distinctive regional and global forces that shape it in different ways. Austin's dramatic growth, from the sixty-seventh-largest U.S. city in 1970 to the thirteenth-largest in 2010, owes much to regional and global shifts that have drawn capital from older industrial centers in the Northeast and the Midwest. Owing to the New Deal, World War II, and, later, to policies that subsidized industrial decentralization, the federal government invested a disproportionate amount of capital into the region. As technologies rapidly emerged during the Cold War and businesses became more mobile, lower costs in real estate and labor and lower taxes drew national and multinational corporations to less developed areas. State and municipal governments committed to free market principles offered subsidies and tax breaks and discouraged unions, promising lower costs and higher profits in an effort to attract businesses. They legislated a smaller regulatory apparatus and allowed for more permissive business climates.[46] Federally subsidized air-conditioning, modern roads, and automobiles facilitated these new policies and practices, making life in warmer, less developed areas more comfortable and convenient. Newer, less dense, more suburban landscapes secured the deal, proving attractive and affordable to workers.[47]

Austin certainly benefited from these macroeconomic transformations, both directly and indirectly, and is best understood as a city closely aligned with the Sunbelt's growth trends. The expansion of Texas especially aided Austin. Money and consumers flooded the city as the state government and the university grew to meet the demands of Texas's increasing population. More than any southwestern university, the University of Texas gained from federally sponsored defense spending. By the 1980s, state administrators invested heavily to make Austin the state's technology hub in an effort to diversify away from oil.

Yet Austin was, and in many ways remains, a social and economic anomaly in the region. Austin has shaped its policy and landscape in a way that takes advantage of global flows of people and information and has become a site of global-focused research. Unlike established industrial cities and most Sunbelt cities,[48] Austin sought to capitalize on the natural environment and sustainable urban planning as assets that could potentially drive economic growth from an early date. The University of Texas and the state government allowed Austin's leaders to eschew industrial development and emphasize natural beauty and quality of life. For these cities, heavy industry and its attendant management apparatus, as well as secondary industries, drove not only the urban economy but also migration, including a significant, long-term African American migration extending from 1900 into the 1970s. While the African American population steadily increased in these other cities, at times creating significant friction in the labor market and residential patterns, Austin's African American community steadily declined or grew in a much smaller proportion relative to white residents.[49]

In its broadest conception the story of Austin that I tell, then, is important because it illuminates contradictions and tensions that are vital to understanding how urban life is changing in the early twenty-first century. As environmental problems become more central to American liberalism and policy (including, of course, urban planning), it is of utmost importance to remember that improving and protecting the natural world has often come at the expense of marginalized communities and people of color, even as a variety of liberals have attempted to include or at least account for minority voices. In this sense, the history of discrimination in America is more than simply present; it is continually reconstituted to fit different social and economic imperatives, imbued in the landscape in clandestine ways and always reflective of a racial history that is difficult to square with our broad notions of progress and improvement. Yet this story, like city building itself, is part of ongoing processes that are continually up for interpretation, debate, and transformation. Despite the often sad histories in the pages that follow, the future is open to changes that can make cities more egalitarian, responsive, and vibrant. The first step is to acknowledge historical discrepancies and understand how they continue to foil even the best laid plans.

1 A Mighty Bulwark against the Blind and Raging Forces of Nature

Harnessing the River

The people of Central Texas had a river. It was given to them in the Divine plan of their homeland. It was a fickle, mean, dangerous river. . . . They have built dams to control flood waters and store up water for drought times. In the future they will build more dams and levees. Someday the whole river will obey them.

—Lyndon Johnson, 1939

The year 1893 was pivotal for Austin and for the United States. The financial panic, signaled by the failure of the Philadelphia and Reading Railroad in February of that year, put an expeditious close to the railroad boom of the 1880s and portended a decade of high unemployment and weak currency on Wall Street. Agricultural uprisings mobilized southern and western farmers against eastern moneyed interests and manifested themselves in the People's Party, whose presidential candidate, James Weaver, won nearly 9 percent of the popular vote in the 1892 election. Labor unrest was also on the ascent; violent battles fought in Chicago and in Homestead, Pennsylvania, and a series of other strikes between 1885 and 1893 undercut the supremacy of industrial capitalism and cast yet another shadow on the unchecked growth of the American economy. Perhaps because the more established cities were in such turmoil, *Harper's Weekly* chose to come all the way from panicked Manhattan, bypassing the midwestern heartland's factory towns, to do a story on Austin, the capital city of Texas and home to its new university, in 1893. It was a positive story filled with hopeful news for a young area that the writer saw as unburdened by the problems of the 1890s.

The short piece, titled "Engineering Triumphs in Texas," was packed with much information about Austin as well as civic booster rhetoric common to the Gilded Age. Contrary to popular opinion on the East Coast, Texas was not a frontier outpost, "home of the outlaw and the desperado," with "rough riders, quicker shooters, hard drinkers." Such stereotypes, the article continued, "do a serious injustice to the largest State in the Union, and to the public spirit and intelligence of a people whose efforts have secured to Texas a rate of progress in the accumulation of wealth and population and in advancement towards a high state of civilization second to no other in America." Austin, the article went on, is ready to change places with Boston and become the modern Athens. And what will be the economic engine for this all-but-certain growth? It will be the new high-water dam at Austin, recently completed and a symbol of the "great public spirit in new enterprises," with "a capacity to deal with large public improvements in a large way." The Colorado River, as of yet not contributing anything worth mentioning to the city's wealth and prosperity and, in fact, detracting from the city's value with floods and droughts, must be harnessed and put to work by the large engineering marvel. When finally and fully completed, the $1.4 million structure, financed with municipal bonds, will give the city much-needed flood protection and also be able to deliver over fourteen thousand horsepower for sixty hours a week to Austin. When completed, the dam did provide Austin with consistent power and flood protection. It was also featured on the cover of *Scientific American* in 1896 and widely considered an engineering marvel.[1]

In the several years following the dam's completion in 1893, it became evident that the "engineering triumph" so lauded by *Harper's* was anything but a triumph. The granite and limestone structure, sixty-eight feet high and sixty-six feet wide at its base, was not securely anchored to the rocky soil, which makes up the ground in most of Central Texas. Within months of completion, cracks began to show in the dam's base and were letting through small amounts of water, and water was also passing directly underneath the dam. The leaks became ever more apparent over the following years, until in April 1900, the first major flood since the dam's completion came roaring down the river, destroying the dam and power generating station and leaving over $9 million worth of property damage and forty-seven fatalities in its wake. A large portion of the destroyed dam sat in the river near downtown Austin for many years after the flood, a constant and obvious reminder that the city was still at the mercy of the Colorado, rather than the subjugator of it.[2]

As for many burgeoning western and southern cities in the late nineteenth and early twentieth centuries, the relationship between the urban and the natural worlds was paramount in Austin and its hinterland. While some utopian thinkers dreamed of urban space integrated with nature, the gap between theory and practice was often wide.[3] Many planners and engineers, increasingly forced to deal with human settlements in less than ideal locations, imagined nature as something to be overcome, harnessed, and put to work for humans. Similarly, by the late nineteenth century, conservation efforts were increasingly viewed as part of the engineering realm. Politicians and residents began to view science and technology as reasonable solutions to social problems, and engineers led the way.[4] "Finishing" nature, improving the landscape while still using it, was a social goal. One of the highest expressions of improvement was multiple-use development for waterways that combined flood protection with other social and economic benefits.[5] Central Texans almost universally viewed the harsh yet picturesque landscape from this perspective; controlling the unpredictable floodwaters was of great significance to regional progress, but there were other distinctly important uses for the river. It would transform the lives of rural farmers in the western reaches of the Hill Country by providing electricity and bring manufactories, resorts, and cheap power to Austin. Politicians, businesspeople, and boosters spoke of the dam as a panacea that would provide stability and economic growth in the form of flood control, irrigation, and hydroelectric power.

This chapter looks at the difficult process of achieving that goal from the 1890s to the 1930s in Austin. During this period, damming the river was a central theme when Austinites and other businesspeople and politicians in Central Texas discussed being progressive. To be progressive meant to harness the region's capital, labor power, and engineering expertise to control nature, profit from it, and stabilize society. The dam allowed Austinites to reimagine their city as they wished it to be, and they viewed its failure as a civic shortcoming. After a number of failures and fatalities and the destruction of property, by the 1930s even conservative Austinites recognized that using federal resources to harness the river was the best way to secure the region's future. Dams became a symbol of elusive modernity in Central Texas, technologies that demonstrated the region's collective civic pride as well as the importance of water to society. At the same time, regional leaders yoked their future to New Deal liberalism in the 1930s, sometimes apprehensively, because private outfits and the City of Austin alone could not marshal the funds and expertise necessary to complete even one dam. With the creation of the Lower Colorado River Authority, Austin's New Dealers transformed Central Texas into a progressive political place defined by cooperation with the federal government. Yet the geography of improvements also exacerbated racial differences, demonstrating that even New Deal liberalism stopped short of racial reconciliation.

The Landscape

The site for Austin, from its founding, was associated with natural beauty and a pristine landscape. In 1839 Mirabeau Lamar and a group of five elected commissioners chose the site for Austin as the capital of the Republic of Texas at a small outpost called Waterloo, which had natural beauty and pleasant hills and was located close to what they thought would eventually be the center of the state. Although it was not as good for agriculture and trade as some spots along the Brazos River to the north and east, members of Lamar's party and others agreed that the site "would give delight to every painter and lover of extended landscape." The site was "composed of a chocolate colored sandy loam, intersected by two beautiful streams of permanent and pure water." The beautiful hills to the west, the commission wrote to the government at Houston, are "generally well watered, fertile in a high degree, and has every appearance of health and salubrity of climate." One migrant called the site "a fairy land,"[6] and a young Frederick Law Olmsted claimed in the 1850s that the city was "the pleasantest place we had seen in Texas."[7] From the beautiful territory, Lamar envisioned a garden, closer to wilderness than to the current capital at Houston, from which he wanted to distance the new capital. Although he admired the large river winding down from the mountains and hoped it would be navigable and supply waterpower, he chose the site primarily because its beauty and health benefits would be an attractive possession for all the people of Texas.[8]

Owing to its unusual siting choice and unlike most prominent cities of the nineteenth century, Austin lacked "natural advantages," physical aspects of the landscape that were beneficial for trade, shipping, or defense. This almost always meant proximity to water, usually to sites with natural harbors, or access to large rivers. Indeed, the sites for almost all the major cities of the United States in the nineteenth century were selected because of their natural advantages. The earliest were port cities on the East Coast established in the seventeenth and eighteenth centuries principally for trade. New York City was located at the southern tip of Manhattan Island, in perhaps the best natural harbor on the East Coast and with access to two large rivers to service internal hinterlands. Baltimore and Boston were similarly established in fine natural harbors with inland water access to facilitate trade. Philadelphia was situated as the shortest portage between the Delaware and the Schuylkill Rivers, the former bringing ships from the Atlantic and the latter sending them inland. Newer cities on the West Coast, such as San Francisco and Seattle, were similarly located in natural harbors advantageous to waterborne commerce.[9]

The major inland cities of the nineteenth century were similarly located on water trade routes, either the Great Lakes or large rivers. Chicago was situated at the shortest portage between Lake Michigan and the Illinois River, which provided access to the Mississippi. Buffalo grew into a bustling city as the western terminus of the Erie Canal, where it met Lake Erie. The largest inland cities of the antebellum period were all sited at strategic locations along the major rivers. Cincinnati and Louisville were founded as trading ports on the Ohio River. St. Louis, the largest inland city by 1870, was sited to control all trade flowing from the Missouri River into the Mississippi. New Orleans's location between the mouth of the Mississippi and Lake Pontchartrain was so strategically important for trade that founders ignored its perilously low topography.[10]

Austin lacked these advantages and was subject to the unpredictability of Central Texas's weather and the harshness of its topography. Like all rivers, the Colorado is characteristic of the topography, geology, vegetation, and climate of the regions it drains. The river itself winds an 865-mile course from its headwaters in far eastern New Mexico to the Matagorda Bay on the Gulf of Mexico. It begins on the high plains of northwestern Texas as a small stream. As it reaches the Edwards Plateau, the 36,660-square-mile ecoregion directly to the west of Austin, it becomes more dramatic, carving out canyons and changing its elevation more quickly. Its drainage area is over 42,000 square miles, much of which is flood-prone. Average rainfall increases as the river enters the Balcones Escarpment at the eastern edge of the Edwards Plateau, immediately adjacent to Austin. The three major tributaries all enter within 150 miles upstream of Austin. The Colorado exits the escarpment at Austin, and downstream it becomes a coastal plain river.[11] Although humans had been using the river's resources for millennia and many imagined different uses for it with modifications, little about the river was altered. It maintained its natural state well into the late nineteenth century save a smattering of small dams used to power grist mills and cotton gins upriver from Austin.[12] Lamar hoped that the Colorado River would prove navigable, but the decision to locate Austin near it was made quickly and in the absence of scientific hydrology. It did not, largely because of its unpredictable flow. Just four years after Austin was founded, the river rose thirty-six feet above its normal level, destroying much of the town and punishing the small agricultural settlements downriver from Austin. The Colorado Navigation Company, founded in 1851, used a small federal grant to improve the river for navigation in 1854, but by 1860 the river was no longer navigable due to lack of upkeep.[13]

Lamar also could not have imagined how perilous and dangerous life along the Colorado could be. If someone came to Central Texas during a moderately wet year, the river might look placid and the entire landscape could appear absolutely verdant, similar to areas with higher, more consistent rain totals to the east. The region gets roughly thirty-two inches of rain per year, but the average is not a meaningful number given the drought-flood cycle. While normally arid, Central Texas is prone to severe flooding because it can receive rain from tropical hurricanes coming off the Gulf of Mexico and large, wet air masses arriving from any direction. Texas has six of the twelve largest-recorded forty-eight-hour rain totals in the world, and the Balcones Escarpment, a 400-mile-long uplift immediately west of Austin, is particularly vulnerable to heavy flooding because large storms often stall over it. The rocky, nonporous limestone exacerbates the problem. The lower Colorado basin has been called "flash flood alley" for its propensity to flood quickly and without warning. U.S. Geological Survey hydrologists consider the basin along the escarpment the most flood-prone in the continental United States. For residents, life along the river or its numerous creeks was risky and uncertain and often entailed major damage to property, injury, and death. Conversely, under drought conditions the Colorado routinely slowed to a trickle. Thus navigation for transport and trade via water was unrealistic in Austin, and the city remained largely separated from other Texas regions for most of its early history.[14]

While many people found the landscape beautiful, it was also harsher, isolated, and more difficult to cultivate than Lamar had imagined. Austin is situated on the eastern edge of the Edwards Plateau, which was created millions of years ago when the area immediately east of Austin slid downward by about seven hundred feet, creating the long coastal plain that makes up most of eastern and southern Texas. Thus Austin straddles two very distinct geological landscapes: the rugged hills of the western part of the city (the eastern edge of the plateau) and the flatter, blackland prairie of the eastern section. Over time gravity and water eroded the soft sediment away, exposing the plateau's limestone and dolomite bottoms. Erosion also created the steep canyons, cliff faces, and sharp hills characteristic of the region. Unlike in the lower-lying lands to the east, where most Anglo settlers to Central Texas came from, the lack of deep top soils makes the region difficult for farming, save some areas in valleys where soil and water collect. These areas are also the most flood-prone. Rocky soil and steep gradients mean that water runs off very quickly, in many areas causing flooding and making much rainwater unusable. The terrain was difficult to navigate and travel was challenging, isolating settlers and creating a culture that consisted of small subsistence farming, cutting of cedar, charcoal production, and hunting-gathering in the rural areas to the west of Austin.[15]

"Finishing" Nature and Imagining the Future

In Austin, and in Texas more generally, transforming harsh landscapes into livable spaces necessitated a deep understanding of the natural environment and natural processes in order to properly apply technology to it. Growth in Austin remained slow and sporadic in the three decades following the founding of Texas, which was still at the far reaches of American geography. Water technology, both in terms of capturing water and in terms of flood control, was rudimentary. The state's population growth remained slow through 1870, but it began to rise during the 1870s due to the arrival of the railroad. The population nearly tripled to 2.23 million people from 1870 to 1890. With little attracting residents save the state government, Austin's population remained relatively stagnant through 1870. Much of the growth during the 1860s was due to African Americans escaping to Austin after emancipation was enforced in Texas; in 1870 36.5 percent of Austin residents were African American. The city became a boomtown in the early 1870s because two railroads finished spur lines to the city, allowing for cotton and cattle shipping, but newer regional railroads siphoned off trade by the late 1870s.[16]

In the 1880s two events gave Austin a new sense of stability and guaranteed long-term investment in the city. The 1881 decision to locate the flagship University of Texas campus in Austin engendered the first sustained efforts to increase economic growth in the city. The university, while small at first, brought into Austin white-collar workers whose salaries were paid by the state. The campus and main building provided a peaceful park-like setting on a hill overlooking the downtown area and the river, a sure tourist attraction. Students bought supplies from local businesses. In 1883 the state commissioned a new capitol complex, which was finished and opened in 1888. Like the university campus, the capitol was a symbol of civic pride and modernity for all of Texas and would bring money from around the state to Austin. The twenty-two-acre grounds were meticulously manicured and open to the public; like the university, the complex was pastoral and overlooked downtown Austin and the river. The beautiful statehouse, one outside observer noted, and the university "quite possibl[y] . . . may some day be the greatest seat of learning in the Western world."[17]

In the 1870s, boosters and business clubs began to imagine ways to sell Austin to prospective businesses, tourists, and others who could improve the economic conditions of the city.[18] But the new architecture and attendant "cultured air"[19] provided a model for boosters to create an image of place that drew on a mix of urban amenities and natural qualities. The region, many thought, had beauty and a wonderful climate that improved both physical and emotional health, aspects of human life that cities were often thought to vitiate.[20] In the 1880s and 1890s, members of the Austin Commercial Club[21] envisioned a leisure economy that was based on the city's unique natural characteristics and on the culture of refinement available because of its educated and professional population. They wrote that "the city has recuperative power in its almost magic atmosphere which is generally lacking elsewhere." Health, fine residences, and educational advantages distinguished Austin from other Texas cities, as did Austin's beauty, which possessed "every facility that prodigal Nature can bestow or art produce."[22] Like Lamar before them, Austin boosters saw the natural landscape, unique physical characteristics, and climate as attractive features that warranted growth.

But for Gilded Age boosters, the natural landscape by itself could provide only so much. While early guidebooks were quick to discuss Austin's beauty and charms, no mention was made of the floods that destroyed the city almost every decade or the droughts that routinely caused the city to run out of water.[23] As many as seventeen major floods were recorded along the river basin in Central Texas from 1819 to 1900, and many of the most brutal happened within the lifetime of a number of residents. In 1843 the river rose to thirty-five feet above normal, killing an undisclosed number of residents in Austin. In 1857 a flood along Shoal Creek, Austin's largest urban creek, carried a grist mill into the river. On July 6, 1869, heavy rains upriver came crashing down on Austin, carrying with them the carcasses of hundreds of buffalo. Downtown, water reached forty-three feet above the normal level, still the highest ever recorded. The following year another flood destroyed more property. Yet the arrival of the railroad in 1873 brought a larger regional market for Austin's farmers, and the city experienced a short yet intense upsurge in population and wealth. With modest growth during the 1870s, Austin patricians saw a chance to harness the capital necessary to harness the river.[24]

Although prominent residents began mentioning the possibility of a dam as early as 1854, the region's lack of people and capital made the idea impractical.[25] By 1886, however, the Austin Commercial Club, already a proponent of urban growth as were many such organizations in western towns, gained confidence with the new university opened and the new capitol building under construction. Its support was almost universal, beginning when Alexander P. Wooldridge, president of the City National Bank and of the public school board, wrote a series of articles advocating for the dam's construction in the late 1880s. The series opened on New Year's Day in 1888 with a front-page article in the *Austin Statesman* that linked Austin's future as a progressive, prosperous community to the proposed dam, which would provide flood control, irrigation, power, and consistent water supply. Wooldridge, who went on to become Austin's most influential mayor prior to the 1930s, chastised the city's citizens for indifference and lack of imagination, arguing that the cost of a survey would be minimal. Construction could be financed by bonds. A dam would lead to other improvements, such as sewers and paved roads, as well as industrial opportunities attracted by cheap water and power. Wooldridge skillfully employed the ideas of grandeur and civic pride that promoted large-scale projects such as the Brooklyn Bridge and other emerging technologies that fascinated Americans in the late nineteenth century and became models for other technological improvements. He linked the dam with a greatness that transcended Austin yet also reflected the city's spirit.[26]

In the late nineteenth century, such lofty propositions were not uncommon among urban boosters guided by the emerging idea that technology and science were both civic virtues and potential paths to prosperity and growth. In Texas, though, a public culture that valued individualism and distrust of government projects often trumped civic-minded rhetoric of better days ahead. Wooldridge kept at it, periodically writing in the *Statesman* and forging alliances with other progressive-minded people who saw hope in the dam project.[27] By 1889 the dam had become the central political and social issue for Austinites, who increasingly equated environmental stability with growth and prosperity. As with other technological projects that sought to enhance nature with engineering, the dam system was widely considered the key to regional progress and modernization. Boosters believed that the dam could enhance the region without losing its natural, spiritual qualities, and most voters agreed.[28] John McDonald, who supported the dam, defeated incumbent Joseph Nalle, who opposed it, in the mayoral election of 1889. In December 1889 the first record of the dam appeared in the city council minutes. Public money, never easy to come by in Texas cities, was allocated for a survey and to hire an engineer and buy land that would potentially be flooded along the watershed upriver. Early in 1890 the $1.4 million bond proposal passed by a vote of 1,354 to 50, and a board of public works was established by ordinance to oversee dam construction and operation and the newly acquired public water and sanitary system. The dam was thus not just a physical structure; it was the impetus for city council members and citizens to begin imagining Austin as a modern city with a collective civic vision. It allowed for the city to establish public operations and to plan improvements based on the way the dam changed the landscape. Understanding the natural environment and trying to integrate it with technology—finishing nature to make it more usable—was essential to the process.[29]

Like many Americans of the period, Austinites began to view technological improvement through a lens that was both pragmatic and idealistic. The dam symbolized man conquering nature but also maintaining and improving on an already endowed natural world in and around Austin. Numerous boosters as well as travelers pointed to Austin's particular natural advantages in climate and beauty, but most were equally as critical of the lack of improvements that could help the city realize its full potential.[30] Infrastructure, for example, could allow for easy passage from urban to natural landscapes and provide a source of civic pride for residents and revenue from tourism. One publication reporting on a proposed highway connecting downtown to a scenic river overlook opined that "such things as these knit our hearts together in a common cause—nature's beauty—but we must build with vision and generosity."[31] The dam, boosters hoped, would create a collaboration between nature and technology to improve the landscape and bring jobs, opportunity, and stability to the region. They discussed the placid lake that would emerge above the dam, close to downtown Austin but far away enough for a person "to seclude himself from the hum of busy streets."[32] The engineering prowess and sheer size of the dam (it was thought to be the largest in the world when completed in 1893) were sources of pride for residents and politicians, and the building of the structure "attracted worldwide attention" and discussion in newspapers and technical publications. Many lauded it as a symbol of democracy and greatness because it was publicly owned and operated.[33] Predicting that the dam and its power would bring nineteen thousand new jobs to Austin and allow the city to grow to its proper urban dimensions, an engineer commented, "When Austin grows up to its water power it will be a big place indeed." John

Bogart, the state engineer of New York, envisioned incredible manufacturing growth for Austin because the dam would bring "the finest water power in the United States."[34]

Residents enjoyed modern conveniences, unusual leisure opportunities, and national attention after the final stone was laid in 1893, and nature cooperated with a few continuous years of decent weather. The dam caught minor floodwaters well and created Lake McDonald, the largest contained body of water in Texas at the time. Boating became a pastime for residents and visitors. The enormous steam-driven paddleboat the *Ben Hur* took sightseers around the lake, and Austin established an annual international regatta in 1894, at the time a peculiar event in arid Texas. The dam itself became an attraction for tourists, who marveled at its dimensions and engineering brilliance. At 60 feet high, 68 feet wide, and 1,235 feet long with a 1,160-foot spillway, the dam was huge by the era's standards. Visitors were awestruck by its size and by how it enhanced the landscape, a machine placed auspiciously in the garden to help make it more verdant.[35] One compared the water flowing over it to the iconic symbol of American natural beauty in the nineteenth century: "It was grand and awe-inspiring, and nothing in my opinion could in any measure compare with it, except the falls of Niagara." A streetcar took passengers straight from downtown Austin to the dam site two miles upriver. The city began supplying water from the lake directly to public mains and into many houses. Perhaps nothing changed day-to-day life as dramatically as hydroelectricity. Power, previously provided by a steam station, became more consistent with the use of the dam's hydroelectric turbines. The great powerhouse rose about fifty feet above the riverbed just downstream from the dam. Power rates were cut in half, and the streets were bathed in light at night by moon towers placed throughout the city.[36] Austinites thus enjoyed both easy access to nature and the conveniences associated with the most modern, cosmopolitan cities of the day.

The Intractable River

But problems with the system emerged even before the inevitable drought hit in 1896. The first head project engineer, J. P. Frizell, who resigned his post over worries that the dam was not structurally sound, wrote multiple letters to Austin mayor Lewis Hancock warning him that water was flowing underneath the dam because the riverbed was fracturing it away. The foundation was not properly secured in the hard limestone. Within months of completion the dam's base showed that cracks were letting small amounts of water through, and water was also passing directly underneath the dam. People fishing near the dam noticed that lines were passing underneath the structure. One problem was the topography and geology of the riverbed. In 1898 the U.S. Geological Survey concluded that the original 1893 structure was faulty. Engineers also underestimated the minimum flow that the river could produce, and the long, low-flow periods characteristic of drought in Central Texas meant that power could not be produced, leaving residents to endure lengthy intervals without electricity. During droughts the dam could not even guarantee the availability of the most basic resource, water, to residents. A severe water shortage hit in June 1896, the first since the city passed a large water and sewer bond package in 1887. In 1897 the city council allocated funds to study the availability of other water resources in Barton Creek on the south side of the river and in Bull Creek to the north.[37]

The leaks became ever more apparent over the following years, until in April 1900 the first major flood since the dam's completion came roaring down the river. After several days of widespread rain throughout the area, on the morning of April 7 the river began to rise rapidly at the dam site. Hundreds of people gathered along the eastern bank below the dam to watch the incredible force of a ten-foot wall of water cascading over the structure on an otherwise sunny morning in Austin. The ensuing break, however, changed the mood from amazement at the grandeur of nature's beauty to the horror of nature's power; two roughly 200-foot sections of the dam came loose from the bed at mid-river and immediately jolted downstream. The enormous breach allowed a forty-foot wave of water through that slammed into the power station at the base of the dam, destroying much of the structure in seconds, trapping dozens of workers, and killing seventeen. The wall of water continued through downtown Austin, knocking out bridges and leaving flotsam and debris in its wake, including uprooted trees over forty feet in length. Within twenty minutes the river was nearly a mile wide downtown, about four times its normal size. In all, forty-seven people died and over $9 million worth of property was destroyed, including houses seen "sweeping . . . over like they were tenpins." A newspaper headline the following day coldly summarized the break: "Death and Ruin at Austin." Newspapers from around the region and nation commented on the catastrophe, one calling the dam "the greatest blunder ever made by a Texas city or town." National engineering publications, such as *Scientific American*, which had covered the building and operation of the dam for years with consistent excitement, and *Engineering News*, quickly tried to explain the failures.[38]

Over the next two decades technical reports and professional opinions on the dam breach and failure blamed human engineering decisions rather than the natural power of the river. Thomas U. Taylor, a University of Texas hydrologist, produced a major report on the dam failure in 1900. He found that no hydrological data had been collected in Lake McDonald and that a fine silt characteristic of limestone erosion had built up quickly in the reservoir. As a result, in 1900 the lake's storage capacity was roughly half of its total in 1893, leaving the dam much more prone to floods than most engineers believed. From 1891 through 1908, the Army Corps of Engineers produced a series of reports recommending that the federal government avoid undertaking improvements on the Colorado because harnessing the river was not feasible. Daniel W. Mead's official report, published in 1917, did not list a specific cause for the dam breach but did indicate that the limestone riverbed at the dam site was poorly chosen because it was already prone to cracks and fissures.[39]

The failure of the Austin Dam project in 1900 was only the first in a series of disappointments over the next thirty years along the Colorado.[40] Just months after the failure Austin businesspeople, hoping to quickly rebuild the dam to furnish them with cheap power and attract other business, began inquiries with private engineering firms.[41] In 1903 the city council initiated a committee to investigate rebuilding the dam, and in 1908 voters passed an ordinance allowing the city to sell the rights to the dam. Several private firms made preliminary reports on the dam in 1908 and 1909 before the contract was given in 1910 to William M. Johnson, who again sold bonds to finance construction. Nature failed to cooperate. The Colorado flooded twenty times between 1900 and 1915, causing over $23 million in property damage and taking at least forty-nine lives. Against the suggestions of a number of engineers, Johnson's contractor, the Carmichael Company, began rebuilding the dam on its original site in 1912, this time using a system of gates rather than gravity to let water through. City officials were optimistic that the second dam would be sturdier, but dozens of gates were damaged and blocked by debris during another flood in 1915, rendering the dam once again unusable.[42]

The failure of the second Austin Dam and the high occurrence of river-related problems did have the effect of galvanizing different interests into larger associations whose main purpose was to control the river. Concentrating enough capital and technical expertise was still an enormous issue,

The first Austin Dam, destroyed by a flood in 1900, and the picturesque Hill Country in the background. Photographed in 1910. PICA 18022, Austin History Center, Austin Public Library.

however, yet residents still expressed that the river could be harnessed and the region's future secured.[43] Federal interest in dam projects during the early twentieth century focused more on desert states farther west, and most boosters still remained skeptical of a large federal presence in the region.[44] Local people, though, were more apt to cooperate with one another and, if deemed necessary, ask federal entities for assistance. The Colorado River Improvement Association, formed after the second dam was destroyed in 1915, was a group of landowners and farmers living along the river or near it. The group was able to win an appropriation from the U.S. government to undertake a survey of the Colorado watershed as part of the Rivers and Harbors Act of 1916. In its final 1919 report, the Army Corps of Engineers provided the basic geological and technical framework that informed all future work on the river, although navigation was its primary focus. The report detailed the river's propensity to flood and the long drought cycles endemic to the region. The final report recommended a series of multipurpose dams on three specific sites above Austin as the only lasting solution to the flooding problem on the river. The report, however, did not recommend federal funding for the project, and the dramatic increase in railroad transportation linking Austin to other parts of the state since 1901 indicated a greatly diminished need for a navigable river.[45]

The State of Texas also took some initial action by approving a conservation amendment to the Texas constitution in 1917. The amendment allowed for the state to legally create agencies and authorities that provided for the development and protection of natural resources, while placing rivers and waters in the public domain. After more floods battered the state in 1919, 1921, and 1922, the Texas legislature funded a statewide study of rivers undertaken by the Texas Board of Engineers. The board worked with the U.S. Geological Survey to amass more data, but little was done in practice.

During the 1920s, private contractors attempted to build dams upriver from Austin at sites chosen by the Army Corps of Engineers and also endeavored to rebuild the Austin Dam. C. H. Alexander, who owned water rights to extensive territory along the river, tried to build a dam at Marble Falls, but was only able to amass enough capital to complete a low-water dam about a third of the way across the river. In 1922 William M. Johnson defaulted on the bonds he had issued to build the dam a decade earlier and the dam was foreclosed. A group called Austin Dam, Incorporated (ADI), took over control of the project and funded an extensive engineering survey of the site that found the dam would have to be completely rebuilt and would be unable to generate as much power as previously thought. ADI tried to sell the rights to the project in 1926 to Texas Power and Light, the largest energy company in Texas and a subsidiary of Middle West Utilities, but could not. In 1928, after nearly a decade of inactivity, the city filed suit to reclaim the dam site from ADI; after a three-year court case and two more years of inactivity from ADI, the city reclaimed the site in 1933.[46]

The power industry also saw the Colorado as a potential source for huge hydroelectric generating capacity, but development was stalled by the financial uncertainties of the era. Middle West Utilities, which was one of the largest energy companies in the United States and led by Chicago industrialist Martin Insull, began surveying the river above Austin in 1927 with the hope of building a series of dams whose principal function would be power generation; Texas conservationists hoped that the dams would also be able to protect the watershed from floods and perhaps irrigate crops along the coastal plains downriver from Austin. Middle West surveyed a site about sixty miles northwest of Austin and spent close to $1 million on water rights, but a suit brought against the company by upriver ranchers compounded by broad financial issues forced it out of the project in 1928. Insull's withdrawal and the continued failures at Austin Dam were yet another disappointment to conservation advocates and citizens in Central Texas. Water law expert and lobbyist Alvin Wirtz, a longtime advocate of river development, got the Emery, Peck, and Rockwood Development Company of Chicago to purchase the rights-of-way and take over the project. The company hired a contractor to begin building the dam in 1931, but then sold its holdings back to Insull's new subsidiary, Central Texas Hydroelectric Company. The new owners spent $3.5 million to complete half of the two-mile-long dam before Insull's holdings collapsed in April 1932 and Insull himself was indicted for fraud and embezzlement and deported to Canada. A frustrated Wirtz took the half-completed structure and the water rights into receivership.[47]

For forty years, the Colorado River and its half-built, partially destroyed dams stood simultaneously as a symbol of hope and as a monument to human failure and natural disaster for Austinites and Central Texans. As in many western watersheds, engineers and private contractors constantly underestimated both the river's power to wreak havoc and its propensity to run dry. They treated the landscape as more malleable than it was. In arid Texas, where surface water is the most valuable of scarce resources, residents saw controlling it as the key to growth and the key to improving their own local fortunes. Thus from a social and political perspective, river development was plagued by disparate interests and fractured geography. The City of Austin lost millions of dollars trying to build its own dam, without considering that dams above Austin could mitigate floods. The energy industries and other private companies lost millions trying to generate power upriver. The Army Corps of Engineers, West Texas farmers, and the Colorado River Improvement Association all saw the river through a specific lens and with their own particular interests in mind.

Harnessing the River: The Landscape of Modernity

The federal government provided the final hope for Central Texans in their quest to develop the river, and it was the last place most Austin businesspeople wanted to turn. The federal government began its involvement in western conservation and water development programs in 1902 by founding the Bureau of Reclamation and passing the Newlands Reclamation Act, which provided for small damming and irrigation projects throughout the West. Through the next two decades, the bureau grew slowly and generally supported only small projects. The late 1920s were pivotal for flood control and reclamation projects in Texas and in the United States more broadly. A series of destructive floods damaged river basins throughout the country in 1927. After torrential rains in April, over eighteen million acres on the Mississippi River floodplain were inundated along twelve hundred miles of the river, from north of St. Louis to the river's mouth south of New Orleans. It was by far the largest flood recorded in American history. Floods destroyed much land and property in the Tennessee Valley, in the San Joaquin Valley, and along the Colorado River basin in Colorado and Arizona. Congress passed a $325 million relief bill for the Mississippi River flood in 1928, but did little else for the rest of the country. The floods of 1927 and the beginning of the Boulder Dam project in 1928, the largest federally funded infrastructural project to date, garnered the interest of large construction and utilities companies looking to expand markets and increased interest in federally supported dam projects. But, by and large, the bureau remained underfunded and the federal presence in conservation projects in the West continued to be small.

The beginning of Franklin D. Roosevelt's administration and the widespread programs created by his New Deal in 1933 signaled an astounding and provocative change in the function and authority of the federal government. The New Deal created a wide range of agencies and directed funds to states to mitigate the severe unemployment that was crippling the U.S. economy. The idea was driven by demand-side economic theory: using federal dollars to hire workers would inject capital into the economy by allowing laborers to become consumers. Infrastructure development was central to the program because it would provide employment opportunities and simultaneously modernize lagging regions, especially in the South and West. Interior secretary and Public Works Administration (PWA) director Harold L. Ickes titled his history of the PWA *Back to Work*, indicating what he felt to be the key aim of the programs. The main method used to combat unemployment would be public works, largely paid for with federal dollars but implemented on a local level using unemployed, mostly unskilled and semiskilled workers to carry out construction projects. The magnitude and breadth of the programs, along with the dollars spent on them, were indeed fantastic. The federal government spent on average 1,650 percent more money on construction between 1933 and 1939 than between 1925 and 1929. Roughly two-thirds of that money was spent on public works. At the New Deal's height in 1935, the U.S. government appropriated an incredible 6.7 percent of the nation's gross domestic product for public works and in total built over seventy-eight thousand bridges and forty thousand public buildings.[48]

By 1934, Ickes had concluded that water resource development ought to be a major piece of broader New Deal infrastructural investment. He received a report from two consultants recommending that the federal government should radically shift its policy from funding only small projects to funding only large ones. Ickes was initially skeptical, but once he realized how much hydroelectric power dams could generate, he began to see that long-term solutions to the labor, infrastructural, and energy issues plaguing much of the South and West during the Depression were multipurpose.[49]

In Central Texas, the federal funds made available by the PWA were too enticing for even the staunchest small government conservative to resist. Texans marshaled a number of resources in Washington in an effort to secure funds, and Central Texans were at the forefront. Congressman J. J. Mansfield was named chairman of the House Committee on Rivers and Harbors in 1931. Congressman James P. Buchanan was appointed as the chairman of the Appropriations Committee in 1933, a powerful position for directing federal funds. Austin citizens elected Tom Miller as mayor on his New Deal platform in 1933. Lawyer Alvin Wirtz, one of the region's most avid dam enthusiasts, brought private interests from the newly formed Colorado River Company and representatives from the state board of water engineers to work on securing funding. The entire group met with Ickes in early 1934. Ickes, though unwilling to discuss funding to complete Hamilton Dam, a key flood control structure about 60 miles upriver from Austin, did recommend an engineer to investigate the problem. More importantly, he told Wirtz that the State of Texas would have to create a public agency to administer funds and employ contractors to build the dams and generators because it was illegal for the federal government to loan money to private businesses or directly to states.[50]

The formation of the Lower Colorado River Authority (LCRA) is a story in itself; it took Wirtz over a year from the meeting with Ickes to persuade the Texas legislature to create a public authority to take over development on the Colorado, which it accomplished on September 17, 1934, amid protest from a handful of private Texas power companies and a group of West Texas ranchers who feared that the LCRA would impinge on their water rights.[51] Support for the dams in Central Texas was almost universal, even among the heavily free market–oriented Austin Chamber of Commerce, which repeatedly opposed federal intervention into the city's business affairs. As in most smaller southern and western cities, the chamber of commerce in Austin was a primary political and economic force in the community, usually made up of wealthy businesspeople who were also involved in politics. In the late 1920s and into the early 1930s, most chamber members adamantly opposed the New Deal or any kind of federal assistance. In 1927 the chamber went on record "as opposing any action of the federal government at this time towards increasing cotton and farm products"; it was "also opposed to federal operation and ownership of any local enterprise especially . . . which contemplates the construction and operation of an electric and irrigation project as Boulder Canyon."[52] Even chamber member Walter E. Long, who would later write booster pieces on behalf of Austin that focused on the relationship between infrastructural improvements and urban growth, was skeptical of any federal interference into local business.[53]

The influence of Mayor Miller, a Democrat, and the universal need for water control and power influenced the chamber to support Roosevelt for a brief period during the 1930s. The Highland Lakes dam project was foremost on the chamber's agenda. In early 1934, the chamber's president, A. C. Bull, declared its support for the completion of Hamilton Dam as its major goal for the year. The chamber unanimously supported the creation of the LCRA later that year, despite the widespread protests of Texas's business community, which deemed the LCRA unconstitutional and unfair to Texas's private utility providers. The chamber continued to vociferously support the dam building project throughout the 1930s. On January 1, 1935, the *Austin American* ran an advertisement, signed by the chamber, encouraging citizens to actively support federal measures that would help the city grow. Six months later the chamber was cited in an editorial suggesting that Austin use its waterpower (when the dams were completed) to attract new manufactories to the city. The *American* continued to be an unabashed supporter of the dams throughout the 1930s, writing multiple pieces every year linking the dams with regional economic prosperity and modernization.[54]

On May 7, 1935, after months of work from Wirtz, Buchanan, Miller, and other dam advocates, the PWA resolved to loan the LCRA $15 million to complete the dam, reservoir, and "other works" at the Hamilton Dam site. An additional $5 million grant would provide the flood control portion of the project and would not need to be paid back. The language provided that the LCRA could also complete a hydroelectric generator and power lines, which would electrify rural areas in Central Texas and provide funds to pay back the $15 million loan to the PWA. The $20 million in total allocated funds was the third-largest sum apportioned for multipurpose dams, behind only the Hoover and the Grand Coulee Dams.[55]

The two main upriver structures, Buchanan Dam and Mansfield Dam, were completed in 1937 and 1939, respectively, just after intense flooding destroyed over $16 million in property and killed dozens of people in 1935, 1936, and 1938.[56] The 1935 flood, coming just a month after the initial loan was secured, was the second-largest Colorado River flood on record.[57] While private utilities claimed the floods were proof that the dams did not work, LCRA advocates argued that the completed system would have corralled the destructive floods. After a prolonged battle, in October 1937, the City of Austin agreed to amend its charter to authorize the city council to lease the old dam site to the LCRA, which would secure funding to build a new structure and then sell power back to the city at a reduced rate; at the end of the lease, the city had the option to buy back the dam.[58] The contract was signed in February 1938. When the first round of four dams was finished in 1940, the Highland chain was the most extensive multi-dam complex west of the Mississippi River. The dams did have an immediate impact on both everyday life and the regional economic and social growth ideologies. Flood control, irrigation for downstream farmers, electricity, and pristine nature available for recreation were all made possible by the dams. At the height of construction, nearly fifteen hundred laborers were employed in building Buchanan Dam; thousands of others worked on Mansfield, Inks, and Austin Dams.

In 1937 the first of the Highland Lakes dams was dedicated about sixty miles northwest of Austin. Originally called Hamilton Dam, it was officially renamed Buchanan Dam on October 16, after the late James P. Buchanan, the longtime U.S. congressman from Austin's district and the LCRA's most ambitious Washington advocate. Buchanan had died three days after ground was broken on Mansfield Dam. Lyndon Johnson won a special election to take his seat, his first elected office, after his support of Roosevelt, the New Deal, and the LCRA. Johnson gave a short speech at the dedication for the purpose of introducing Secretary Ickes, but his comments also demonstrated what the dam meant to Central Texas. It was broadcast across the entire region on KNOW radio. After almost exactly one century of Anglo settlement, the region finally had a technology to harness the river. "Today we are gathered here before this magnificent structure," Johnson began, "a mighty Bulwark against the blind and raging forces of nature, better to make it do our will. We are not gazing upon a vast pile of steel and concrete, of towering abutments and staunch piers. . . . We must peer beyond the concrete, beyond the steel, beyond the placid waters of this rising lake. . . . We must look to the minds of the men from which they sprang into being."[59]

The hydroelectric generators promised that residents would never again suffer "uncontrolled horsepower in Nature squandering its fury." Electricity, irrigation, and flood prevention were the most important functions of the dams. The metaphors Johnson used consistently spoke of man's triumph over nature. Only as an afterthought did Johnson mention the recreational possibilities that the lakes would provide, and even then largely as a place for residents to relax and bring their children. The entire event was a somber occasion filled with speeches by government leaders and polite applause, representing a collective yet restrained catharsis, a constant source of misery and tension finally controlled by the fortitude of visionary leaders. The speech not only reflected the economic and social turmoil of the time; it also reflected the difficult lives led by the mostly rural, agricultural citizens who lived in the area. Their dreams were not of urban growth or making money through tourism. Electricity and flood control were thought of as modern conveniences that would make life more predictable and controllable.[60]

The quest to harness the river demonstrates that progressivism and modernity were linked to controlling nature using political, technological, and civic power. After nearly fifty years of trying, Central Texans had turned the rough wilderness of the river basin into a garden, a natural area that was safe for humans and would produce for them while retaining its majestic beauty and grandeur. The dams and reservoirs obviously had practical uses that made citizens' lives better. But the dams also allowed them to imagine and build new landscapes, celebrate the completion of a collective vision, and reimagine the federal government as friend rather than foe (these topics are discussed in chapter 4). And for Austinites, controlling the Colorado was one factor in a larger vision of reorganizing the city based on its natural qualities as socioeconomic structures.

Yet as Johnson's speech articulated the benefits spread evenly throughout the region, in reality racial geography and increasing racial discrimination cut a deep chasm between those who benefited and those who did not. The dams changed the landscape, but they did little for urbanites living away from the river. No Africans Americans or Latinos participated in the planning or even got to articulate a vision for the improvement of

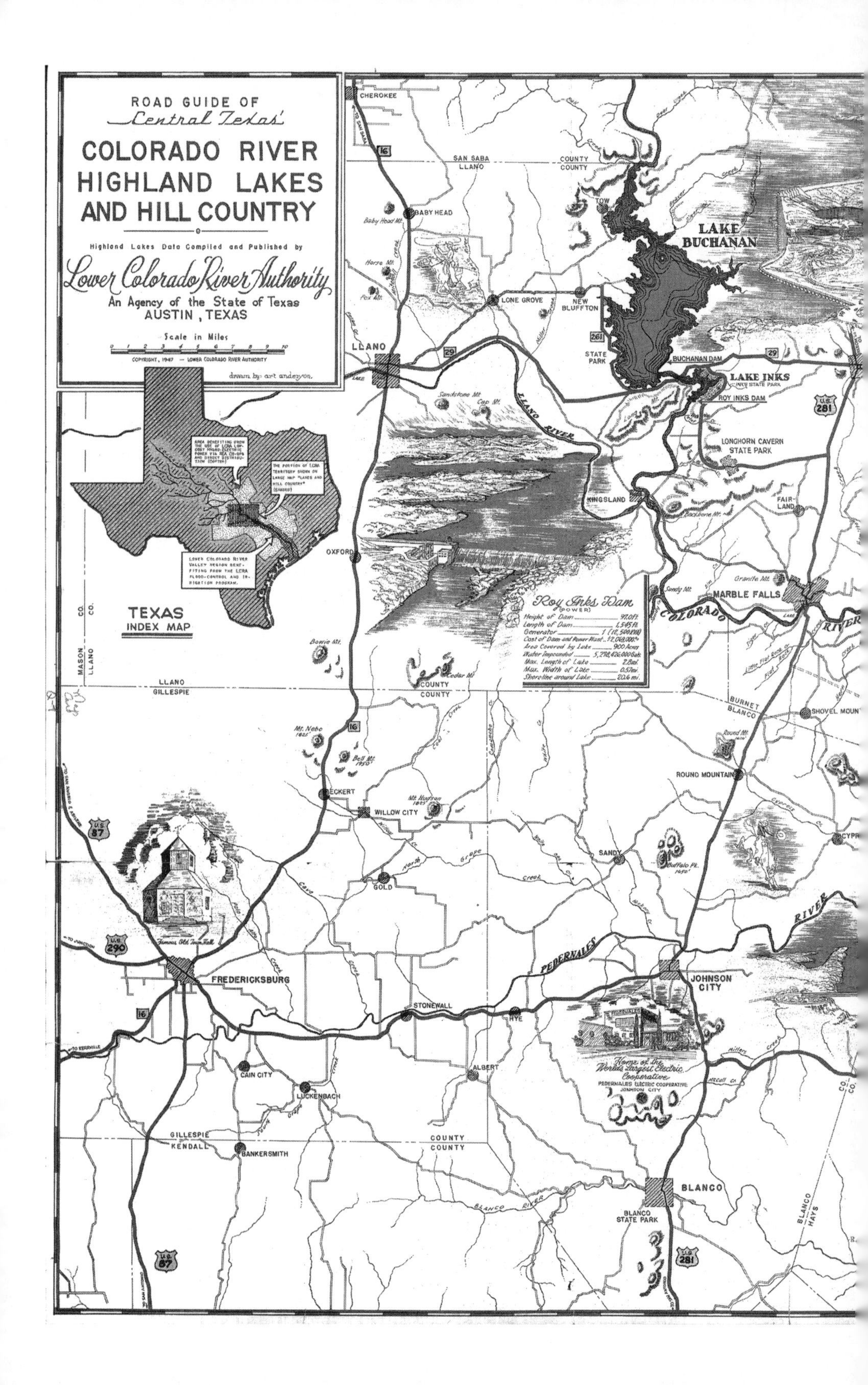

ROAD GUIDE OF
Central Texas
COLORADO RIVER
HIGHLAND LAKES
AND HILL COUNTRY
Highland Lakes Data Compiled and Published by
Lower Colorado River Authority
An Agency of the State of Texas
AUSTIN, TEXAS
Scale in Miles
COPYRIGHT, 1947 — LOWER COLORADO RIVER AUTHORITY
drawn by: art anderson
TEXAS
INDEX MAP
Roy Inks Dam
(POWER)
Height of Dam 97.0 Ft.
Length of Dam 1,545 Ft.
Generator 1 (10,500 KVA)
Area Covered by Lake 900 Acres
Water Impounded 5,292,426,000 Gals.
Max. Length of Lake 2.8 mi.
Max. Width of Lake 0.57 mi.
Shoreline around Lake 20.6 mi.
LAKE BUCHANAN
LAKE INKS
BUCHANAN DAM
ROY INKS DAM
LONGHORN CAVERN STATE PARK
LLANO RIVER
COLORADO RIVER
PEDERNALES RIVER
BLANCO RIVER
CHEROKEE
BABY HEAD
LONE GROVE
NEW BLUFFTON
LLANO
KINGSLAND
OXFORD
MARBLE FALLS
FAIRLAND
ROUND MOUNTAIN
ECKERT
WILLOW CITY
SANDY
GOLD
FREDERICKSBURG
STONEWALL
HYE
JOHNSON CITY
CAIN CITY
LUCKENBACH
ALBERT
BANKERSMITH
BLANCO
BLANCO STATE PARK
SAN SABA COUNTY
LLANO COUNTY
LLANO COUNTY
GILLESPIE COUNTY
BURNET
BLANCO
GILLESPIE
KENDALL
BLANCO
HAYS
MASON CO.
LLANO CO.
Famous Old Town Hall
Home of the World's Largest Electric Cooperative
PEDERNALES ELECTRIC COOPERATIVE
JOHNSON CITY

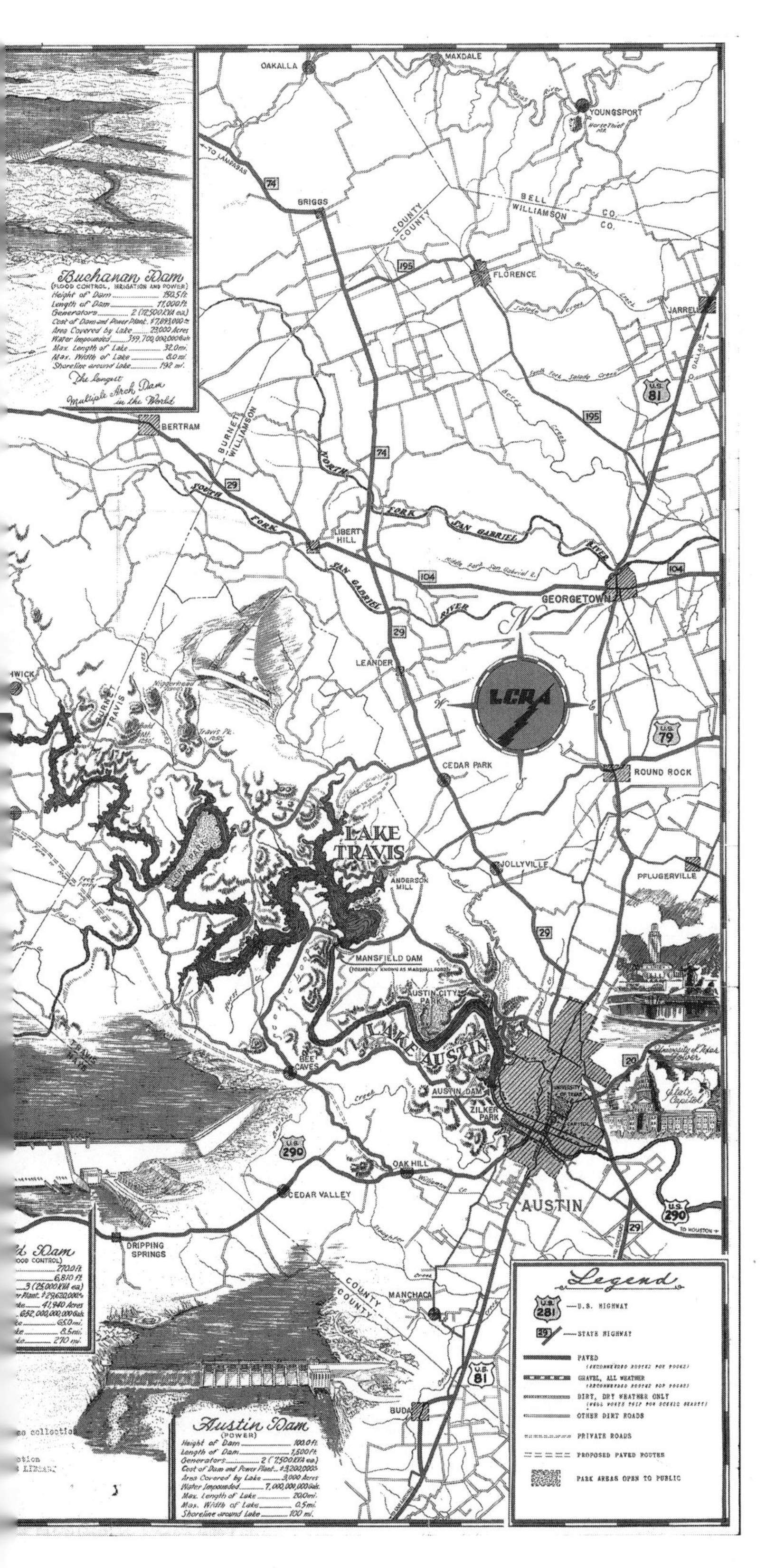

Lower Colorado River Authority map of the Highland Lakes reservoir system and the Hill Country of Central Texas, with Austin pictured in the lower right, circa 1940. LCRA Corporate Archives.

minority communities. White construction workers threatened to strike if nonlocal or nonwhite labor was used. And minorities were barred from many of the improved spaces and the outdoor leisure sites and denied the cheap water and power that the dams and reservoirs provided. Increasingly, their lives were lived in the city and by the 1930s in specific areas of the city set away from technological improvements and white residential neighborhoods. As the river transformed the region, urban planning transformed the city. Despite the collective discourse generated by harnessing nature in the early decades of the twentieth century, in the city minorities bore the burden of the human-environment interaction as well as environmental improvements.

2 A Distinct Color Line Mutually Conceded

Race, Natural Hazards, and the Geography of Austin before World War I

The Colorado River and its terrific power and potential occupied the imagination and energy of both engineers and boosters in Austin and its environs from the 1880s into the 1930s. Yet life in the city carried on despite setbacks on the river and a paucity of industries to attract workers and investment from other places. In the years preceding World War I, Austin looked like many small cities throughout the South, minimally industrialized with an isolated, largely agricultural labor market, and its geography reflected urban, social, and economic relations.[1] There was no zoning, and land uses were determined by private preferences. Race was predominant among social divisions and the division of labor, although residential segregation was not intense. Rather, African Americans lived in scattered pockets throughout the city, often close enough to white neighborhoods to easily get there for work as domestic laborers or near the city's small decentralized industrial or production facilities, often in activities related to agriculture or geared toward local markets. Natural topography, as well as man-made hazards, often determined where African Americans' small colonies were located, and they were frequently near the dangerous beds or mouths of creeks, close to places where refuse was collected, or on the outskirts of town, where land was inexpensive. The city was fairly isolated in a region that was mostly rural and agricultural; railroads linked the city and region to national markets beginning in the 1870s, and the University of Texas was established in the city shortly thereafter, yet growth was slow during the late nineteenth and early twentieth centuries.[2]

This chapter sketches racial and natural geography in Austin in the years before World War I prompted Austin's leaders to explore options to regulate land use and fund infrastructural improvements in the city. In the portrait that emerges, natural hazards, health risks, and proximity to waste, rather than the strict physical separation of races into distinct districts, are what demarcated space. Early on, more casual attitudes regarding racial geography indicated that Anglo Austinites were not threatened by proximity to blacks, an outcome of both the division of labor, where blacks had long worked in Anglo homes and lived near Anglos since slavery, and the rigid social division of race present throughout the South. Yet African Americans and increasingly Mexicans bore the brunt of natural hazards and garbage and waste disposal and suffered far more health problems related to disease and pollution. While some of this discrepancy was due to poverty and social custom, it is also clear that race was an important factor in the city's geography. Although Austin was not an industrial city where production was central to the economy, it still generated waste that needed to be stored somewhere.[3] Additionally, certain locations in the city were extraordinarily hazardous. The unimproved urban creeks, much like the Colorado River, proved extremely susceptible to flash flooding and carried all types of waste through the city to the river. The creeks also took away waste free of charge in an era when sewerage systems were practically nonexistent, and their floodplains were some of the cheapest land in the city. Thus in the early twentieth century, Austin's proximity to urban creeks could be treacherous on numerous fronts. Before World War I, then, the unimproved natural landscape of the city provided a possessive investment in whiteness, along with a strongly unequal educational and labor system that made it practically impossible for minorities to move up the economic or social ladder.[4]

Geography, Labor, and Social Relations in Turn-of-the-Century Austin

In 1900 Austin was a relatively small city with a total population of just 22,258 residents and a developed area of only about six square miles. While the city was home to the University of Texas and the state government, which both provided consistent sources of revenue and employment, no large manufactories or industries existed to attract migrants. The city had small brick-making and food processing outfits, but neither employed more than about thirty people.[5] Railroads, construction, and small factories also employed semi- and low-skilled workers. Labor unions were likewise very weak in the city compared even to other Texas cities. The dam, long seen as a boon for local businesses, did not increase employment opportunities or entice businesses to relocate, and its destruction in 1900 curtailed discussion about Austin's industrial growth for the next decade. Agriculture was still the main occupation for most people in the region and especially for Latinos and to a lesser degree African Americans. The improvement in railroads around the turn of the century, however, favored larger agricultural operations, and many smaller farmers and tenants were forced to the city to seek employment. Thus rural-to-urban migration generated much of the population growth in Austin because the need for agricultural labor was slowly contracting in Central Texas, especially among whites. Large rural farmers increasingly looked to short-term laborers and hired help from farther away.[6]

As in most cities at the time, land uses in Austin were haphazard, but they were oriented around the river at the south end of downtown, the capitol at the north end, and the university to the north. Rugged hills provided a geographic barrier to the west, while the eastern hinterland was composed of flatter agricultural prairie. The city was laid out with the downtown facing the Colorado River on the south and the capitol complex to the north. Residences were interspersed with businesses in the downtown area, with predominantly residential districts to the east and west of downtown, although a lack of zoning meant that there was rarely a perfect separation among uses. North of the capitol complex the University of Texas dominated the landscape, and beyond that to the north was the exclusive community of Hyde Park, a racially restricted residential suburb serviced by a streetcar line, opened as a bedroom community for white-collar professionals in 1904. A few other small, new, racially restricted subdivisions that catered to university employees and other white-collar workers emerged before World War I around the university area. Between 1900 and 1930 upper-middle-class neighborhoods appeared around the university campus and to the north and west.[7] A group of small settlements emerged south of the Colorado River in the late nineteenth century after the major bridge along Congress Avenue was reconstructed out of iron in 1884. Barton Springs, a spring-fed pool and grove located south of the river, was a principal site of recreation and civic activity for white Austinites.[8]

In Austin, as in Central Texas more broadly, the need for water and water's inherent dangers were central to people's ideas about the relationship between the natural and the urban worlds. While residents were often quick to point to the city's natural beauty as a point of civic pride and as a marker of quality of life, Austin's rivers and creeks were both vital sources of precious surface water and dangerous locations to live near. The university campus, the capitol complex, and most of downtown were located between two creeks, Shoal Creek on the western side and Waller Creek on the eastern side. Both funneled rainwater, waste, and debris southward from the city's northern periphery to the Colorado River and were subject to flooding, drainage problems, and clogging.

People view destruction from Waller Creek flooding, April 1915. The flood was among the deadliest in Austin history and disproportionately affected minorities. PICA 27261, Austin History Center, Austin Public Library.

What few manufacturing and other labor-intensive industries existed often operated away from population centers or close to undesirable areas near train tracks. All eight of the city's privately owned slaughterhouses, for example, were located on the outskirts of town, with seven in the eastern part of the city.[9] Many of the city's smaller industries, such as a brick manufacturing plant, a mattress assembly facility, and an auto body shop, were located near the terminus of the two railroad lines, one east of East Avenue at Fourth Street and the other just west of Shoal Creek near the Colorado River.[10] Walker Manufacturing, the largest food processing outfit in Central Texas, had small factories located at the mouth of Shoal Creek and in the far eastern part of the city.[11] Many unskilled positions open to minorities existed at small businesses and shops located near Waller Creek on the east side of downtown. There minorities could sometimes find work in bottling facilities, lumberyards, horse and mule lots, and wagon yards.[12] The Missouri Pacific and the International and Great Northern railroads also provided sites for Austin manufactories looking to minimize shipping costs.

Like much of the South and many eastern parts of Texas, Austin conformed to social and spatial patterns that reflected spurious "separate but equal" ideology but did not reflect the rigid spatial segregation that would come to dominate residential life by the mid-twentieth century. As in the previous slavery system and in the contemporary Jim Crow South (Texas adopted its Jim Crow policies in 1893), inequality was pervasive and did not need a particular residential spatial pattern to justify itself. Most African Americans were barred from voting and public facilities were strictly segregated, largely to keep African Americans from gaining access to power. Austin had a large number of African American residents in the late nineteenth century, topping out at 36.5 percent of the overall population in 1870, because a high percentage of early settlers in Austin were wealthy and thus owned slaves. The city also housed the first freedmen's colony west of the Mississippi River, Clarksville, on the western edge of the settlement, as well as at least four other small freedmen's colonies by 1880. Thus Austin, like most cities in the South, had a far higher percentage of African Americans than any large industrial cities of the nineteenth century. Although growth among blacks was slower than among whites, in 1900 African Americans still represented 26 percent of Austin's population.[13] That percentage remained virtually unchanged in 1910. Significantly, very few Mexicans lived in Austin at the turn of the century, with only nine households reported in 1880 and eighty in 1910. Most came to Central Texas as migrant agricultural workers and stayed for only a short period to harvest spinach, cotton, or the other agricultural products so vital to the region's overwhelmingly agrarian economy.[14]

Before 1920, residential segregation between blacks and whites was relatively fluid compared with later periods, but social relations were extremely separated into distinct worlds. That is to say, racism was highly institutionalized, but racial groups were not segregated into particular areas of town; rather, minorities tended to live in less desirable and more dangerous pockets throughout the city. Owing to the proximity between races dating back to slavery and the rigid social structure inherited from the Jim Crow South, most middle- and upper-class whites did not perceive blacks as threats to their property, community, or status, although that slowly changed as time passed. African Americans held a low position in the social and economic hierarchy, but most whites considered them an important component of the community. According to newspapers of the day, whites saw blacks as essential to the existence of Southern society and critical to daily life there, yet the distance between the races was enormous. There was little social contact between the races outside of employer-employee relationships, and a long record of racism and discrimination existed; whites often depicted blacks as less than human, alternatively lazy and dangerous, shiftless and shifty. As was the case throughout the South and in much of the North as well, whites ascribed negative characteristics to blackness and viewed blackness as a quality that always indicated the lowest social position.[15]

The social and economic bifurcation was most visible in the labor market and in housing conditions. Blacks were consigned to a small variety of jobs, most of which were difficult and low paying. Middle-class whites understood them as a lower class whose presence was necessary to conduct daily functions, such as cooking, maintenance, and yard work, or to toil in unskilled and semiskilled manual labor jobs that were often also unsteady. The employment disparity between white and black women was especially dramatic; even in 1930 black women were almost twice as likely to work as white women were.[16] Alternatively, because of the presence of the state government, the University of Texas, and state agencies, Austin's population overall was much more educated and white-collar than that of most cities in Texas or the South. The labor market was almost completely segregated; Jason McDonald found that roughly 60 percent of the white labor force was employed in white-collar occupations in 1930. Whites held virtually all the white-collar jobs in the government, at the University of Texas, and at the various state agencies or commissions located in Austin. Whites also dominated the skilled trades in Austin, as demonstrated by their higher levels of membership in local trade unions. While blacks had trade unions, they were generally much smaller and were often excluded from white local unions and prohibited from bidding for contracts. Most whites imagined race in terms of the division of labor. According to one observation published in 1891, "women don't want to do negro work" and "the feeling that contact with colored people is degrading still rules" among Austin's whites. The same publication justified building a dam in Austin because it would create new jobs for white women, who would then be less likely to "be put in the same place as coloreds through labor."[17]

Housing also demonstrated a wide gap between whites, blacks, and, increasingly, Mexicans; one commentator described these groups as three "classes of population," thereby linking race and class status in terms of home quality. Before 1910, African Americans lived in pockets scattered throughout the city, usually in areas that were inexpensive and provided easy access to work. There were multiple African American colonies in Austin as early as the 1870s; Clarksville and Wheatsville on the western side were joined in the 1890s by Masontown and Gregorytown on the eastern side. While they almost always lived among other African Americans, they were not segregated into a particular area of the city and often lived near white neighborhoods, especially on the eastern side, which housed numerous small ethnic enclaves, including Germans, Italians, and Swedes, close to small industries and the railroads tracks between Fourth and Fifth Streets. The main occupation for African Americans was in domestic service, where as much as 40 percent of the African American workforce toiled. This was especially true of African American women, who were employed far more than white women in Austin and almost always as domestics. Many blacks lived with their employers, in their houses or on their property. The two largest African American neighborhoods on the Westside, Clarksville, about one mile directly west of downtown, and Wheatsville, between Shoal Creek and the University of Texas and northwest of downtown, afforded easy access to wealthier white areas. Blacks were also scattered throughout the western part of downtown between the river and Nineteenth Street.[18]

In 1880 the largest African American agglomeration was centered from Red River Street and Waller Creek in the eastern section of downtown to past East Avenue between Sixth and Twelfth Streets.[19] By 1910 the African American population had grown by about twenty-five hundred people and had moved east and south, taking up residence along East Sixth Street and to the north of Eleventh Street to the city's eastern edge. These residences were in proximity to a number of Austin's small factories, slaughterhouses, mills, and lumberyards, where blacks had access to jobs and other unskilled and semiskilled positions. In the late nineteenth century, the East End district was Austin's most eclectic, housing blacks as well as a number of white ethnics, many of them first- or second-generation immigrants, who were less affluent and had less access to full white status among Austin's upper classes and established Anglo communities. The census of 1900 counted 5,447 whites who were either foreign-born or the children of foreign-born parents, roughly half the number of native whites with native parents in Austin at the time. Swedes and Germans, by far Austin's two largest white ethnic groups, had neighborhoods in the East End, and smaller Italian and Syrian neighborhoods existed there as well.[20] As we will see, in the first decades of the twentieth century the East End lost a significant portion of its white population while gaining black and, eventually, Latino residents.

Whereas white neighborhoods varied in quality and often had very nice houses intermixed with more dilapidated stock, minority housing was almost always substandard and usually described as "shanties" or "shotguns," long, thin, single-story wooden structures set on pier foundations. Clarksville and the Eastside suffered high levels of congestion, while Wheatsville was less dense. Even so, crowded conditions were the norm, with two or more residents living per room on average among African Americans citywide. Even in the nicest neighborhood, streets were not paved or graded, leading to instances of flooding. While Wheatsville had access to city water, many residents there found it too expensive. Clarksville and East Austin did not have access to municipal water or sewerage services. No African American houses had access to sewerage systems, and none had plumbing. African Americans, though, owned their homes at a rate comparable to whites and at a far higher percentage than Mexicans. Yet housing value was an obvious indicator of difference; white-owned residential properties had an average value close to five times greater than black-owned properties. An indoor toilet or bathtub, one observer noted, is "entirely out of the question" for minorities. In some of the more impoverished areas, mainly along Waller Creek and in the East End, "practically all facilities for ordinary cleanliness are lacking."[21]

Commercial life was largely segregated as well, ensuring that at least a small black business and professional class catered to the needs of the entire black community and increasingly to Mexicans. The East End, and East Sixth Street on both sides of East Avenue in particular, also made up the heart of Austin's African American commercial, professional, and entertainment districts. While African Americans might be able to shop for daily necessities at a small store in their neighborhood, most would need to go to East Sixth for the majority of their goods and services because they could not shop in white-owned stores downtown. Black-owned grocery stores, hardware stores, dry goods and clothing stores, restaurants, saloons, pool halls, and other shops served black clientele, and sometimes Mexicans and even whites, in and around East Sixth. The vast majority of professional services for African Americans were located in the vicinity of Sixth Street as well. Doctors, dentists, and lawyers made up a significant portion of the African American professional class, along with skilled businesspeople who provided services to African Americans that whites would not: funeral home directors, undertakers, teachers, and pastors. Easy access to the African American commercial district likely drew African Americans to residences on the Eastside as well.[22]

Schools and churches were centers of culture, education, and civic pride for Austin's African American community, and most were again located in the city's East End or in Clarksville. Segregation ensured a high degree of cultural autonomy for minorities. Educational facilities for African Americans in Austin paled in comparison to those provided for whites, but relative to the rest of Texas and the South they were robust. The East End housed two African American colleges, Samuel Huston College and Tillotson College, which provided higher education as well as basic primary and secondary education to the community. Each Westside African American neighborhood housed a basic grammar school; the Eastside contained three more. Churches functioned as the centers of cultural and political life for African Americans and Mexicans, but they also reflected rigid social segregation. The Ebenezer Baptist Church, organized in 1875, was built on the corner of Tenth and San Marcos, adjacent to the East End commercial district on Eleventh Street, in 1885. In Clarksville, the Sweet Home Baptist Church served as the cultural and religious center of the neighborhood from the 1870s onward. Three-quarters of the thirty-two African American churches in Austin were located east of East Avenue by 1929; the rest were located in other black neighborhoods. Five of the six Mexican churches were located in Mexican East Austin, while the sixth was in Little Mexico, a red light district that contained the most concentrated and haphazard neighborhood, near the mouth of Shoal Creek. African Americans also published newspapers that informed citizens of local activities and kept them abreast of news in minority communities. For both African Americans and Mexicans, the calendar was marked with festivals that celebrated cultural heritage. For African Americans, Juneteenth memorialized emancipation and routinely drew thousands of patrons to multiple parties in black neighborhoods. Mexicans celebrated Diez y Seis de Septiembre and Cinco de Mayo in segregated locations as well.[23]

In 1900 native whites, immigrants from Europe, and African Americans made up over 99 percent of Austin's population and roughly 98 percent of Travis County's population.[24] Racial relations thus were similar to those in the South, where one was either white or black, though white ethnics were often perceived as a separate class. But from 1910 into the 1920s, changes in the local economy transformed Austin's ethnic landscape and labor market to include permanent residents from Mexico, whose presence altered the way that Anglos imagined race by the 1920s (discussed in chapter 3). Mexican workers had long been a component of the agricultural economy in Central Texas as migrant laborers, and a small colony of permanent residents had existed in Austin since 1880. But their presence was minimal, as noted above. Two major factors brought Mexicans to

Austin beginning in 1910. First, the Mexican Revolution pushed thousands out of Mexico and into Texas and created migration networks that attracted other potential migrants. Second, the increased availability of railroads in Central Texas after the turn of the century had a significant impact on the region's mode of production. Access to distant markets caused economies of scale to become more pronounced, and larger farms quickly undersold smaller family farms. This process forced former white landowners or their children into cities and towns and created more demand for low-skilled agricultural workers, positions that whites were increasingly unwilling to take. Labor agents began recruiting Mexican workers to fill the void, and, increasingly, white landowners began to favor Mexican laborers over both white and black workers because of their willingness to work for less money and because of other racial stereotypes. By 1920 2.5 percent of Austin residents and 4.5 percent of Travis County residents were of Mexican origin. In the city, the vast majority of Mexicans lived in a district that was formerly the site of the Austin penitentiary and that by the late nineteenth century had industrial uses and also served as the city's main red light area. As more Mexicans settled there it became known as Little Mexico.[25]

In the two decades before World War I Austin's social landscape and division of labor were similar to those in many southern towns, with the exception of a higher level of education and white-collar work among Anglos and the presence of a small but growing Mexican American community. Whites tended to work in white-collar professions or in skilled trades, but many worked as semiskilled laborers as well. While whites' homes were not uniformly of high quality, most white neighborhoods had at worst a mix of dilapidated and standard housing. African Americans occupied a distinct space in the socioeconomic hierarchy and were considered lower yet indispensable by whites because of the labor they provided as domestics and as semiskilled and unskilled production workers. A small percentage of African Americans owned businesses or worked as professionals who catered to the African American community. They lived scattered in pockets throughout the city, often in relatively close proximity to whites, and presented little threat to the engrained social order. The vast majority of their houses were of substandard structure and none had access to Austin's limited water or sewerage lines. The principal way they were segregated was into areas that were more susceptible to environmental hazards, diseases and other health problems, and solid waste exposure.

The Urban Environment: Race, Refuse, and Reform in the Proto-City

Throughout the late nineteenth century, a number of engineering, planning, and health advancements swept across Europe and North America in response to new urban problems that arose due to spectacular industrialization and urbanization. During the antebellum period the vast majority of Americans lived in rural areas or small towns. Cities in the United States were smaller and less dense than those of Europe, and although they were certainly subject to epidemics, uncleanliness, and other health issues, abundant space and natural resources mitigated environmental problems. Some cities were kept cleaner than others, but in almost all cases cleanliness was seen as a personal, not a civic, issue. Garbage and animals in the streets were accepted parts of daily life for most residents, and even when cleanliness laws were passed they were usually not enforced. Only the largest and most crowded American cities, New York, Boston, and to a lesser extent Philadelphia, had established boards of health or regulated the disposal of refuse on the eve of the Civil War.[26]

These largest cities experienced massive industrialization and urbanization in the years following the Civil War. The highly concentrated city became the dominant urban form between 1870 and 1920. While economic growth was often intense and astonishing amounts of capital were generated, a number of new and troubling environmental issues emerged. Smoke from large, spatially concentrated factories fouled the air, dimmed natural light, and choked pedestrians. Coal was used to heat most residences, producing a dirty haze and a black film that coated most city buildings. Residents were forced to endure noises from industry and from increasingly crowded streets filled with horsecars, vendors, and the crush of people. Water supplies were increasingly tainted by the growing amount of refuse produced by both poorer workers and the burgeoning middle class, whose desire and ability to consume was increasing rapidly due to a growing economy facilitated by mass production and scientific management principles. Massive immigration and attendant demographic growth and population density exacerbated the problem and forced city leaders to create a new strategy to better handle the garbage piling up in streets and alleys. Over thirty million people immigrated to the United States between 1850 and 1920, and many settled near entry-level industrial jobs in urban factories. Internal rural-to-urban migration was also robust; perhaps half as many migrants came to cities from the American countryside as from Europe.

Because cleanliness was seen largely as an individual issue, those living in the dirtiest, most dire situations, largely immigrants and the poor, were usually viewed as the cause of the new unsanitary conditions.[27]

The intellectual response to these growing urban problems took many forms and came from many disciplines. The disparate yet often socially and professionally connected group of urban reformers shared the notion that humans would be better off living in cleaner surroundings with access to clean spaces and fresh air. They sought to reimagine sanitation and public health as collective environmental problems that necessitated collective action to combat, and they aimed to mitigate the environmental risks associated with densely populated communities. This meant harnessing the capital and the knowledge necessary to remake the environment and create a discourse that emphasized collective action. In their worldview, human morality could be improved simply by improving the environment humans lived and worked in. They also shared the belief that society's leaders had the responsibility to commit to improvements and to act as role models for the lower classes, who were often too ignorant and enmeshed in preindustrial habits to care for themselves.[28]

Engineers, a newly minted professional class in the late nineteenth century, sought to improve urban sanitation, transportation, and access to water resources through scientific design. The first major engineering innovations of the nineteenth century were municipal sewers, which took waste and wastewater away from heavily populated areas. George E. Waring, the most influential sanitation engineer and urban public health advocate of the late nineteenth century, was one of the first people to articulate the relationship between health and collective action guided by society's betters, the political, economic, technical elite, whose infrastructural improvements and behavior would have a positive influence on the lower classes. He reorganized New York's various sanitation-related departments, attempted to institute a citywide trash collection program, and increased the pay for municipal workers, along with writing dozens of reform tracts aimed at the wealthy. By the early twentieth century, professional sanitary engineers were perhaps the most important figures in regulating refuse and cleanliness in industrial cities. They had invented scientific methods for measuring, collecting, and disposing of garbage and had convinced nearly all urban politicians that environmental problems were best met with collective and professional force. Every major city had a type of public health department, a streets and sanitation department, and a sewerage system by

World War I. By that time most urban residents in industrial cities had access to running water in their homes as well.[29]

A second type of progressive reformer focused on the built environment and the social upheavals brought about by immigration as a primary cause of misery, especially for the rapidly increasing throngs of America's urban poor and laboring classes, most of whom were newly arrived immigrants. Jacob Riis, an immigrant himself, used photography to document the deplorable conditions in New York's immigrant slums and overcrowded tenement houses and called on politicians to upgrade building codes to uplift the masses in *How the Other Half Lives*. Without a suitable environment, he argued, immigrants would likely never assimilate or become productive members of society. Muckraking journalist Upton Sinclair also portrayed the relationship between poverty and environment in *The Jungle*, graphically depicting the inhumane working conditions in Chicago's stockyards and the morally enervating culture that many immigrants and workers succumbed to. Jane Addams, founder of Hull House in Chicago, likewise sought to provide basic accommodations, improved living conditions, and education to the urban underclasses, who she thought were tied to their physical and moral environments. All three spearheaded new progressive regulations that became the rationale for more systematic studies of urban poverty and health that improved conditions in homes and workplaces.[30]

Perhaps the most famous environmental reformers of the late nineteenth and early twentieth centuries were the new city planners and landscape architects who believed urban space should provide health and mental benefits and also reflect participatory American civic culture. While civil engineers concentrated on the physical aspects of cities, such as street patterns and graded roads, and oftentimes moved buildings, planners and landscape architects conceived of rationalizing and improving urban space to fulfill their social visions. Commerce, for example, was often cast as a necessary evil, a fact of urban life that needed to be balanced with spaces designed to mitigate its negative effects.

The most prominent was Frederick Law Olmsted, who designed hundreds of public parks, college campuses, and suburbs over his fifty-year career. A close friend and mentor to Waring, Olmsted shared the belief that collective action, led by professionals, could make life better for all urbanites, and he believed that public open space was a comprehensive solution to the crowded, dirty cities that he loathed. Green space could combat every urban ill. For physical health, trees cleansed the foul air of the city and open space provided respite from the clogged streets, cramped apartment houses, and huge industrial facilities that increasingly dominated urban landscapes. Fields encouraged park goers to engage in beneficial physical activity. For mental health they provided a quiet sanctuary away from the bustle of city life where people could contemplate less mundane ideas and gave people the chance to enjoy music and culture among other citizens. And for emotional health, they encouraged *civitas*, for city streets were places of rudeness, anonymity, and alienation, where urban dwellers "every day of their lives . . . have seen thousands of their fellow-men, have met them face to face . . . and yet have had no experience of anything in common with them." For Olmsted, parks gave people the opportunity to be human again. Perhaps most importantly, parks were the physical manifestation of democratic recreation, places where all were fundamentally equal as citizens but also where newcomers would learn how to be more American.[31]

While Olmstead's utopian rhetoric may have been a bit optimistic, open space proved wildly popular in cities in the early twentieth century, and his views about its value influenced those of other city planners. Daniel Burnham helped to reinvent Chicago in 1909 by integrating a series of public parks using a boulevard system. *The Plan for Chicago* was the first comprehensive project designed to control urban growth in an American city, and it drew on Olmsted's visions of democratic public space. The neighborhood-based park system, which was unlike Olmsted's famous Central Park in New York, was designed to give everyone in Chicago easy access to the advantages provided by open space; he wanted every citizen to be within walking distance of a public park. Ebenezer Howard drew on Olmsted in his creation of a utopian city in his *Garden Cities of To-morrow*, where he envisioned a comprehensively planned city perfectly integrated with nature, which he called the "Town-Country." In his design everyone had access to open space for recreation and for gardens, which would assuage the pressures and mitigate the deleterious effects of life in the industrial city. Though utopian, a number of new settlements that followed his recommendations were built between 1900 and 1930.[32]

In the first decade of the twentieth century, then, a group of professionals from diverse backgrounds, such as landscape architecture, engineering, journalism, and social work, formed a loosely based coalition that saw improved urban environments as intrinsically linked to healthy, vibrant cities. Political and business leaders in large industrial cities turned to their professional expertise in the hopes of improving their cities by giving citizens access to beautiful and serene open spaces and buildings, by creating decent work and living places, and by integrating nature back into cities. While professionals around the country had access to the base of knowledge created by this cohort of urban thinkers, improvements were rarely employed outside the large cities in the Northeast and the Midwest. Progressive urban reform was linked predominantly to industrialization and immigration, which were dramatically reshaping the physical layout of cities as well as their social organization. In the South, urbanization was in general a much slower process owing to dramatically lower levels of industrialization and immigration.[33] Leaders in many northern cities were likewise coming to see immigrants as potential white Americans who could certainly provide necessary labor power and, if assimilated, could contribute to civic improvement. Such was not the case with African Americans in Austin. City leaders in Austin and elsewhere did not perceive the need to enact sweeping changes to clean up their cities; advocates of reform were often businesspeople attempting to improve real estate values or infrastructure for profit. As a result, planning and infrastructural upgrades lagged far behind, and few environmental improvements (outside the attempts to dam the Colorado) were made through 1910.

Yet even in Austin a small number of academics familiar with urban reform were realizing that a small, nonindustrial city could likely have major problems generated by a lack of infrastructure and services. In 1913 University of Texas sociologist William B. Hamilton published a report titled *A Social Survey of Austin, Texas*, after studying social and economic conditions in the city. Writing as a reformer and social scientist, Hamilton had much contemporary literature to draw on as Progressives were beginning to link urban environmental problems with urban social problems in cities around the country.[34] He was interested in the health and well-being of the community and used his report to advocate for civic improvements, such as modern sewers and water delivery systems, and siting decisions that would produce a cleaner, healthier, more modern city. Like many academics of his day, he argued that poor environmental conditions fostered poor social conditions and that it was the duty of Austin's better-off to engender changes that would improve the community. Waste disposal and cleanliness were thus more than personal issues to be taken care of by each household. Hamilton envisioned the city as a collective whole that needed to be improved publicly for the benefit of all citizens. Yet he also remained tied to the idea that the failures in public health were moral failings on the part of the city's lower-class citizens, who needed to be improved by their betters and by an augmented environment. Like education, infrastructure and public health programs could have a civilizing influence on Austin's lower classes and racial minorities and make the city a better place. Much of his report attempted to alert the city's middle- and upper-class whites to the conditions and activities of the poor, and especially minorities, while at the same time frighten them into action.[35]

Although the city had begun the process of some basic infrastructural improvements, such as paving major streets, installing sewers, and providing public lighting, its health and sanitary conditions remained deplorable even by the standard of southern cities. Hamilton spent months observing environmental, sanitary, and health conditions in all sections of the city. His descriptions, however, also provided a valuable portrait of inequality and environmental injustice in 1910s Austin. His report was followed up four years later by another similarly titled study that shows how reticent Austin's leaders were to institute improvements that would benefit minorities and impoverished whites in the years before 1920. In the first two decades of the twentieth century, urban Austin's infrastructure and development of the natural environment left a great deal to be desired, and minorities were often prone to the lack of development as well as to both obvious and subtle environmental racism. Although not strictly segregated, minorities were disproportionally subject to health hazards, pollution, and the effects of waste disposal, conditions that white people rarely had to endure. Before urban space was organized and minorities were segregated with dirty but necessary urban functions and environmental peril, those functions and hazards still affected life profoundly for minority residents even before large-scale industrial production became commonplace in Austin.[36]

Hamilton noted the relationship between water and disease as a primary problem for all Austinites in 1913. Because the City of Austin did not deliver water outside its municipal jurisdiction (or to many areas within the municipality), much of the developed region did not receive water and people had to dig wells without engineering knowledge, leading to stagnating water sources. Even in better-off areas, unsanitary conditions due to the amount of privy vaults were common. Typhoid epidemics traced to infected water from private wells hit the Anglo middle class in Hyde Park and a poorer white South Austin neighborhood in 1912. Hamilton estimated that 35 percent of Hyde Park residents had privy vaults that ran into water sources during storms. About five hundred open cisterns spoiled the Austin landscape and emanated noxious odors across the city, especially in the hot, dry summer months. The city had no regulations regarding manure handling, and Hamilton found that alleys throughout the city were particularly subject to piles of rotting manure and the insects that lived on them and often carried disease. Throughout the area east of East Avenue, a heavily black neighborhood that was poorly graded and lacked access to sewers, Hamilton observed “large cesspools” where human waste and other garbage collected openly. All wells dug in East Austin, which also lacked access to municipal water sources, risked being inundated with waste after rainstorms. Less than half the population had access to the city’s sewer system despite the miles of new sewer lines built since the turn of the century. The subsequent study in 1917 found that few improvements were made; pumps used to draw water from the river were placed in areas where water was stagnant, and as a result drinking water was filled with bacteria.[37]

Life along Austin’s creeks could often be miserable and unhealthy, and during floods areas adjacent to the creeks were the most dangerous in the entire city. By the 1910s both blacks and Mexicans were much more likely to live near Waller and Shoal Creeks, often because the creeks provided a ready-made sewerage system that carried off waste and garbage free of charge. The land on unimproved creeks was also extremely cheap or free because it was so vulnerable to flooding and offered no municipal services. Many of the shacks along the creeks backed right up to them to make waste disposal an easy task. Hamilton referred to these houses as some of the worst in the entire city, and he described the soil in the black and Mexican neighborhoods around Waller Creek and in Little Mexico as “polluted,” “reeking with hookworms,” and “in its present state unfit for human habitation.” He called Waller Creek “a large open sewer” that brought all sorts of refuse from the university through downtown Austin before discharging it into the Colorado River. In the 106 houses along the roughly two miles of Waller Creek south of Nineteenth Street, there were 122 privies that emptied directly into the creek along with twenty-three private drains. Both creeks discharged into the river near heavy concentrations of minorities, blacks around Waller Creek and Mexicans in highly concentrated Little Mexico at the mouth of Shoal Creek. At Shoal and Waller Creeks, the Colorado was a popular spot for fishing among minority residents, even though they were near untreated sewage. Hamilton reported that fish caught along the Colorado were often sold at the markets on East Sixth Street.[38]

Flooding on Waller and Shoal Creeks could be even more deadly than in the Colorado River basin because they had far less capacity to carry water. Thus some of the biggest catastrophes in Austin history are due to creek flooding. One such flood devastated Austin in April 1915, shortly after Hamilton’s study was completed. On the night of April 22, a stalled thunderstorm dropped upwards of eight inches of rain around the area in four hours, with some localized reports of over ten inches of rain in less than three hours. The flood, the worst in the city's history according to the *Austin American*, carried away houses and even streetcars along with thirty-two bodies. Almost one thousand people were left homeless, roughly 4 percent of Austin's entire population. The vast majority were poor people who lived along the creek. The Austin chief of police was straightforward about what type of residents were most affected by the flooding: "If the wealthy people of Austin could have seen what I saw last night . . . they would liberally subscribe for the relief of the people" because of the generosity and mutual support demonstrated by the "really poor" who suffered the most. Large floods, however, were only the most spectacular and harrowing of the creek hazards. Dozens more died through the late nineteenth and early twentieth centuries in smaller floods, and property was often damaged as well.[39]

Perhaps nothing illustrates white attitudes about the relationship between race and environmental hazards more than the use of garbage and the siting of garbage dumps in Austin. Because the city had no zoning laws, the removal and placement of the city's trash was unregulated and thus reflected power relationships. Both municipal authorities and private garbage haulers picked up and disposed of trash and also determined where the garbage would be dumped. Hamilton considered the system archaic even for a southern city, claiming that "Austin is in the Dark Ages of civic sanitation. In no one thing is the city farther behind than in the disposal of garbage and waste."[40] While the wealthier whites with more money to spend produced far more garbage than minorities did, it was the minorities who bore the physical and psychological brunt of Austin's waste disposal system. Minorities had to endure trash in their neighborhoods and also had to rely on trash for sustenance and the necessary goods they often could not afford. They were likewise often demonized for their substandard living conditions and lack of cleanliness even though most of the refuse that wound up in their communities was not theirs.[41]

The casual disposal of refuse illuminates the dire conditions that most minorities were subject to in 1913. Much of the garbage produced by businesses never made it to the haulers because it was picked through by poor residents to feed animals or even themselves. "Slop wagons" owned by whites picked up garbage from residences and businesses and sold usable materials to other residents, usually minorities. Many poor minority residents raised chickens, pigs, and goats, and collecting food scraps was a cheap way to feed them. Hamilton reported that "negroes" picked up garbage from restaurants, hotels, and residences to feed hogs. He likewise described an incident of a Mexican family picking through trash and consuming food waste on the spot. These practices were common to minorities, who even the progressive Hamilton likened to animals. "Just as the chickens follow the farmer's plow to pick up fresh earthworms, so do Mexicans, negroes, and poor whites follow the city wagons to the dumps to pick out rags, boxes, and decaying food." The practice, he argued, "has undoubtedly caused much disease among a class which is very hard to reach." By including only poor whites but all minorities, Hamilton makes plain how the use of garbage reflected the relationship between race and economic class in Austin. He recommended that the city outlaw the removal of anything from dumps, failing to see that eating garbage was likely a desperate act caused by severe poverty.[42]

Food processing spaces likewise produced garbage and other unsanitary conditions that Hamilton argued were largely used by minorities and hence minority problems. The city's slaughterhouses in particular drew his ire; he claimed that the conditions inside were "the dirtiest in Texas," and for this he blamed the black workers' "sloppy methods." Black and Mexican workers were also chastised for taking offal home to feed hogs and for "carrying these [trimmings] back to their shanties, where they are used in making soup." Bakeries were also oftentimes unsanitary to Hamilton. He found that all but five of the city's commercial bakeries were located in residences and that in four the baking was done in the same room where people slept. Often food made in minority residences was seen as a health hazard for whites. Candy and tamales, which were produced by Mexicans but largely consumed by whites, were likewise made and sold out of shanties that Hamilton found crowded and dirty. A message was clearly being sent to Austin's Anglos: regulate industries where minorities produce food, because you and your families are consuming it.[43] Hamilton used the same argument for why whites needed to regulate cleanliness when he was discussing minorities' high levels of tuberculosis. "Who is surprised at the moral and physical inefficiency of the 'modern negro,'" he wrote, "when from early childhood to old age they live under such bad housing conditions?" But the real problem was their contact with the whites who employed them as "nurses, cooks, and launderers."[44]

The siting of garbage dumps and the collection and disposal practices of both private and municipal trash haulers had devastating effects for minority communities in early twentieth-century Austin. Both private haulers and the city's collectors used a large city dump that was located "almost in the very center of Wheatsville," one of Austin's largest African American neighborhoods. Hamilton observed that public and private haulers were not always careful that the trash made its way to the actual dump; many would dump their wagons on streets or alleys and in residents' yards to save time. Manure, produced by horses at an individual rate of thirteen to thirty-five pounds daily, also fell from carts and from horses on the way to the dump. The psychological effects were obviously negative. Blacks residents rarely complained, Hamilton believed, out of fear that the city would force them to remove their hog pens from their property. Hamilton also speculated that after years of enduring unsanitary conditions at the hands of the city, many black residents took garbage-filled streets as "a matter of course." While haulers would bring trash from around the city to the Wheatsville dump, they refused to take trash out of the neighborhood or from residences to the dump. "Why should [residents] collect it," Hamilton argued, "when the city wagons rarely, if ever, cover this section?" Because the city wagons only went to Wheatsville to dump trash, the refuse had "almost turned the streets and alleys of these people into a dumping ground." The location of the dump also caused negative ecological impacts that affected African American residents disproportionately. The dump was sited adjacent to a draw that took refuse into a nearby ravine during storms.[45]

The dump in Wheatsville and another smaller trash collection area farther south on Shoal Creek were both located near African American schools and caused health problems for students. The Wheatsville School's outdoor grounds, located less than a block from the city dump, served both as a recreation area for students and as a community gathering space. The garbage from wagons often fell or was dumped directly into the school yard by drivers. The ground, Hamilton claimed, was infested with hookworms. Another ward school sat on a small bluff where a stream entered Waller Creek at Ninth Street amid open privies that drained into it and produced "foul air."[46]

Other dumps were found in pockets throughout the city, and most if not all that were located among residences were in minority communities. The largest Mexican neighborhood, Little Mexico, near the mouth of Shoal Creek and at the end of the Missouri Pacific Railroad, contained over half of Austin's Mexican population in 1910 and was one of the densest and least developed areas in the city. Along with the city's red light district, where ten houses of prostitution and five saloons served men of all colors and backgrounds, Little Mexico contained the city's second-largest dump, just above the mouth of the creek where haulers and locals alike dumped trash. Hamilton was quick to write this off, opining that "it is true that a national trait of the Mexican is to live in shanties surrounded by rubbish and tin cans."[47] But he also argued that improved living conditions would lead to improved morals among the city's minority populations. Unfortunately, surveys conducted four years later in 1917 found that waste collection was still highly uneven and many minority areas were still neglected, some to the point where collection was still nonexistent.[48]

Minorities also suffered from health problems associated with unsanitary conditions and malnutrition far more than white Austinites did. Tuberculosis was a particularly serious problem in Austin owing to its relationship to overcrowding, malnourishment, and poverty. In 1916 Austin's rate of tuberculosis deaths was twice that of the United States, with ninety-five deaths that year (close to 3 percent of the entire population). Due to a lack of drainage and sewers, Austin's rate of typhoid fever was triple the national average as well. African Americans, who were poorly nourished, confined to dirtier and more congested neighborhoods, and often lacked access to professional health care, were far more likely to suffer or die from communicable diseases. Blacks had a 50 percent higher mortality rate from tuberculosis than whites did. Conditions had particularly deleterious effects on black infants, whose mortality rate was 300 percent higher than that of white infants.[49]

Through the first two decades of the twentieth century, Austin's minorities were subject to both environmental and man-made hazards that stemmed from white discrimination as well as from poverty and the lack of better housing choices. Newer subdivisions with professionally built homes employed racially restrictive covenants to keep the few African Americans who could afford such conveniences out.[50] Socially, whites generally did not feel threatened by blacks because of existing social customs and Jim Crow laws, yet it was also extremely rare to find whites living in the same area as blacks or Mexicans, except for the still common practice of female domestics living on their employer's property. But by the 1910s there were some signs that whites were aware of black mobility and increasingly perceived space as a racial battleground. In 1915 politician Henry Haynes warned residents that African American encroachment in East Austin had displaced his family and that, left unchecked, the same plight was possible for neighborhoods in West Austin. He led an unsuccessful campaign to stop the construction of an African American school in East Austin the following year.[51]

So when the Commercial Club wrote that in Austin "there is a distinct color line, mutually conceded" in an 1891 booster publication, the idea was likely not one based in geography or entirely mutually conceded by both races.[52] Instead, it was the product of unequal social and economic relations and racially discriminatory customs in Austin, which reflected an obvious and ubiquitous possessive investment in whiteness that largely determined economic and social possibilities and defined the health and quality of life for Austin residents. While race was increasingly perceived spatially, especially after 1910, the city lacked a formal mechanism to separate races as well as a plan to improve the city by making it a healthier and more modern place to live. Increasingly Austin's leaders agreed with Hamilton, that creating a better city necessitated both civic vision and professional urban planning and engineering. Yet improving the city did not mean improving it for all residents equally or taking specific measures to eradicate the social problems or the glaring inequities that existed. Instead, urban improvements were increasingly defined by removing minorities from improved spaces. By the end of World War I Austin's business and political leaders, known as Business Progressives, reacted to changes in racial ideology as well as to changes in urban planning ideology in an effort to modernize and beautify Austin. By the 1920s their vision for an improved city linked environmental improvements with a new racial geography and set in motion the policies that would define Austin's socioeconomic and racial landscape for the rest of the twentieth century.

3 A Mecca for the Cultivated and Wealthy

Progressivism, Race, and Geography after World War I

> One of our visitors some years ago paid us the doubtful compliment of saying that Austin was unique in being "the city that God has done the most for and man less for than any other city in the world."
>
> —President of the Austin Exchange Club, 1926

From 1918 to the 1930s, Austin underwent significant social changes that led to the radical reorganization of urban space. City leaders created a land use policy that simultaneously created rigid racial segregation and funded improvements that would control and augment the natural landscape, making it more usable for residents and more attractive for economic growth. The Austin City Plan of 1928, the city's first official planning document, used contemporary urban planning ideas, such as incorporating green space, improving sanitation and waste disposal, and implementing land use zoning, to order spatial production in a way that reflected progressive attitudes of the day. Despite strong anti–federal government rhetoric, by the 1920s Austin's politicians and business community saw planning as a way to create a more desirable, livable city. Because of long-standing beliefs about race, they saw no contradiction in improving the city while simultaneously segregating it; in fact, segregating it was perhaps the most important element of improving it. Ghettoization of minorities was simply the most obvious way to allow Anglo residents to fully benefit from urban improvements and ensure a place to sequester Austin's urban elements and unskilled labor force, including many unskilled whites. So to understand Austin's spatial changes, it is essential to understand the social and economic changes that transformed Austin from 1900 to 1930 as well as the ideas that Austin's business leaders, known as the "Business Progressives," employed to transform the city.

The Austin City Plan of 1928, as well as zoning and other formal restrictions on land use in Progressive Era cities, embodied the idea that the physical design of the city and its attractiveness reflected its moral order and civic virtue.[1] Yet race and class relations were also reflected in planning, indicating that the urban moral order was strongly related to erasing or removing undesirable elements to make cities more usable and more beautiful.[2] Planners used zoning, the story goes, to encourage people to move from one area to another, to increase real estate values in one place and diminish them in another, and to make existing land uses legal or illegal. The way that zoning was implemented reflects power relationships and often racism. There is a great deal of truth to this argument, but often the role of environmental improvements in augmenting socio-spatial change is overlooked in discussions of Progressive Era planning. In Austin, urban planning and zoning were simultaneously used to improve natural spaces and encourage minorities to move out of those improved spaces. Indeed, businesspeople in Austin thought that enhancing the city's natural beauty should be the main goal of the city plan. Segregation was also a main goal, albeit one that was kept largely out of the city's newspapers and public record; yet as Eliot M. Tretter and M. Anwar Sounny-Slitine have demonstrated, private market forces and racially restrictive covenants were already rapidly solidifying residential segregation in Austin in the 1920s.[3] As elsewhere, restrictive covenants were widely viewed as essential to securing high property values in subdivisions; by the 1930s the federal government was actively promoting restrictive covenants as a stabilizing influence in the housing market.

Understood in this framework, whiteness became an asset that increased the monetary value of real estate, similar to infrastructural and natural improvements and the ordering of spatial production via zoning and professional land use planning. Proponents of racial segregation in Austin argued that real estate appreciation was perhaps the most important economic benefit of segregation.[4] I argue that this functions in a similar way that infrastructural and natural improvements do. In Texas, private restrictive covenants could not be voided by public zoning measures. Also, because racial zoning had been repeatedly outlawed by a number of federal courts beginning in 1917, planners needed to figure out a way to circumvent these laws. They devised a plan to institutionalize segregation using service delivery and implemented a zoning code that allowed for nonrestrictive land uses in exclusively minority locations. The Home Owners' Loan Corporation institutionalized racial difference in real estate markets by creating redlining maps of Austin (and 234 other cities), which made attaining loans almost impossible for minorities in East Austin. Thus by the 1930s and especially by the 1940s, Austin was heavily segregated by both race and land use. Minority residential and social spaces were not only segregated from white residential and social spaces, but they were also located in close proximity to, and often intermixed with, industry and other environmental hazards and pollution. While their areas were sometimes improved, they lagged far behind white areas of the city in terms of access to attractive natural spaces as well as basic infrastructure and other amenities that the city began providing in the 1920s. While the normal pattern of white suburbanization became common after World War II, Austin's changing geography after World War I demonstrates how the removal of urban elements from white residential areas and the simultaneous improvement of those white areas also radically altered urban geography.

World War I and the New Urban Landscape

If Hamilton's study of Austin reflected the Progressives' good intentions inflected with contemporary ideologies of racial superiority and paternalism, Austin's postwar conceptions of race and urban development took a turn toward a brutal retrenchment of the color line in response to a perceived African American threat and a sharp increase in the size of the Latino population in Austin. Race and segregation became pressing concerns for Austin's Anglo community in the immediate aftermath of World War I, and in the 1920s the relationship between race, geography, urban planning, and nonindustrial growth became most visible. World War I had dramatic effects on American urban life. As immigration from Europe was curtailed, production increased, and men were pulled out of the labor force to join the military, industrialists faced severe labor shortages. For the first time they looked to African American labor as a viable option, and roughly five hundred thousand African Americans migrated from the rural South to midwestern and eastern cities during the First Great Migration. Many more made the shorter journey to southern cities, including Houston and Atlanta. From 1916 to 1920, Austin, a city with almost no heavy industry, lost roughly 20 percent of its African American labor force, mostly due to relocation for jobs. Northern labor agents were in Austin recruiting blacks in the summer of 1917, and African American newspapers reported that many were leaving for Houston of their own accord after word arrived that industrial jobs were available.[5]

In part because of the loss of African American laborers, Austin's business community began actively recruiting Mexicans to the labor force during World War I, a shift that had a dramatic effect on the city's demographic makeup and racial outlook in the 1920s. While Mexicans had been performing agricultural labor in Central Texas for decades, before 1910 most were

Pemberton Heights, a new West Austin neighborhood, in 1938. Unlike many Depression-era cities, Austin thrived in the 1930s due to environmental improvements, urban planning, and a strong relationship with the New Deal. PICA 02562, Austin History Center, Austin Public Library.

migrants who stayed only for brief periods and were thus not viewed as permanent components of Austin's social structure. Jason McDonald convincingly argues that up until the 1910s Mexicans were viewed by Anglo Austinites as lower-class whites, not members of a separate racial group but as foreign national guest workers who were marginally white. A small number of wealthier and more educated Mexicans had access to Anglo status and also organized and sponsored cultural institutions for Mexicans in Austin, including newspapers and festivals. Their presence in Austin was minimal before the war; Mexicans made up less than 2 percent of the city's total population in 1910, even after political upheavals in Mexico forced numerous migrants into Texas. During the war, shortages in agricultural labor led the Austin Chamber of Commerce to request special permission from the federal government to import laborers from Mexico. The permanent population of Mexicans living in Austin increased dramatically to at least 6 percent by 1930.[6]

The end of World War I also had dramatic and devastating consequences for race relations and social stability in American cities. White soldiers returning from the war found glutted labor markets in cities as production slowed and the removal of wartime price controls sent inflation spiraling. Unemployment spiked. During the war urban geography had changed to accommodate much larger African American populations; many of the newly arrived blacks expanded traditional African American boundaries into adjacent neighborhoods that were often immigrant ethnic enclaves. Servicemen and members of white working-class communities, many themselves first- or second-generation immigrants, lashed out. African Americans, many of whom served in the war and felt a new sense of justification for equality, fought back in many cities. Xenophobia and anti-immigrant sentiment increased as anti-Bolshevik hysteria became common after Communists took over Russia. In 1919 urban African American communities felt the greatest brunt of this social turmoil. Race riots erupted in thirty-eight U.S. cities during what writer James Weldon Johnson dubbed the "Red Summer."[7]

While the most notable disturbances occurred in larger cities, Texas suffered riots in Longview, Port Arthur, and Texarkana that scared white leaders throughout the state. Fearful of the "New Negro" who was less willing to accept social and economic subservience, Southern leaders began to intensify race prejudice, and racial violence spiked throughout the nation. Despite intense support for the American cause during the war and a higher level of draft registration, many whites felt that blacks were more prone to communist agitation because of their low socioeconomic position and gullibility. In Austin as elsewhere, the Ku Klux Klan became more active and vociferous in maintaining traditional racial and ethnic relationships in the face of demographic change and social upheaval. Most notably, Austinites demonstrated their resistance to African Americans and to Northerners they perceived as integrationists or outside agitators. In response to the Longview riots and amid rumors that blacks were buying firearms in Austin, John R. Shillady, a National Association for the Advancement of Colored People (NAACP) representative who came to Austin to help organize a local branch of the organization, was severely beaten on a public street by three men, including a police constable and a judge, after he was released on charges of inciting racial violence. Shillady died shortly after leaving Austin, and the national press condemned the incident as an example of racial brutality and vigilante justice gone too far. Texas governor William P. Hobby responded by defending the beating and authorizing the suppression and harassment of the NAACP throughout the state. Austin's NAACP branch was disbanded, and at least one prominent member was forced to leave town by militant whites.[8]

Austin's growing Mexican population also felt the pressure of racial retrenchment and hostility during and after the war, and Mexicans' status as white foreign nationals was put in jeopardy by their increasing numbers and larger permanent enclaves in the city. During the war whites questioned the allegiance of Mexicans and saw them, as they did Austin's Germans, as dangerous foreign nationals and, increasingly, as racial others. As with blacks, whites cast Mexicans as more susceptible to bolshevism and other anti-American ideologies despite their being outwardly patriotic and amenable to military service during the war. School programs were designed to teach Mexican children English and "the fundamentals of American citizenship" rather than knowledge that would encourage social mobility, and educators often blamed them for not mixing well with white children.[9] While the business community sought their labor power, most argued that it was because Mexicans were more docile, worked for lower wages, and were more adapted than blacks or whites to living in deplorable conditions. Racial stereotypes also provided reasons for why whites thought that Mexicans should be segregated and why they increasingly occupied a social position between whites and blacks in Austin. On the eve of World War I, the predominant Latino neighborhood in the city was one of Austin's most dangerous, unhealthy, and derelict. The half-mile-square area contained ten houses of prostitution as well as the largest manufacturing agglomeration in the city and a large municipal garbage dump. While Latinos appear to have lived in conditions similar to those of African Americans, during the war their status also changed from foreign to domestic underclass. Anglo perceptions of Austin's racial hierarchy also changed from dyadic (whites and blacks) to triadic (whites, Mexicans, and blacks) between 1910 and the 1920s.[10]

The Spatial Is Social: Separating City and Garden

Partly in response to growing concerns over race and partly in response to calls for economic improvements, the Business Progressives, represented by the chamber of commerce, imagined urban planning as a tool that would simultaneously solve two problems. First, a city plan could incorporate progressive ideas about ordering and cleaning urban space and integrating natural space for the benefit of the community and especially for the city's economic growth. Second, planning could solve the city's mounting race problem by codifying segregation and removing minorities from white spaces, effectively finishing the social segregation that Jim Crow began. Both solutions, they argued, would make Austin a more desirable city for businesses and people of means to relocate to. In answering the first issue, they sought to raise capital to make improvements and use professional urban planning techniques and laws to produce the pastoral, residential space that they thought best reflected the city as "a mecca for the cultivated and the wealthy."[11] The second issue was more problematic but was equally as vital to Austin's growth in their minds. For as bad as an unordered, haphazard, and even environmentally dangerous city might be for growth, it paled in comparison to the prospect of a city with no obvious color line. It was in the radical reordering of urban space, a geographic fix to a social problem, where the Business Progressives found their most suitable answer.

Even before the 1920s, economic forces and social mores were slowly segregating blacks into the city's Eastside. Whites began to voice concerns about African American encroachment on their residential neighborhoods in the early part of the twentieth century; methods for maintaining segregation were both causal and legal. Property agents for both renters and buyers observed an informal color line, where integration was perceived as damaging to business. They almost always steered clients to sections populated by their same ethnicity. Most African Americans and Latinos who owned property built their own homes because it was difficult to acquire a loan. The rental market was similar. Rental units were only open to blacks in black areas. In the northern and western peripheries, new housing subdivisions, such as Hyde Park in the 1900s and University Park and Pemberton Heights in the 1920s, had racially restrictive covenants that forbade property sales to African Americans.

As race relations slowly declined and the spatial division of labor became more pronounced, fewer blacks worked as domestics for whites, and domestic workers were less likely to live on their employer's property. Social services were often at the heart of arguments about segregation. West Austin whites protested a proposal to build a new school in Clarksville in 1915 because they thought it would increase the African American presence there. In Wheatsville there was evidently no need for protest. The African American school's enrollment dropped from 177 in 1914 to 44 in 1924, in large part because university leaders were growing wary of the large African American area so close to their expanding student residences. Some African Americans left because they wanted their children to attend Anderson High School, the only one open to African Americans, located across town in East Austin. One family moved because their son was racially harassed walking across the University of Texas campus to get to school.[12] The section of Austin east of East Avenue was also undergoing slow demographic changes related to property values. One white resident complained at a city council meeting that he had been driven out of East Austin by minorities and that other whites had left because of fears over declining property values. While African Americans had long had a presence in East Austin, the growing Mexican community and the dwindling white population in East Austin in the 1920s indicate that the in-migration of African Americans may have caused a decline in property values there that made the area more accessible to poorer, often Mexican, households.[13]

By the 1920s, Austin elites had begun to expedite the process of spatial reorganization and consolidated political power among whites in the city using new legal instruments and contemporary urban planning philosophies. Legal changes during the Progressive Era dramatically increased the power and scope of municipal governments. Foremost among them was "home rule," an effort that sought to give municipalities more autonomy from states and that grew along with rapid urban expansion in the late nineteenth century. Home rule allowed localities to create charters and establish their own policy priorities and led to an increase in spending for services that middle-class urbanites were increasingly demanding.[14]

Home rule also encouraged experimentation with alternative forms of municipal government that proved especially attractive to cities in the Southwest. One of the most important changes was the turn in 1924 to a form of government with open seats on the city council, rather than geographic districts that each elected a representative. In Austin this process came in two steps. The ward system, a relic from the nineteenth century, was considered an offshoot of the machine politicians and pro-labor politics that dominated industrial cities in the late nineteenth century and, for many prominent Austinites, hindered growth because council members only looked out for their wards.[15] Walter E. Long, president of Austin's chamber of commerce and a growth advocate, was singularly concerned with abolishing the ward system in the early twentieth century and replacing it with a commission system, where a commission elected via citywide vote would have legislative and executive power over the city. Long considered the commission style more modern and more amenable to business interests, and the Commission Charter was adopted in 1909 despite protests that it was a less direct form of municipal government that would fail to be sufficiently democratic.[16]

Yet by 1917, he and other chamber members had come to also see the commission system as inefficient because power was decentralized and commissioners still catered to special interests. Specifically, the commission had repeatedly failed to pass a bond package to supplement and extend utility services. In the city manager form of government, which was gaining popularity in the early decades of the twentieth century, the city council was elected and then hired a city manager to carry out daily operations. Long argued that the city manager system would get more for the taxpayers' money and allow the "best businessmen in the city" to have more power over decision making.[17] Long was unabashed in his argument that city government should first and foremost provide for "a sound business community" that would then engender and manage economic growth.[18] His friend and Beaumont city manager, George Roark, likened the proposed government to a corporation, where the city manager form is absolutely the same as the operating plan of all large corporations, "which makes it unpolitical" and takes politics out of city government. In this system, Roark summarized, "the voters are stakeholders."[19]

But the city manager form of government had the added benefit of giving almost unlimited power to small groups that could mobilize interests across a city, which meant that business elites could essentially take over city government in a smaller city such as Austin. The debate over the change to city manager government was heated even after the city charter was amended to provide for it in 1924. Austin commissioners called the system undemocratic because the city manager was appointed and thus not open to recall, and they complained that the job was too much for one person. Commissioner Henry Haynes compared the power afforded citizens by the manager system to the power "the negroes have to say about the Democratic Party."[20] The *Austin Statesman*, the city's mouthpiece for the chamber of commerce, publicly described the battle over the change "as the most hectic in Austin in many years," and, privately, Long characterized opponents to the city manager government as unprogressive and self-interested people who cared more for their personal wealth than for the public good.[21] Yet in 1926, Austin became the first city in Texas to adopt the city manager system. Lost in the discussion, however, was the quiet erasure of geographic districts in electing city council members under the new system. Under the older systems, commissioners were elected from five districts throughout the city, where residents of each district voted for one representative. Under the new system, all council members were elected by citywide votes, so each resident voted in all five council races. While blacks were legally barred from voting in primaries, some could vote in general elections. The erasure of geography from voting was thus facilitated by the spatial segregation of minorities, who would not be able to elect a representative despite their geographic concentration.[22]

The Business Progressives also initiated the process to create the first Austin City Plan by forming a City Plan Commission in the 1920s, a chamber goal since 1915 and the institutionalization of the Anglo business community's vision for the city's future. For the Business Progressives, urban planning was a way to organize and improve space, but collecting the ideas generated by the entire business community made that planning truly progressive and, to Long and other prominent members, more democratic. Their attitudes reflect the very narrow conception of whose ideas counted in remaking the city, but they also provide a strikingly comprehensive and coherent record of the collective consciousness of the city's business class and the importance its members placed on finishing nature to improve the city. Like most booster publications from that era, theirs characterized Austin as a naturally endowed place that simply needed minor adjustments from man to reach its full potential. After hiring the Dallas firm Koch and Fowler to develop the plan, chamber president Long surveyed the group's 317 members in an effort to gain a comprehensive civic assessment of what the new plan should accomplish. The results, which Long cataloged and delivered to Koch and Fowler, provided the basis for Austin's first planning document.[23]

Most members of the chamber viewed infrastructural, natural, and social improvements as inexorably linked, and they considered beauty to be Austin's most important marketable asset. Like many Progressive Era urbanites, they usually saw open space as a valuable part of quality of life assessment and encouraged more orderly urban development, a separation of land uses, and better transportation options to get back to nature more often.[24] Industry was almost never discussed as a practical option for growth; only one of the 317 suggestions expressed a desire for industrial growth in Austin. The creation and extension of parks was seen as a far more effective means of improving the city and as a better growth magnet. Before the 1920s, Austin had almost no park space; as one letter writer complained, practically nothing "had ever been done here towards beautifying the city." Even a traveler passing through Austin on route from Los Angeles to Miami took the time to comment on Austin's unimproved beauty. "You have the most beautiful site for a city that I have seen between the coasts," he wrote, "but you are negligent in not beautifying and parking river banks and valleys." Only about five square blocks of municipal open space existed within the city, and four of those blocks were parkway along East Avenue.[25] An unpaved dirt road took tourists and leisure seekers to the secluded spots along the bluffs adjacent to the river just west of the city limits. Barton Springs, purchased by the city in 1917 and largely unimproved as well, was a popular spot for picnicking, swimming, and recreation and already a symbol of Austin's outdoor culture, but it was also difficult to access for residents who did not have a car. Businessman Andrew J. Zilker, who sold the springs and the parkland around them to the city, opined that "Austin can be made the most beautiful city in America" with improvements.[26]

Guidebooks and other booster literature also imagined picturesque landscapes augmented by infrastructural improvements as Austin's greatest potential asset, an advantage that set it apart from other Texas cities. Austin was seen as "one of the garden spots of Texas" in its natural beauty but was lacking in access to nature.[27] For many Austinites, access to nature was both a social and an economic benefit. The option to easily leave a city was imperative in creating a livable, vibrant place. A resident could quickly "seclude himself from the hum of busy streets" and easily access water for leisure activities.[28] Roads literally connected urbanites with access to nature and therefore better lives. Many boosters, encouraged by the success of Daniel Burnham's boulevard system in Chicago, envisioned a similar system for Austin. Cleaning and organizing the city, along with adding modern roads, would allow the natural world around Austin to shine. The city's "rugged hills and flowing river ask no aid from man," but the city itself needed "the vision of an artist in landscape" to design a plan that would complement the natural setting. Warmth and sunshine were considered important aspects of the environment and writers argued that Austin's dry, warm air was a "magnificent climatic advantage" that, when combined with water resources, provided the optimal environment for those suffering the maladies associated with northern winters and southern humidity.[29]

The most imaginative progressive travel literature linked natural beauty with tourism but also with civic democracy and cultural progressivism in an era when technological improvements to the natural landscape were often considered patriotic and civically virtuous. A system of paved scenic roads would bring "tourists here from the remotest countries to play and spend their money," as well as "open up a natural playground at our very door." But the civic spirit demonstrated by public improvements could be just as important. The University of Texas campus was central to this vision. Well manicured and designed to promote intellectual contemplation and thoughtful reflection, the campus was already the city's largest and best open space and a symbol of Austin's importance to the state of Texas. State money was used to beautify grounds on the campus. University Avenue and Lavaca Street were parked, paved, and lighted, for example, before 1913.[30] The problem, as the chamber of commerce saw it, was that "its beauty is lost unless it is part of a setting that is in itself beautiful." Like nature, the pristine campus grounds could only flourish alongside a city improved by a rational plan. Boosters imagined that natural beauty reflected social and cultural refinement and that improvements demonstrated the "highest citizenship" possibilities for residents. "Such things as [improvements]," one publication mused in 1922, "knit our hearts together in a common cause—natural beauty—but we must build with vision and generosity."[31] It is difficult to gauge how realistic such grandiose ideas were considered among Austinites, but, clearly, access to picturesque landscapes was an important aspect of positive urbanity for many people.

Accessible park space in the city served a variety of functions for chamber members, and their recommendations about park space varied. The most common suggestion was to continue park development along East Avenue because it would best showcase the city's outdoor space to travelers passing through from out of town. For some, widening East Avenue would also provide for a more pleasant experience driving though Austin as well as make the two sides of town more distinct. Many noted the relationship between quality of life, outdoor space, and water in Austin. Some encouraged the city to make the entire riverfront into a park area, and many saw the floodplains around Waller and Shoal Creeks as attractive lands for public parks, if they could be improved. One resident wrote that the creeks' improvement would make Austin "the most beautiful city of Texas and the South and this is our greatest asset." To some chamber members, parks were also a way to demonstrate their support of segregation and their ideologies of limited government. George Heflybower encouraged the city to "perfect two parks, one for white and one for colored, and no more."[32]

Chamber members also saw improved streets and sewers as assets that would allow residents and tourists to access Austin's natural spaces and also curb air pollution, disease, and illness attributed to dirt roads and sewage. As the capital of Texas and therefore the symbolic center of the state, Austin needed to be improved for the benefit of the entire state's population; at the same time, however, natural beauty was a unique asset that separated Austin from other Texas cities. Businessman A. D. Boone articulated this relationship clearly when he wrote, "The creator has furnished Austin and vicinity a natural endowment for one of the most beautiful cities in America. Let's beautify and provide good roads so the nation may enjoy with us the view which other close by cities would spend millions to possess." Numerous members thought that paving specific scenic roads as well as major thoroughfares would lead to improved social and economic conditions in Austin. Paved scenic drives were suggested for both banks of the Colorado

River downtown, boulevards running adjacent to Shoal Creek, and a scenic highway running from downtown to Mount Bonnell, the city's highest point overlooking the river west of Austin.[33]

Perhaps only the most engaged chamber members referenced the emerging legal framework that could institutionalize nature and separate it from urban elements in the city: land use zoning. Broadly, zoning allows planners to determine how certain parcels of land are used by designating them for different levels of intensity. The three classic designations are residential, for homes and apartments; commercial, for retail businesses and shopping; and industrial, for more intense applications including manufacturing, processing, and other heavier uses. So, theoretically, zoning allowed for more organized use of space by legally separating different uses into different areas of the city, and planners and politicians lauded it as yet another progressive governmental intervention that would help alleviate crowding, congestion, and unsanitary conditions in cities. Zoning would also lead to uniform land uses and, theoretically, higher levels of stability in real estate markets. Although in the United States cities adopted policies to control and organize land use in the mid-nineteenth century, comprehensive zoning was not adopted in any U.S. city until New York instituted building restrictions in 1916. The New York statutes became the blueprint for the federal Standard State Zoning Enabling Act of 1924, which legalized land use zoning; the Supreme Court upheld zoning ordinances in 1926, paving the way for states to allow cities to implement zoning laws. The Texas legislature legalized zoning in 1927. Over the next few years urban planning grew from a practice employed by only the largest cities to a common means of ordering urban space for hundreds of American cities.[34]

While in some southern cities zoning was disregarded as government intrusion on private property rights,[35] in Austin zoning was seen as another tool to make the city more progressive, livable, and business-friendly. Chamber president Long understood that zoning necessarily undermined property rights, but he viewed it as necessary for communities that were growing and also trying to preserve their natural beauty. Chamber members vehemently rejected federal intrusion into local affairs but were more amenable to the local implementation of what they called "wise practices." As early as 1915, Long was advocating for zoning, claiming that it allowed municipalities much more control over private property and led to an increased tax base, increased property values, and a more attractive and stable city. The Progressive Era emphasis on trusting professionals was ubiquitous in zoning rhetoric, where experts could "direct growth of [the] city along the best lines." Long and his friend John E. Suratt wrote to one another praising Texas for quickly legalizing zoning statutes in 1926 and planning to enact further legislation enforcing building codes and regulating platting outside municipal jurisdiction in 1927. Another writer to Long argued that zoning legislation and comprehensive city planning meant that Texas "has placed itself among the most progressive states in the country." Long himself saw zoning as something of a panacea that would mitigate "irregularities . . . which are tending to mar the attractiveness of Austin." To Long, zoning would also allow Austin to be more aggressive in attracting businesses because segregating intense industrial functions would help preserve property values and natural beauty for the rest of the city. Quietly, city leaders also discussed how racial segregation might have the same effect.[36]

While discussions of incorporating, preserving, and improving nature were common in Austin during the mid-1920s, talk of the city's vexing "race problem" was far less overt. Yet the issue of race certainly lurked in the background of planning and political discussions and, no doubt, in the parlors and offices of the city's Business Progressives. Ideas about improving and organizing the city were closely bound to ideas about racial geography and the prospect of creating a minority city within the emergent Anglo garden. Where Long wished to segregate urban functions, he also hoped to segregate blacks. As early as 1891, booster publications advertised the "color line" in Austin, calling it "mutually conceded, but without friction or race antagonism."[37] The color line, however, was a social rather than a spatial demarcation, a matter of practice, occupation, and custom rather than geography. By the 1920s the old color line was in jeopardy as racial tensions between blacks and whites hardened and, in Austin, the number of Mexicans grew rapidly and their status as minorities became solidified.

Institutionalizing Spatial Difference

The Austin City Plan of 1928 institutionalized racial geography in Austin and, with it, split the city into an urban-industrial section and a residential-natural section. As with public discussions of planning, race was always present in the geographical imagination but only emerged sporadically in the actual plan. As consultants, Koch and Fowler followed the chamber's suggestions, citing the "civic survey" throughout the plan and advocating for a number of improvements that would maximize the city's beauty and enhance outdoor spaces to make Austin "an ideal residential city" that reflects its function as "a cultural and educational center," and minimize industry which "will not be the controlling element that will build future Austin."[38] Parks and boulevards were central to this vision, and Koch and Fowler imagined them as intrinsically linked by giving them a section of the plan together. Parks and access to them were cast as indispensable to recreation and therefore to quality of life.[39] The plan suggested implementing a system of boulevards along the riverbanks that would carry traffic away from residential districts and also provide access to natural areas and views of downtown. Koch and Fowler encouraged the city to buy space for parks along the boulevards, linking outdoor recreation with improved quality of life and boulevards with access to that lifestyle. They also discussed a regional plan that would link Austin to the region's most famous and desirable natural spaces, many of which were far from central Austin. Bull Creek and Mount Bonnell were close and easy to access within the framework of the city plan, but more distant outdoor attractions, such as the Pedernales River, Hamilton Pool, and the West Caves, necessitated far more advanced infrastructure and the coordination among Austin and many other cities in the western hinterland.[40]

Perhaps the most radical transformation of natural space that Koch and Fowler imagined was the plan for Austin's major creeks and waterways. Where areas along creek banks had been home to Austin's most destitute minorities, sites of terrible destruction during floods, and some of the most unsanitary places in the city, improvements would transform this land into winding parks serviced by scenic roads. The plan stated that the ground was not suitable for residences, but parks could allow the land to function as a repository for nature and the precious water flowing in the creeks. Koch and Fowler wanted Barton Springs and its environs to be the city's largest and most important park, accessible by Barton Springs Drive and a new bridge over the Colorado. The entire north bank of the river should also become a park, they argued, bisected by a new boulevard along First Street, running west from East Avenue through downtown. Pease Park, a long-planned tract along Shoal Creek, could be realized by buying inexpensive, flood-prone land along the creek and below the bluff; the plan also called for Waller Creek to become a greenbelt running from the university campus south to downtown. To transform Waller and Shoal Creeks, however, would necessitate removing from their banks the "unsightly and unsanitary shacks inhabited by negroes." Koch and Fowler laid out a plan for a comprehensive system of parks that included smaller playgrounds and neighborhood parks, accessible by foot for most residents, and several larger outlying parks that "should preserve natural topography and scenery" for residents and tourists to enjoy.

Improving urban creeks and removing minority residents would also have the obvious advantage of bolstering real estate values in neighborhoods near creeks; not surprisingly, many of the newer subdivisions close to Shoal Creek experienced dramatic growth in the 1920s and 1930s.[41]

Removing minorities from the creeks' banks, though, was only part of a larger aspect of spatial transformation through segregation. There is little record of a public debate over institutionalizing racial segregation in Austin, yet in letters and discussions the desire to segregate the city was talked about casually. Long corresponded with two people about the possibility of a "Race Segregation Enabling Act," similar to a law passed in Louisiana, but worried that the law would be deemed unconstitutional.[42] The Dallas firm Kessler Planning Associates circulated a memo in 1927 to Texas chambers of commerce arguing that "whites who have bought homes are entitled to protection from encroachment of negroes moving into their neighborhood."[43] Mayor P. W. McFadden quietly wanted to use racial zoning measures to mandate segregation. And in the city plan, Koch and Fowler casually stated that there had "been considerable talk in Austin, as well as in other cities, in regard to the race segregation problem."[44] But when the Louisiana law was struck down by the U.S. Supreme Court in March 1927, Koch and Fowler used their experience sidestepping constitutional bans on legal segregation in Dallas to effectively complete racial segregation in Austin. While private property rights were inviolable and "the problem" could not "be solved legally under any zoning laws" known to them at the time, separate but equal public facilities were still legal and customary in Texas. Koch and Fowler argued that segregation could be accomplished simply and effectively by providing or enhancing public services—parks, schools, libraries—to African Americans in only one part of the city, a small price to pay for racial exclusion. They suggested East Austin, where the population was already mostly black. This, Koch and Fowler agreed, would also be cost-effective by eliminating the need for duplicate facilities throughout the city. Moving African American facilities to the Eastside would allow for the Westside and more rural environs to become more natural.[45]

While the plan to provide services for minorities in East Austin expedited the process of segregation already under way,[46] zoning suggestions, which were turned into laws when Austinites adopted a comprehensive zoning plan in 1927, radically altered the landscape and yoked minority space to urban-industrial space in Austin. The zoning proposal included five primary designations ranging from the least intensive A zoning to the unrestricted E zoning. The entire block of land abutting the proposed black district to the south was zoned industrial; to the immediate south was the area of highest Mexican concentration and in-migration. And while much of the African American district was zoned residential, zoning regulations were rarely enforced there. The boulevard system was absent for East Austin, while East Avenue would be widened, paved, and parked in an effort to accentuate the line of racial demarcation. Instead of boulevards, Koch and Fowler suggested "trafficways" that would allow workers to access industrial facilities on the Eastside without entering the central business district. Comal and Chicon Streets, both running parallel to East Avenue on the Eastside, were designated as trafficways.[47]

At the same time, Koch and Fowler advised the city to eliminate the industrial district west of Congress Avenue and south of Fifth Street, near what was formerly Little Mexico. Railroad tracks coming from the west along Fifth Street cut off the central business district running along Congress Avenue. Koch and Fowler found that real estate values dropped precipitously south of Fifth and that south of Third was "practically a blighted district."[48] Their suggestion was to remove the tracks between East Avenue and West Avenue and to zone the area for less intensive activity, which would increase property values and thereby force the remaining Mexicans out of the area.[49] The industrial district west of there, along the International and Great Northern Railroad tracks, should also be removed via zoning, they argued, because "the area being absorbed at the present time is a high-class residential area."[50] All train traffic could likewise be routed around downtown and to the Eastside by building two new stations and a bridge at Pedernales Street.

While industry was segregated on the Eastside, new suggestions for subdivisions would work to further enhance real estate values on the Westside. Private racially restrictive covenants, which promised socioeconomic stability through racial exclusivity, had been part and parcel of new Westside subdivisions since at least the development of Hyde Park just after 1900 and likely before. Yet Koch and Fowler agreed that standardizing developments and controlling density could have equally beneficial effects, and they pointed to new state legislation that allowed cities more say in the character of residential developments. Due to a lack of planning, much of the existing street pattern was poorly graded and inefficient.[51] As part of the move toward home rule, the Texas legislature adopted a bill that gave city plan commissions authority to approve plats before they could be filed. Therefore, municipal government could create guidelines that developers must follow to have new subdivisions approved or, more commonly, annexed into the city. The plan recommended procedures needed to make subdivisions specific and standardized; for example, streets should be uniform in width, alleys should facilitate drainage, lots should have minimum size requirements to stabilize values, and, above all, lots should not be irregular. The bill, however, did not mandate that municipalities regulate developments, and since no developers were interested in building subdivisions or having them annexed by the city on the Eastside anyway, guidelines were only enforced on the Westside.[52]

The Austin City Plan's and the chamber of commerce's suggestions for improvement paint a portrait of a city that was increasingly interested in improving both property values and quality of life for white citizens. These two aims included removing minorities and industries from white residential areas and improving infrastructure and open space throughout the city. The new city manager form of government allowed elites to essentially reserve political decision making for themselves; even in 1933 only 23 percent of Austinites were eligible to vote, and only roughly a third of those eligible did vote. In 1927 less than 5 percent of Austinites voted.[53] Residents who could and did vote supported the efforts to remake the city by passing a $4.25 million bond package via referendum in May 1928, by far the largest in the city's history. The bond was designed to provide the capital to remake the landscape to conform to the city plan. It gave the city $700,000 to buy and develop 112 acres of land for boulevards, parks, and playgrounds and over $2 million to implement infrastructural improvements. By 1931 Austin had forty-seven new miles of sanitary sewers, twenty-six new miles of storm sewers, five new bridges, and new schools, fire stations, and a hospital and multiple creek improvements had been made.[54] An augmented road network connected new Westside subdivisions to downtown and to natural areas, including Mount Bonnell, and money was poured into improving creeks and making them safer.

As promised, bonds also paid for minimal but important improvements in East Austin. Some of East Austin had sewers by 1930, Rosewood Park was completed and opened, and improvements were made to black schools and libraries in East Austin. Zaragosa Park, an almost completely unimproved field in the barrio area south of East Sixth Street, was dedicated as the city's Mexican Park in 1931. However, African Americans who chose not to move east were severely penalized. Residents of Clarksville did not receive any municipal upgrades, and the area lacked running water, sewers, and paved streets into the 1970s, even as some of the most expensive white neighborhoods in the city were developed around them. Wheatsville School, which had been shrinking for two decades, closed in 1932. Landlords began refusing to renew leases to blacks in Wheatsville shortly thereafter in favor of University of Texas students. Census Tract 6, which contained Wheatsville and other Anglo neighborhoods, dropped from 17 percent black in 1915 to 3.6 percent black by 1940. The only school for blacks in more rural South Austin, Breckenridge School, saw enrollment decline precipitously between 1924 and 1934, and the school was shuttered shortly after that as residents left for the Eastside.[55]

The improvements and planning initiatives set the stage for dramatic growth in Austin during the 1930s, despite the Depression. The city saw a population increase of 65 percent during the decade, its largest percentage increase since 1880 or any time thereafter. New subdivisions in Bryker Woods, Enfield, Tarrytown, and Pemberton Heights, all located west of Shoal Creek, were developed rapidly during the 1930s, in accordance with the plan's recommendations for restrictive land uses: uniform, low-density, residential buildouts with small commercial districts, no industry, easy access to open space and downtown, and no minorities. Conversely, the two districts east of East Avenue lacked the uniform, controlled qualities of the new Westside subdivisions. As early as 1931, the black and Mexican neighborhoods on the Eastside contained 2,700 residents per square mile, more than double the city average of 1,250.[56]

Perhaps the most deleterious for Eastside real estate prospects were the Home Owners' Loan Corporation residential security maps and housing survey completed in 1935. As the Depression worsened in the early 1930s, Roosevelt created the corporation to help stabilize the real estate market by providing loans, insurance, or other types of financial help; the corporation worked with local appraisers to create maps that showed where investment was deemed safe and where it was considered dangerous (or "redlined," because of the red indication on the maps). In Austin, real estate professionals redlined the entire section of the city east of East Avenue and south of Manor Road, effectively the predominant black and Mexican areas, along with other areas they considered hazardous in South Austin and downtown, although the Eastside was the largest district marked red.[57] The maps meant that securing a federally backed loan, real estate insurance, or other federal subsidies in the area would be next to impossible for individuals or developers. Essentially, the areas that minorities were forced into in the 1920s and 1930s were devalued by the federal government almost immediately. Even if developers wanted to build in subdivisions in East Austin, the risk would have been too high, especially given the availability of federally

HYDE PARK

The most beautiful, healthful, and practical place for homes in the city of Austin. It's the safest place for investment. The terms offered are remarkably easy. The prices are very reasonable. Any person buying two lots WILL BE GIVEN ONE LOT FREE OF COST. There are six miles of beautiful graded streets in HYDE PARK, and a magnificent

SPEEDWAY FROM THE PARK TO THE CITY.

THE FINEST DRIVE IN TEXAS.

HYDE PARK IS EXCLUSIVELY FOR WHITE PEOPLE.

The main line of Electric Street Cars run into and around a belt in the Park. Free Mail Delivery twice a day. There is no limestone dust. The soil is the best for Fruits, Flowers and Lawns. No one thinks of taking a carriage drive without going to Hyde Park. The drives are free from mud and dust. The scenery is interesting. The altitude of Hyde Park is 185 feet above the river. Hyde Park is Cool, Clean and Restful. Invest while YOU CAN SELECT, and SECURE ONE LOT FREE. If you wish to buy on the installment plan the terms are $3.00 per month on each lot. If you pay all cash a discount of 8 per cent will be allowed. If you wish to invest and do not live in Austin, we will pay your fare both ways, if the distance is not over 300 miles. Strangers who wish to see the city can have a Free Carriage by calling at our office.

Extraordinary Inducements Are Offered

To persons who will agree to erect good houses. If parties wish to build in Hyde Park we will trade lots for other Austin property on a fair basis, and DONATE ONE LOT as a Premium. Beautiful Views of Hyde Park, and of THE SPEEDWAY sent free upon application. Write to us, or call at 721 CONGRESS AVENUE, AUSTIN, TEXAS.

M. K. & T. LAND AND TOWN CO.

M. M. SHIPE, General Manager

Advertisement for the Hyde Park subdivision in North Austin, circa 1900. Note the environmental amenities offered and the promise of racial homogeneity. PICA 25419, Austin History Center, Austin Public Library.

secured mortgages and higher real estate values on the Westside. Eastsiders were left with little choice but to endure haphazard street patterns, poorly graded properties, and a lack of opportunities to refurbish their homes, a problem that would have long-lasting ramifications for socioeconomic stability well into the 1960s and 1970s (discussed in chapter 6). A study in 1939 found that well over 60 percent of the more than thirty-two hundred residential properties surveyed were deemed substandard in East Austin. Alternatively, the safest areas on the maps, all in white areas in West Austin, saw rapid growth.[58]

On the eve of World War II, Austin's racial and natural geographies had changed radically, although not all of the proposed planning changes had come to fruition. Zoning maps from 1939 show that the entire area bounded by East Avenue, Third Street, Pleasant Valley, and Seventh Street was zoned as either industrial or unrestricted. Train tracks coming from the northwest had been routed around downtown. A small industrial area remained adjacent to the rail lines west of West Avenue, but it was separate from residential West Austin.[59] Austin's zoning system was also cumulative, allowing for more uses in higher-intensity areas rather than totally separate uses, which frustrated residents for decades. This meant that areas zoned for industry could legally include residences. Often cheap residential properties in areas zoned for industry enticed less affluent residents who were displaced from changing neighborhoods downtown and on the Westside.[60] Trash and waste disposal and other unseemly urban functions also followed in the exodus from west to east.

Urban planning, zoning, and service delivery thus hastened the spatial transformation already under way that tied minorities to industry and haphazard and confusing land use and curtailed their ability to improve their property, acquire loans, and avoid pollution. Yet these interventions simultaneously guaranteed economic stability and access to desirable outdoor space for most of Austin's whites. By the 1930s, professional planning and infrastructural improvements had fulfilled a grandiose vision for the city articulated by boosters in 1909:

> Austin is pre-eminently a city of residences and a seat for education . . . statesmen, lawyers, judges, government officials and wealthy men have sought the social and educational environments of the place as the proper atmosphere in which to raise their families. While it is true that the position of the city as capital of the State has had much to do with this attraction of wealth and talent, still it must be admitted that the natural charm and beauty of the location has made it doubly attractive . . . the whole country surrounding Austin may be described as a *grand public park*. Nature has done more for it than human energy and skill could ever have accomplished.[61]

Yet the last sentence belied the difference in environments. It was specifically human energy and skill along with the will and political capital to separate races and functions that had transformed Austin into the grand public park that reflected the proper atmosphere for cultured citizens and their families. Finishing nature was part and parcel of creating a physical landscape that reflected the social imperatives, economic dreams, and racial

Lenders used residential security maps, like this one, to determine where the federal government would guarantee mortgages and thus where it was safe to lend. In Austin, the African American–designated area east of East Avenue, as well as all other areas that still contained high concentrations of African Americans, was deemed "hazardous," and thus federal guarantees were not available. Home Owners' Loan Corporation, residential security map of Austin, 1934 (National Archives, Record Group 145, Austin Texas Folder). Outlines added by author.

hostilities of Austin's elites. The augmented landscape would shortly pay dividends. As World War II approached, Austin's Business Progressives and growth-oriented politicians used the improved landscape to promote the city and express a new vision driven less by civic pride than by economic profit.

4 The Playground of the Southwest

Water, Consumption, and Natural Abundance in Postwar Austin

In 1952, Granite Shoals Dam, about twenty miles closer than Buchanan Dam to Austin, was rededicated in honor of LCRA general counselor and conservation advocate Alvin Wirtz, who died unexpectedly in 1951. Similar to the dedication of Buchanan Dam in 1937, many leaders spoke at the ceremony; LCRA general manager Max Starke presided over a series of speeches by Lyndon Johnson, Congressman Homer Thornberry, Governor Allan Shivers, and former Austin mayor Tom Miller, all of whom lauded Wirtz as the single most important figure in the development of the region's new water resources. Shivers boisterously claimed that "water is of more value to the economy and future prosperity of the state" than oil and gas are, and for Central Texans he may have been correct. But the 1952 speeches and celebration were palpably different from the somber, pragmatic tone of the 1937 dedication. No speakers addressed the new possibilities engendered by the technological advances along the river; they did not have to. Instead, the focus of the day was on leisure activities. Only the first twenty minutes of the daylong festivities were given to speeches. For the rest of the day the enormous crowd of over six thousand people, one of the largest gatherings ever in Central Texas, had fun. The state paid for a giant barbecue. The LCRA encouraged guests to swim and boat, and there were sightseeing tours all around the area. Hundreds of people fished and swam in Lake Granite Shoals (later Lake LBJ), and boat races began at 10:30 A.M. and lasted all day. Numerous newspapers reported on the event. Most hailed the celebration as a symbol of the economic success of Central Texas, the growth of Austin, and the enhanced quality of life that the dams provided for residents.[1]

The differences between the two ceremonies indicate the changes that took place in Central Texas and Austin between 1937 and 1952, largely because of the dams. Journalist Raymond Brooks wrote that the dams had not only saved millions of dollars in flood control and irrigation; they also brought unforeseen economic expansion to Central Texas and Austin. Over

$100 million worth of new wealth was created in the Colorado's watershed by 1952, much of it coming through tourism to the quickly growing resort areas now dotting the reservoirs. Plans were in the works to develop a $1 million luxury resort on the north bank of Lake Travis on the outskirts of Austin. Power supplied by the dams not only modernized rural areas in Austin's hinterland; it also allowed Austin to attract some small industries, which encouraged the city's steady growth throughout the 1940s and 1950s. And picturesque real estate along the river, now safe from flooding, was some of the most valuable in the region. As a result, Brooks wrote, "Austin grew; and with the completion of the dams [the city] has grown tremendously, both in population and in diversity of business and industry."[2]

This demographic and economic growth was accompanied by a less obvious but equally important change: the dams also allowed Austin politicians, citizens, and businesspeople to reimagine, and thus further remake, the region in a way that was highly entrepreneurial and driven largely by a leisure economy. From the creation of the LCRA in 1934, public and private interests used the lakes and dams as tools to attract both capital and tourist money into Central Texas. The dam system allowed Austinites and other Central Texans to create uses for the region that went far beyond the structural concerns voiced by early reclamation advocates. Ironically, Wirtz was one of the most pragmatic of the early dam advocates. As late as 1944, he scoffed at the idea that the lakes would become primarily sites of recreation and real estate a significant engine of growth. He thought the next big project should be improving the river for navigation.[3] But by the 1950s it was clear that water-based recreation could be an incredible economic engine for Central Texas. Imagining and implementing growth occurred in a variety of ways, from the City of Austin using its new hydroelectric power to attract a magnesium plant during World War II to the LCRA planning an entire recreational region catering to upper-class Texans. Earlier conservation and reclamation enthusiasts in Texas imagined a modernized rural, agricultural hinterland freed from drought and flooding and able to take advantage of electricity. Quickly, however, it became clear that the change in the landscape would have far more dramatic consequences for Central Texas.

This chapter looks at how water became a primary growth paradigm for Austin after World War II. Between the 1940s and the 1970s, new types of cities with more suburban geographies emerged and grew rapidly in the modernized South and West. While many emphasized leisure opportunities, climate, and low taxes, Austin's growth advocates advertised pristine outdoor space, with natural areas woven into urban fabric and hinterlands, and celebrated unique environmental characteristics and their recreational possibilities. Americans, with more leisure time, more money, and higher expectations for a life that included recreation and access to nature, began to see nature as something to be enjoyed as part of everyday life as well as part of vacations.[4] At the same time, Keynesian economic policy promoted consumption as a patriotic act that stimulated economic growth and as an ideology that material abundance was part of a postwar American lifestyle.[5] Consumption, however, was not limited to material goods, such as houses, autos, and televisions. People also expected to consume new experiences, more quality time with family and friends, and even nature itself. In Austin, water became the central symbol of the area's abundance of natural spaces; the lakes produced social and cultural meanings that tied residents' sense of the "good life" to natural landscapes and provided a theme for boosters marketing a nonindustrial city for future economic and urban expansion. Abundant water and water recreation were thus vital components of the urban forms—suburban, low-density, integrated with nature—which growth promoters advertised in the postwar decades.[6]

Abundant water also differentiated Austin from drier regions to the north, west, and south and provided a sense of collective regional pride.[7] The LCRA's struggle against private power interests, its widespread support from across the region, and its ability to cheaply electrify even the most remote hinterland property formed a collective consciousness in Central Texas. The system of lakes and dams linked the region together physically and symbolically. The landscape along the watershed after the dams were built was an envirotechnical system, a blend of ecological and technological systems that connected city and hinterland and blurred boundaries between natural and man-made.[8] Although humans had been augmenting the river and natural landscape in the watershed for centuries, the dams and reservoirs provided an entirely new system that radically altered social and economic life for Central Texans. In Austin, city and watershed evolved as related forms of settlement space that were created by technological augmentation of the watershed. Thus the polarization between rural and urban, developed and undeveloped, was not definite.[9] The seamless blurring of technology and nature also emphasized the duality of Austin's attractiveness: it was both a beautiful natural place and an improved place that reflected a civically engaged, vibrant social and cultural life.

The new physical landscape and the newfound stability promoted new strategies for regional growth that commodified land and water that were previously susceptible to natural cycles of drought and flood.[10] In the 1930s, Lyndon Johnson and Alvin Wirtz, two of the most vociferous proponents of the LCRA and regional modernization, considered rural farmers the principal subjects of the watershed's new benefits. But while rhetoric surrounding the New Deal focused on improving and modernizing the lives of rural Central Texans, it was Austin businesspeople and the other townsfolk along the watershed who saw their vision materialize. They viewed rural land in terms of its economic benefits as real estate and recreational space and as the natural part of a dynamic growing region with Austin at its center.[11] The system of reservoirs, linked to one another and to Austin by highways, became the geographic and symbolic connection between downtown and the recreational hinterland, and water evolved into the feature that defined Austin as a natural city that cared about its environmental qualities, much different from its sprawling Texas neighbors who sought accumulation via resource extraction, manufacturing, finance, and energy.[12] Residents believed technology could make nature better, and their hopes and dreams for the future were closely tied to hydro-technology.[13] Water and power were imagined principally as tools to attract business, increase local consumption in the summer months, showcase the region's amenities to prospective newcomers and businesses, and provide a motif for civic pride.

Commodifying the Landscape

From the late 1930s to the early 1950s, efforts to promote economic development in Austin focused on attracting tourists and creating a high quality-of-life image for Austin. Boosters promoted growth by constructing Austin as the center of a pristine natural world rather than as a city in any contemporary popular sense of that term. As the institutional center of Texas, Austin already had plentiful well-manicured, pastoral, and natural space right in the city and, as mentioned earlier, had little industry.[14] The University of Texas campus, which was a painstakingly manicured pastoral space designed to promote intellectual reflection and peacefulness, and the capitol complex, with its parkland outdoor space, were symbols of the region's nonindustrial urban qualities. Barton Springs, a spring-fed pool south of downtown, was the city's outdoor social center and most famous natural urban space. The 1928 city plan set aside many acres for parks and greenbelts, and subsequent plans and ordinances ensured West Austin ample green space. Upon completion of the first few dams and reservoirs, however, guidebooks and promotional literature began to cast the city as the hub of a vast regional environment perfect for vacations, retirement communities, and eventually suburban development. They consciously linked city to hinterland by focusing on regional environmental characteristics and urban amenities. Tourists first engage with images and writing pertaining to the places they visit and events they attend and then engage with visual cues in order to read and assess landscapes.[15] The guidebooks and other promotional materials were also silent about the city's urban elements, which they erased from images of Austin and geographically sequestered from the physical landscape.[16]

In the waning years of the New Deal, public entities began to imagine a new landscape for the lakes. One of the first and most comprehensive attempts to chart the region's course was a brochure produced jointly by the National Park Service, the LCRA, and the Texas State Parks Board in 1941 titled *The Highland Lakes of Texas*. Created from a joint survey conducted by the LCRA and the Department of the Interior, the brochure was intended to provide an outline for potential recreational development around the lakes.[17] The authors based the report on a three-year survey that used histories, maps, interviews, and personal observations to devise a land use proposal suitable to both the new watershed and the growth-minded outlook of the region. Most of the river was harnessed during the time that the survey was put together; Lake Buchanan and Inks Lake were fully filled during the period, while Mansfield Dam and Austin Dam were completed shortly before the report was published.[18] *The Highland Lakes of Texas* is thus an early working vision of what the dams and lakes symbolized for the region's future.

Unsurprisingly, the report considered the region as primarily public land and advocated public cost and public uses for the sprawling state-run parkland, which would include nearly eight hundred thousand acres of land, including over 670 miles of shoreline (more than the length of the Texas coast), stretching from West Austin to Lake Buchanan, roughly 60 miles to the northwest. While the plan allowed for private rights to be protected in the park, it would not allow further rights of private property to be granted within park boundaries. Grazing rights and other additional commercial rights, such as mining, would be slowly curtailed and eventually eliminated. The state would retain developmental rights over all lands, indicating that private uses would be almost entirely absent from the park. Accommodations, however, would not be duplicated if private enterprises either within or outside the park satisfied demands. While the plan encouraged charging a park entrance and usage fee, it was not really profit-minded.[19]

The brochure also imagined Texas as a symbol of a potential postwar growth pole in the United States; the Highland Lakes Park would be a cornerstone of recreational enjoyment that would serve an energetic population.[20] Texas was cast as the nation's "sum-total of forward moving energies" and "a pendulum to keep other parts of the Nation in productive operation." In sum, Texas produced "accomplishments that have poured forth from a melting pot of resources and human energy to create a State of wealth, a State of power, a State of dignity, and of widely diversified interests." The state's cities, moreover, grew from log-cabin villages to centers of modern industrial power, almost overnight. Texas finally had the capacity to move past "just living" and begin enjoying its abundance and natural wealth. " 'Scenic resources,' 'recreation,' and 'conservation' " today meant something to Texans. These resources had to be viewed as a primary form of wealth, right along with other commodities, such as oil and cotton. Recreation, finally, could make the economy grow.[21]

Most of the text outlined specific recreational uses for the park. Along with obvious outdoor activities, such as hiking, camping, swimming, and boating, the brochure also offered decidedly upper-class pursuits. Plans to build two championship golf courses in the park accompanied images of two couples teeing off under a tree. A young man was pictured lawn bowling in a suit with a woman encouraging him. Most interestingly, the authors framed polo both as a sport and as entertainment for large crowds, in a region "already committed to polo."[22] Recreation was imagined as far more specific than people simply having a good time or randomly enjoying leisure time. Activities were marketed to an upper-class audience, or at least those who aspired to polo, lawn bowling, and golf during an economically unstable time in a rural, historically poor area that had received electricity only in the previous few years. The new, controlled waterways allowed planners to imagine a radically new socioeconomic landscape that accompanied natural transformation.

For Clarence McDonough, general manager of the LCRA in 1941, *The Highland Lakes of Texas* was part of a larger vision for the new lakes where the area would be turned into a 350,000-acre national park. In 1940 he invited Harry T. Thompson of the National Parks Service to come view the area and assess its potential as a national park; McDonough claimed that Thompson preferred the Highland Lakes to other areas of the Southwest, most notably Big Bend in southwestern Texas, which had recently been granted national park status, because of its accessibility and range of potential activities. McDonough and *The Highland Lakes of Texas* exemplify the public-oriented imagination surrounding early visions of how to employ the lakes and dams for regional benefit. Because of the project's deep ties to public capital and the public positions of early advocates of the area, these discourses are unsurprising. While the mode of production for the area would become driven much more by the private sector over the coming decades, the rhetoric remained largely focused on circulating capital.[23]

By 1940, private businesses, led by the Austin Chamber of Commerce, began to recognize the region's new commercial possibilities, and public entities such as the LCRA and the City of Austin worked on their behalf to stimulate growth. Capturing as much of the tourist market as possible was the goal of Austin's early promotional efforts because the watershed was unique to the arid Southwest and could attract circulating revenue.[24] This strategy included using the lakes for recreation as well as catering to conferences, trade shows, and conventions, all of which provided fast, non-intensive, and nonindustrial injections of capital into the local economy. Much of the discourse surrounding the lakes during and shortly after World War II focused on bringing vacation and retirement income to the area by portraying it as a vacation "wonderland" for families and building retirement communities that would use the lakes and the warm climate as their main attractions. Texas House Joint Resolution No. 43, passed in August 1945, was essential to marketing the area. The joint resolution amended an antiquated section of the Texas Constitution, which formerly prohibited the state from allocating any money or promoting Texas's resources to outsiders. The law, referred to colloquially as the "carpetbagger law," was written into the Texas Constitution in 1876 in an effort to discourage outsiders from moving to the state. The resolution removed the prohibition on state funding for advertising, and promotion began soon after the war ended.[25]

Central Texans took advantage of the new law by focusing advertising resources on circulating capital, especially conventions, vacationers, and retirees. The campaign coincided well with the postwar economic boom as disposable income, free time, and demand for leisure and travel increased markedly after 1945. By the early 1950s roughly half of Americans took family vacations, and many saw vacationing as a way to demonstrate class status through consumption.[26] Disposable income increased by 108 percent in the United States between 1945 and 1958, generating all kinds of new markets for leisure products and services. During roughly that same period, the average weeks of vacation per year for an American worker doubled due to increased production per work hour, better transportation, and increased workplace efficiency. By 1960 the leisure market was absorbing $44 billion a year, indicating that vacationing and tourism were already two of the most lucrative industries in the United States. Since tourist dollars also represented the highest level of capital mobility, they were subject to intense regional competition.

Among Texans, economic growth was even more striking. Between 1946 and 1956, total personal income in Texas grew from $7.4 billion to over $15 billion, an increase of over 100 percent in just one decade. Per capita increases in personal income were smaller due to a 25 percent gain in total population, but they still measured a robust 64 percent growth in the decade after World War II ended, from $1,028 in 1946 to $1,685 in 1956. Tourist revenues were substantial enough in the 1950s to support fairly centralized patterns of growth in areas that offered attractive destinations, natural or otherwise. Tourism also does not require a great deal of initial capital, except for advertising, provided there are attractive features in a particular place. The lakes were thus obvious sources of potential tourist revenue that Austin businesspeople and political leaders sought to exploit as the city's first economic growth engine after World War II.[27]

Guidebooks and promotional literature were the first medium Austin boosters used to introduce the city to potential tourists and investors. Booster literature can help to form a regional consciousness that focuses on the positive aspects of a place.[28] In Austin, water attractions and the ability to generate power became idioms for the city's improvement and images that differentiated Central Texas from other less endowed southwestern regions. The publications *The Austin Area Lakes* and *The Highland Lakes of Texas*, which imagined the watershed as an upscale playground complete with polo grounds and a national park, served as the model for later, more descriptive imagery that is especially apparent in the myriad guidebooks, pamphlets, and other booster materials that became commonplace in Austin after World War II.[29] Starting in 1942, the Austin Chamber of Commerce published a guidebook called *Austin in a Nutshell*, which brought together images of pastoral urbanism, natural leisure pursuits, and cultural amenities that increasingly defined suburban-style living in the 1940s. Water provided the central motif. The "mighty Colorado River, cutting its way through this range for centuries . . . has created the 'Palisades of the Colorado,' where high cliffs and peaks offer breath taking views."[30] In some places along the river, pristine natural areas needed no improvement from man.

At the other end of the spectrum, complex technologies that harnessed nature's power were attractive in their own right. The dams were objects of tourism and functioned in a highly symbolic context that became central to the region's identity.[31] As both public works and symbols, dams ushered in modernity to residents of Central Texas. Just like the Hoover Dam and Las Vegas, they were also the objects of both leisure pursuits, in the form of tours, informal sightseeing, and ancillary activities, such as going to a restaurant at the Buchanan Dam, and technical pamphlets and demonstrations designed to create what David E. Nye calls a "technological sublime."[32] The technological sublime creates awe in the power of large technology to reshape the landscape for social benefit and fosters the related belief that this power is linked to national greatness.

The spectacle that the dams created became a major tourist attraction during an era when technology and modernity promised better lives for millions of Americans. When completed in 1937, Buchanan Dam was the largest multiple-arch dam in the world, spanning over two miles. The thirty-seven floodgates, one under each arch, opened independently of one another, allowing specific increments of water through the dam. Visitors could walk along the top of its massive arches and view the 22,000-acre Lake Buchanan, the eleventh-largest freshwater lake in the United States. The dam's ability to change the landscape was immediate and striking, as it provided a perfect vantage point from which to view how completely the river had been harnessed. Inside the dam, tourists could marvel at the three enormous turbines that were able to generate over fifty-one megawatts of hydroelectric power if necessary. The LCRA provided guided tours of the facility and also published yearly newsletters that included technical information and photographs of Buchanan Dam. In the 1940s the LCRA built a dam observation building that could accommodate close to five thousand visitors a day, the average daily weekend attendance during peak season.[33]

The key flood control structure and the most impressive of the Highland dams was Mansfield Dam, about fifteen miles northwest of downtown Austin. When fully completed in 1941, Mansfield Dam rose 278 feet above the riverbed and its base spanned 7,089 feet, making it the fourth-largest dam in the United States. Texas Highway 620 ran right across the top, affording majestic views of Lake Travis, the primary recreation reservoir and the center of high-end development along the watershed. Like Buchanan, Mansfield attracted thousands of tourists each year. Both dams reinforced the idea that the natural could be improved through human ingenuity, technology, and investment.[34]

In Austin, tourism offered a natural experience that blended harmoniously with the built environment and the suburban landscape, though. Nature tourism has long been an important facet of national identity and has functioned as a respite from the ills of the urban-industrial state. Envirotechnical systems combined the two symbols of American national greatness in a way that de-emphasized the urban-industrial aspect of technological innovation. Brochures distributed by the LCRA treated the dams in a highly technical manner, describing the style of their gates, their capacities for hydroelectric production, and their cost. But promotional material was more visual, focusing on the immense size of the dams and their relationship to the environment through the medium of water and to humans through electricity. These were not just machines in the garden, but machines that created a garden out of wilderness. Both narratives were linked to the region's future prospects and imagined prosperity at a watershed moment of public and domestic improvements through technological innovation.[35]

Austin boosters used the lakes and water to market the city nationwide, beginning as early as 1939 when the LCRA sent a twelve-foot-long model of the dam and electrification system to the World's Fair in New York City. The 1939 fair, with its utopian-themed "The World of Tomorrow" slogan and in sharp contrast to the socioeconomic malaise and rising totalitarianism of the 1930s, was an appropriate place to showcase conservation and hydroelectric technology that could jumpstart an entire region. The notion of progress embedded in rural electrification and consistent water supply had the potential to reinvent derelict American landscapes, especially in the South and West, and bring the comforts of modernity to struggling people. Technology was often viewed as a remedy that could transcend the troubled world of free market capitalism and simultaneously avoid the perils of communism—a panacea for a weary nation.[36]

Marketing of the Highland Lakes focused more on leisure and less on practical concerns in the postwar era, but opportunities arose to combine the two while again showcasing Austin's water supply in New York City. In 1950, in a clever, lighthearted marketing ploy called "Operation Waterlift," the City of Austin, working with the LCRA, the Austin Area Economic Development Foundation (AAEDF), and the Texas Motor Transportation Association, sent three thousand gallons of Highland Lakes water by truck to New York City, which was in the midst of a drought. Central Texas went on to survive the brutal drought of the early 1950s far better than other Texas cities because of the region's ample supply of water. The city actively courted small manufacturing outfits in conjunction with a federal plan that encouraged "general decentralization" as a defense measure and offered some of the lowest water rates in the nation. Water thus provided jobs and publicity for the city from a very early date.[37]

Guidebooks cast Central Texas as the best vacation spot the state had to offer, and most promotional tourist literature aimed to attract Texans and people from surrounding states. Brochures and booster articles gave Austin a central location as part of the natural environment rather than as something differentiated from it. Referred to by one writer as "the apogee of vacational wonderland,"[38] the lakes offered a variety of sports and leisure pursuits, from boating and fishing on Lake Travis to more cultural pursuits on the banks of Town Lake in downtown Austin. Hundreds of small resorts and campgrounds, catering to vacationers as well as fishermen, hunters, and campers, had opened for business along the coastline by 1964.[39] Along with festival activities that brought money to the city, the reservoirs generated revenue and created businesses that took advantage of tourist and leisure dollars almost year-round. In the early 1950s the Greater Austin Chamber of Commerce developed the Highland Lakes Committee to oversee and encourage entrepreneurial development along the river corridor in West Austin and in the burgeoning western and northwestern suburbs. Bill Glaston, the committee's chairman, wrote to Mayor Tom Miller in 1958 thanking him for "having great understanding of the lakes' importance to Austin's economy." Glaston, owner of the region's largest manufacturer of fiberglass watercrafts, had recently completed a survey that concluded that boating was a multimillion-dollar business in the city and that Austin's tourist and vacation trade brought in upwards of $5 million each year. While most boosters considered the lakes system a public asset, in reality most economic development in the area filled private coffers but nonetheless stimulated the Central Texas economy.[40]

In the early 1950s the LCRA and the Highland Lakes Committee developed a promotional campaign for the region in *Top Spot for Fun* and other publications. A cartoon spokesman known as Tex McLoch was central to efforts to portray the region as both fun and western. The figure was a peculiar mix of the friendly Texan persona and a frontier-oriented Scotsman who wore both cowboy hat and kilt, smoked a pipe, and recalled the hardscrabble farmers who populated the Hill Country from the 1870s onward. Always smiling and ready to make a joke, Tex extolled the simple virtues of Central Texans and the leisurely activities on the lakes while symbolizing a mythological westerner. The character also recreated the frontier image, ubiquitous in popular culture during the 1950s, and packaged the excitement of the frontier in a safe, wholesome form that appealed to all members of a nuclear family. Women were also central figures in *Top Spot for Fun* promotional photos, especially those emphasizing fishing, boating,

The LCRA created the cartoon character Tex McLoch, a blend of Texas cowboy and Scottish backwoodsman, to advertise the Highland Lakes in the 1950s. Image number PB066845, LCRA Corporate Archives.

and waterskiing. Although some images portrayed women as objects of the male gaze, others cast females as able-bodied, salt-of-the-earth providers in the mold of Rosie the Riveter. In the 1950s *Top Spot for Fun* appealed to a wide variety of middle-class whites seeking an affordable tourist destination that signaled the egalitarian frontier culture of the American West, but it also appealed to those interested in all types of water-based leisurely pursuits.[41]

By the late 1950s the Highland Lakes had helped to completely modernize the region. Reservoirs provided water for five times as many people as lived in Austin and its hinterlands, and the LCRA was able to sell surplus electricity to regions throughout the Southwest. In the prolonged drought of the early 1950s, for example, Austin was the only major city in the state of Texas that did not have to ration water. In 1957 the City of Austin estimated that the dams saved residents at least $13 million in damages when a severe flood on the Colorado was contained. Property along the river, particularly in South Austin, was able to be developed and taxed because the area was no longer prone to inevitable destruction from flooding; much of Austin's civic infrastructure, which was used to attract circulating capital and host cultural events, was built just south of the river in areas that were previously very dangerous to build on. The lakes allowed for enough water to cool new power plants, providing extremely cheap electricity to city residents and businesses. The Highland Lakes Tourist Association estimated annual tourist revenue from the lakes at between $5 million and $7 million in 1960.[42] By 1960 over 215 resorts dotted the "stairway of lakes" that stretched over one hundred miles from Lake Austin to Lake Buchanan. Journalist Raymond Brooks opined that "perhaps the City of Austin is the best example of the direct benefits created by this state flood control and storage project" in the State of Texas.[43]

For both tourists and potential home buyers and retirees, developers cast the Highland Lakes region as a series of prototypical suburban developments, offering municipal amenities and conveniences while promising the relaxation and leisure of rural living. Although the City of Austin was largely suburban in form, new suburbs emerged along the watershed west of Austin in the 1950s and 1960s. Most were filled with comfortable ranch houses on large properties that focused outdoor leisure activities on private backyard spaces and serene water landscapes.[44] The undeveloped hills to the west of the city offered numerous secluded spots, which were usually enclosed and often offered private water access. In 1953 attorney Emmett Shelton developed Westlake Hills, an upscale suburb across from Lake Austin, on four thousand acres, to escape municipal regulation. An *Austin in*

Action article from 1961 referred to Lake Austin as "an all-purpose lake" located in an area that "has become almost a residential suburb of the Capital City."[45] Jonestown, developed in the early 1960s on Lake Travis, ran ads in the same publication calling itself "Vacationland U.S.A.," advocating for its "leisurely, pleasure-filled way of life," and encouraging people to move there because their investment would be protected by "appropriate zoning."[46] Lakeway, just west of Austin on the south bank of Lake Travis, also developed directly out of a resort in the 1960s. Although much undeveloped property remained public, developments along the lake system generally conformed to the suburban pattern of the day by eliminating most public open space from designs. Larger public areas, mostly state parks, were often just a short trip by automobile.

The catalyst for many hinterland communities was Barnes and Jones, a developer that built over fifteen resort suburbs. Far from Austin on Lake Buchanan they created Island Village, after building a causeway to a small, secluded island on the northern part of the lake. On the far reaches of the lake system, affordable rural vacation communities were most common. Closer to Austin, suburbs were usually marketed to a more upscale clientele, oftentimes as more traditional suburban developments. Lakewood Village offered paved, curbed, and guttered streets, as well as water and electricity, and was only steps from a free boat launch. Lago Vista became one of the most exclusive resort suburbs and was marketed as a blend of rural leisure and the most modern of cosmopolitan convenience. It was close enough to Austin for commuters but could also attract retirees and seasonal residents looking to avoid bigger cities and enjoy "recreational living at its best" in a "distinctive resort city." One advertisement asked consumers to "picture [themselves] in one of the delightfully different all electric homes on paved streets, complete with public water supply, garbage collection service, street lights and telephones," while also ensuring homogeneity with "highly restrictive" policies. Restrictive covenants were not just offered to residents of the more economically exclusive communities. Every Barnes and Jones development included protective zoning, even for modest tracts that could be purchased with no money down and only $10 per month.[47]

Austin planners and politicians also viewed the area's outdoor culture and quality of life as draws for both white-collar laborers and the firms that employed them and saw water as part of a growth program that emphasized skilled white-collar labor. Professional planning documents from the late 1940s stressed the importance of attracting an educated labor force and focused attention on overall quality of life as a key in locational choice among skilled workers. As research and development became prioritized in the wake of World War II, cities began to see knowledge work as a way for them to grow while not becoming more urban.[48]

Well into the 1960s the chamber of commerce's *Austin in Action* published monthly articles that viewed the lakes and the natural environment as central to Austin's social and cultural life while encouraging development for economic gain. The retail potential for items related to boating, fishing, hiking, and other outdoor activities was obvious, and outfits catering to the recreational needs of tourists and, increasingly, leisure-minded residents were among the first to open along the watershed. In 1958 Austin boat sales increased by 500 percent. By 1959 an estimated ten thousand boats cruised the lakes, many of these crafts owned by new lake residents.[49]

Articles focused on the natural beauty and leisurely pace around the lakes, while continuing to envision new uses for them. "The Lure of the Lakes" is a particularly descriptive and colorful article that demonstrates the importance of the controlled water system in Austin's collective consciousness as well as the importance of natural beauty and privacy to the city's marketing campaign and lifestyle. The article begins with a landscape photograph from iconic Mount Bonnell, which overlooks Lake Austin and the pastoral, undeveloped panorama bifurcated by the river, toward which a solitary couple is seen strolling. The article blends images of Austin's indefinable natural characteristics, referring to the chain of lakes as the "glamour girls" of all resorts and "paradise for an enchanted land." It continues, "They're not sophisticated, and they're not seductive—but they have an aura of charm and excitement which is captivating." This rhetoric both differentiates the languid, leisurely environs of Austin from more fast-paced, "sophisticated" areas and encourages the reader to concentrate on the lakes' "aura," an indescribable, unspoken quality similar to a sense of place. The article goes on to contrast the natural, rustic beauty of the lakes with the region's determination to harness that nature, but in a gentle manner: "Man, with his engineering ability and far-sightedness, has joined hands with Nature, who so richly endowed this Colorado River area with rugged and picturesque setting, to form one of the most fascinating stairways of lakes ever assembled."[50] The ability to enjoy nature's beauty was an important component of good urban living in Austin, and access to nature was constructed as limitless, despite the growth of development, much of it private, along the water on the western edge of the city and in the suburbs.

Even in downtown Austin, water was showcased as pastoral and surrounded by manicured public green spaces after Longhorn Dam, the final dam in the system, was erected in 1960. A small reservoir, called Town Lake, formed the southern boundary of downtown and extended into the Eastside of the city under the interregional highway bridge. The dam's completion ensured that the reservoir would always be filled to the same level and would never again destroy property in the central part of the city. The last infrastructural improvements on the lakes, right in the heart of Austin, had dramatic effects on the city and demonstrate the extent to which water and natural space dominated Austin's image, both as a perceived public good and as an economic engine for the city. By the late 1960s the reservoirs and public parks surrounding them in the urban core defined residents' vision of Austin's quality of life as pastoral urbanism.

Showcasing a Region: Austin Aqua Festival

Perhaps nothing illustrates the importance of natural images and the turn toward water as a major cultural attraction more than the Austin Aqua Festival (AAF), which debuted in 1962 and became an annual event through the 1990s. The AAF was an example of festival tourism, a spatially localized, temporally contiguous event designed to celebrate a place's unique, authentic characteristics. Festival tourism represents a quick injection of capital into the economy and a form of place marketing, but not something the economy relies on or something that negatively affects local culture. Festivals also must create, through entrepreneurial vision, a sense that what they are celebrating is meaningful and worthy of spending money on for consumers.[51] In Austin the AAF was an obvious symbol of what was unique: abundant water, especially dramatic during the intense heat and dryness of typical Texas summers. It was also a celebration of civic pride for many Austinites; in the festival's first year nearly four thousand citizens volunteered. Most of all, the event allowed the city to show itself off. From the new municipal auditorium to the far reaches of the Highland Lakes, visitors and locals alike were encouraged to look at what the city and region had to offer and to envision Austin as one small part of a much larger natural region, focusing on the seamless transition from growing city to rustic hinterland.[52]

The festival was created shortly after Longhorn Dam was finished and Town Lake's supply of water was finally regulated. In 1961 the Austin Ski and Boat Club sponsored a local waterskiing meet on Town Lake. Shortly after, Tom Perkins, manager of the Austin Chamber of Commerce Tourism and Recreation Department, attempted to organize a water festival. The idea was accepted almost immediately by the chamber and then quickly by

Miss Aquafest contestants pose in front of Lake Travis in a convertible, circa 1966. PICA 15991, Austin History Center, Austin Public Library.

the city council. The two groups cosponsored the festival, providing a major public-private partnership intent on expanding the local economy through the use of water. In 1962 the city and chamber hosted the first of the more than thirty AAFs that took place. The festival's success is indicative of the popularity of water recreation in Austin and of effective event marketing and practices.[53]

The specific features of the AAF reveal ideas and themes that are of great importance to understanding how the new envirotechnical system became the dominant image of Austin's landscape and how appealing this image was to both residents and visitors. The chamber and city were not at all shy about publicizing their intentions for the festival, and they had them published in all sorts of media, from newspapers to flyers and magazines. The goals were fairly simple. The first and most obvious was to publicize the water and recreational resources of the Highland Lakes to both regional and national markets in the hope of drawing long-term business revenue, along with potential newcomers to the region. As with most other promotional opportunities, the chamber and the city worked together in an entrepreneurial capacity, actively seeking investment and long-term capital for the region.

The second goal was the more immediate economic benefit of a yearly festival that would bring tourist and vacation dollars to local businesses, especially during the hot Austin summers, when consumption usually plummeted and businesses often closed or curtailed their hours. This "stimulus," as the chamber called it, provided short-term relief and seasonal stability to a host of small businesses in the area. The AAF was imagined as a source of civic pride and community building that emphasized cultural history and contemporary uses of urban and rural spaces. It offered residents and outsiders alike a chance to create a shared regional history and culture that illuminated positive qualities—from the environment to the high-tech industry to the region's cultural diversity and ethnic heritages. Austin residents were encouraged to take ownership of the festival by volunteering time and by enjoying the myriad entertainments provided by the city. Through 1964 about 80 percent of AAF attendees were from Central Texas, while thousands also came from other areas of Texas and from surrounding states. In the mid-1960s organizers began national marketing campaigns that focused on water and recreation, and boat racing and waterskiing competitions were televised to national audiences.[54]

The ten-day event was structured in a way that emphasized the entire region rather than just the city or the downtown, although most of the eclectic events were situated near the downtown area around the city's new municipal auditorium and on green spaces on the south bank of Town Lake. Each year the AAF kicked off with a four-day canoe race starting at the far reaches of the uppermost lake, Lake Buchanan, and ending in Town Lake near downtown Austin. Canoe teams from all over Texas assembled into relay groups that took turns maneuvering their crafts down the lakes and intermittent stretches of river. The event emphasized natural continuity, focusing viewers' attention on the unbroken string of water extending well over one hundred miles. It also emphasized the continuity between the vast rural hinterlands to the northwest of the city, the upscale resorts and new suburban neighborhoods along the river in western Austin, and the downtown core, as well as displaying Austin as a city rich in natural beauty. As a natural extension of the lakes, Austin provided the central location from which to enjoy them. The festival officially began as the competitors finished the test of endurance amid large crowds downtown. Throughout the festival, events were held in towns all along the lakes, drawing people to outlying areas as well as to the city, emphasizing both the functionality of the dams and the form of the river and its environs.[55]

Other AAF events focused on Austin's water resources, but they blended these with other forms of cultural production and outdoor competition. All the AAFs throughout the 1960s and early 1970s featured competitive water sports geared toward visual consumption and excitement for crowds. The large lake system was perfect for water sports, which drew competitors from all over the country. Professional waterskiing was a main attraction, and by 1966 the festival hosted the National Water Ski Kite Flying Championship as well as other regional waterskiing competitions. Sailing regattas were held annually along with amateur fishing contests, and by the mid-1960s professional drag-boat races were held on Town Lake and shown on ABC's *Wide World of Sports*. Sporting events illustrated the multiple recreational possibilities for the lakes, but they were not the only water-based events. Aquacade, a floating parade and concert on Town Lake near downtown, emphasized civic pride and economic activity right on the water. At the first AAF in 1962, television personality Art Linkletter crowned the "Queen of the Lake" as part of the pageant of the same name. The queen and her court provided many photo opportunities on the water and were one of Aquacade's main draws. The Rio Noche Parade, a nighttime event on Town Lake that showcased Austin's cultural attractions and Tejano distinctiveness, was one of the single biggest draws of the festival, entertaining about 150,000 spectators in 1968. Tennis and golf tournaments were also held in and around the city, and in 1966 the first Carrera de la Capital Auto Race, held at a track north of Austin, and the Commodore's Auto Show demonstrated a growing interest in land-based races and automotive ingenuity. The earlier festivals concluded with a public gospel "singsong" event in Zilker Park, the city's largest.[56]

As the AAF evolved over its first decade new events were created, many of which did not directly involve water. These forms of cultural production were, however, linked to Austin's water-based form of place making simply by being promoted as part of the AAF. One of the most important and interesting of these new events was the series of Hollywood movie premieres at the festival, beginning with the world premiere of *Batman* in 1966. With Adam West and Burt Ward on hand, the Paramount Theater in downtown Austin hosted two screenings of the film, both of which sold out quickly. Where the earlier AAFs had multiple musical concerts, by 1966 the most popular concert was a "Battle of the Bands" held each year at Municipal Auditorium. By 1972 the battle had become so popular that individual tickets needed to be purchased for the two-day event, rather than just having the doors open to the public as was done previously. Film premieres and concerts, begun at the AAF, soon evolved into Austin's premiere tourist attractions and engendered other festivals.[57]

The festival also began offering events related to Bergstrom Air Force Base and the Texas-based National Aeronautics and Space Administration (NASA), demonstrating the growing importance of technology industries to Austin, many of which focused on aeronautics. Throughout the 1950s and 1960s, nascent private research and development and light manufacturing firms and the University of Texas established Austin as one of the Southwest's premiere high-technology centers. Begun in 1966, the NASA Space Exhibit promoted interest in space travel and held educational demonstrations for children. By 1969 the festival included Austin Aerofest, which took place at Bergstrom Air Force Base just southeast of the city. Highlighted each year by parachuting servicemen and an air show put on by pilots, Aerofest rapidly became one of the AAF's most popular events, drawing eighty thousand spectators in 1970. Crowds were encouraged to physically engage with airplanes and to marvel at their jet engines, control panels, and sleek exteriors. Demonstrations of military prowess also played to Cold War crowds eager to feel secure in the United States' technological dominance, cultivated national pride as the Vietnam War became more of a reality, and highlighted the region's growing significance as a center of high-technology research. They also linked high-technology work, mostly carried out in new office parks or university facilities on the urban perimeter, with suburban residential and recreational styles of living.

As the technological aspects of the AAF flourished, so too did entertainment that emphasized and also created regional and local traditions. The binary between technology and tradition collapsed during festivals such as the AAF, which provided imagined spaces where sanitized regional history and modern technologies could comfortably coexist. The technical, scientific, futuristic aerospace events were complemented and balanced by events designed to evoke nostalgia for a more rustic vision of frontier Texas. By the 1970s, ethnic-based components emerged at the AAF, where evening activities were focused on a particular form of ethnic cuisine, dance, and culture usually based on Central Texas's prominent ethnic heritages at the time, German, Mexican, Czech, and African American. Rather than complicate narratives of ethnic encounter on the Texas frontier, the AAF's representations of traditional cultures evoked a sense of local pride and western egalitarianism set against a suburban landscape of increasing homogeneity. Water was presented as a democratizing force, but in reality access to water and natural spaces accentuated segregation.[58]

Although the AAF promoted ethnic diversity, its leadership structure and planning committees came from the upper reaches of Austin society. Their AAF was centered on the Admirals' Ball, an invitation-only dinner that was hosted by that year's festival president, commodore, and vice commodores and where each year's Queen of the Lake was crowned. The leadership of the festival planning committee was set up as a fairly rigid hierarchy with the president and commodore doling out responsibilities to vice commodores, who oversaw specific aspects of the festival. The Admirals Club threw the event and essentially planned the entire festival throughout the year; each ball had a geographic and ethnic theme, usually associated with an exotic port from around the world. The choice of navy ranks for the leaders of the AAF is an obvious indication of the special place water had in Austin, even among business and political elites. AAF leaders and special guests were given badges to wear to demonstrate their positions in the AAF hierarchy.[59]

The AAF was a type of place making that blurred the boundaries between city and hinterland while reinforcing social hierarchies, incorporating images of ethnic diversity, and emphasizing leisure pursuits. It was also a celebration of a resource that differentiated Austin from surrounding cities while simultaneously locating the city in the decentralized, naturally integrated image that separated less concentrated western urban landscapes from eastern cities and from larger, sprawling Texas cities. The diversity that the AAF celebrated in performance turned ethnicity into something to be viewed, performed, and consumed. The festival created an egalitarian image for the city that reflected western myths of egalitarian heritage and the declining importance of racial and ethnic difference, yet daily life was still largely lived in the shadow of the South.

Water and Nature as Possessive Accumulation Strategy

The success of tourism, suburban development, and the water-based festivals in Central Texas allowed the region to resist extensive industrial development while still promoting and encouraging regional growth. Anti-industry efforts, leisurely lifestyle, and reputation paid off in March 1965, when *U.S. News and World Report* ranked Austin as one of the fourteen most desirable places to live in the United States. The same publication later published a Bureau of Labor Statistics report that found that Austin was also the most inexpensive metropolitan area in the country. By the late 1960s the nonindustrial growth imagined by boosters in the early 1940s had become a reality and the region had developed a natural yet cosmopolitan image.[60]

During the first four decades of the twentieth century, progressivism in Austin was defined by rational, scientific attempts to harness nature, organize space, and modernize infrastructure. In the post–World War II decades Austin's political and business elites sought to enhance economic and demographic growth using these improvements. They were uniformly entrepreneurial and imagined that Austin was in competition with other cities for resources and investment; they sought to differentiate Austin from other cities by highlighting its distinct advantages and channeling resources into nonindustrial modes of production. During this era, being progressive in Austin entailed supporting strong ties to the federal government as a source of funding and as a way to avoid industrial development.

By the 1960s, Austin residents expressed a new urban vision by integrating their daily lives into natural spaces, reimagining the region as linked by the river, and celebrating Austin's natural abundance. It was traditional antiurban motifs—the sanctity of nature, open space, wholesome fun, without the loss of amenities—that functioned so powerfully in Austin and made it attractive. Essentially, however, Austin elites viewed the AAF as an economic development strategy. Boosters could claim that they had transformed Austin into the "Playground of the Southwest," a blend of leisure, knowledge labor, and environmental amenities that generated wealth and growth without undermining quality of life.[61] Recreation and outdoor fun, however, was just part of the broad strategy for accumulation that Austin elites devised in the years following World War II. Taking full advantage of the city's potential, they argued, meant more than just transforming the natural landscape. The city's other great resource, its university and ready-made skilled-labor pool, remained essentially untapped. Thus Austin's political, business, and intellectual leaders began constructing a social and physical landscape that would keep knowledge workers in Austin and attract knowledge-based industries to the city. Their efforts are explored in chapter 5.

5 Industry without Smokestacks

Knowledge Labor, the University of Texas, and Suburban Austin

In the fall of 1947, amid fear of a potential return to depression and uncertainty over the mounting housing crisis, businessman C. B. Smith and the Austin Chamber of Commerce hired New York–based planner Richardson Wood, Smith's friend, to chart the city's course for economic development. Rather than plan the physical design of the city, the chamber charged Wood with assessing Austin's socioeconomic infrastructure. How can the city take advantage of its natural advantages and grow in the best way possible? Wood, a seasoned planner who specialized in small cities, prepared his report in less than a week. In early 1948 he submitted his study, titled "Outline for a Plan for the Future Economic Development of Austin, Texas," to the city council. While Wood summarized the macroeconomic changes bringing capital and people to the South, the mounting hostility toward labor unions, and the growing independence of southern financial institutions, his message for the council was straightforward. In Austin the desirability of businesses should be judged by "the character of their personnel rather than the material field of interest. . . . Businesses that have a large professional and skilled element [are preferable] to businesses that have a large unskilled or semi-skilled element." This simple, yet novel, idea reinforced dominant economic ideology in Austin and became the foundation for Austin's postwar planning regime: create an economy and lifestyle centered on skilled labor. While many southern cities were beginning to enter the industrial and manufacturing market after World War II, Austin's focus was increasingly on skilled personnel. In the emerging atomic culture of the postwar period, the knowledge economy, economic and industrial decentralization, and place production became increasingly intertwined. Cities with an abundance of well-coordinated technological and human resources were thus primed to prosper—particularly cities that were not considered major targets for attack.[1]

From the end of World War II to the 1960s the relationship among labor, politics, and spatial organization changed dramatically in American cities.

Owing to the Cold War, the federal government began to prioritize defense-oriented research and development, and politicians increasingly saw research universities as primary sites for technology and science, which meant lucrative investment. At the same time, federal policy incentivized suburbanization, dramatically altering urban geography and decentralizing population and production in nearly every American city. As heavy industry, traditionally the leading sector in urban economies, decentralized from urban cores and eventually fled American cities altogether, cities with the ability to produce and develop technology-based industries stood poised to flourish.[2] Research universities became central sites of knowledge production and also increasingly centers of political and economic power in cities.[3] The federal government's role in capital and other investment favored less economically mature and less urbanized regions such as Texas where flows of capital migrated to areas and cities with higher rates of return.[4] Fear of nuclear attack during the Cold War prompted the federal government to incentivize industrial dispersal, which drove federally sponsored investment for defense-related industries to less concentrated areas in the Sunbelt.[5] But investment in the South was also facilitated by wide-ranging changes in the southern economy: entrepreneurial urban and regional marketing with the goal of bringing in new industry. In many southern cities business coalitions that emphasized economic growth replaced old guard regimes more interested in maintaining rigid socioeconomic hierarchies.[6] In Austin, leaders actively sought economic growth through a variety of models focused on distinct local advantages: low utility rates and cost of living, a friendly business climate, unique place-based characteristics (such as the Highland Lakes region and modern suburban homes),[7] a highly skilled labor market, and especially the University of Texas, which became the central feature in Austin's attempts to attract knowledge capital in the 1950s and 1960s.

During this same period, Austin's political and economic elites became decidedly entrepreneurial; with the city physically modernized and racially segregated, they began to turn to attracting and generating business. They identified research and development and knowledge labor as leading sectors and then transformed the city into a place that would lure and retain technology firms and their knowledge workers. They understood that competition for lucrative federal contracts was intense, and they envisioned their role as facilitators in engendering and attracting knowledge-based businesses. The chamber of commerce developed national advertising campaigns that showcased Austin's attractive amenities and lifestyle for knowledge workers. Chamber members worked closely with university administrators and professors, some of whom held dual roles with the university and the business apparatus, to generate private growth and secure federal funding. The city's relatively small size and the high degree of coordination among business, political, and academic leaders were advantages in attracting capital. That Austin was the site of the flagship university and the state government helped ensure high levels of employment and a well-educated population, which promoters could point to as examples of the region's high quality of life. They actively eschewed heavy industry and economic sectors that necessitated a large low- or mid-skilled labor force and discouraged residential construction that was affordable for lower-class residents.

By the late 1960s the growing knowledge sector had transformed Austin spatially as well as socially. Spurred on by a large, suburban University of Texas research park and a number of high-tech startups, developers turned formerly agricultural or undeveloped land into modern, upper-middle-class subdivisions that catered to knowledge workers and other white-collar professionals. The spatial division of labor in Austin, which never had a large industrial component, reflected a suburban lifestyle in proximity both to work and to natural space. Most new homes were situated on Austin's hilly western and northwestern periphery, generally considered the most desirable and picturesque land in the area. Federal subsidies aided in the promotion and development of suburbs, highways, and shopping malls.[8] The city used federal funding to build highways that connected suburban knowledge areas to other suburban knowledge areas and to residential subdivisions and eventually to suburban shopping malls. The city rapidly expanded north and west, but it grew little to the east; demographically, whites saw large population gains, while the African American population remained stagnant. Life in the workplace was likewise transformed to meet the needs of knowledge workers. University administrators and private research scientists extolled the virtues of scientific creativity and augmented traditional workplace relations to reflect project-based work that allowed knowledge workers flexibility and nonhierarchical modes of production. Working life for non–knowledge workers remained more difficult, as antiunion sentiment remained strong in Austin after World War II.[9]

The growth of scientific labor and of research and development–oriented firms in postwar Austin was the outcome of both federal policies that favored universities as the primary sites of defense-based research and local efforts to take advantage of federal subsidies. Although the University of

Texas never approached the level of investment garnered by top universities such as Stanford and the Massachusetts Institute of Technology or generated the remarkable number of spin-offs that they did, it became a regional leader in defense contracts, knowledge production, and high-tech startups by the late 1960s. Importantly, the city's economy and geography were also deliberately planned to attract knowledge workers, develop desirable spaces for them to use, and keep them away from the city's industrial functions and minority residents. As with water, the city's growth model yoked suburbanized space, knowledge labor, and environmental amenities to whiteness.

Defining the Region: "Industry without Smokestacks," Federal Investment, and the Origins of Human Capital

Beginning during the New Deal, and accelerating during and after World War II, the federal government took on a new and expanded role in facilitating and directing the economy. Regions with less growth, such as the South and West, attracted higher levels of federal investment during the New Deal because their infrastructure and economies tended to be less developed. During World War II, military investment likewise favored these regions because they were cheaper, less dense, and farther removed from major centers of production. The state of Texas, for example, received over $7.4 million in building contracts from the War Production Board between 1940 and 1944, which paved the way for increased industrialization and economic growth there.[10] Austin was chosen as the site for a large air force base as well as a magnesium plant during World War II, both of which were crucial to the city's wartime economic health because they each employed over one thousand residents. Both were also located far from the city's center, roughly seven miles outside existing city limits, in the midst of agricultural land.[11]

Nationally, wartime technological needs created an emphasis on scientific knowledge and the coordination of that knowledge applied to the research and development of military technology. Scientific knowledge was increasingly viewed as a commodity, and people who commanded that knowledge were increasingly attractive as forms of human capital. Austin business and political leaders understood this new focus on knowledge labor as an opportunity to diversify the city's economy away from agriculture.[12] In 1943 G. S. Moore, Austin's city planning engineer, created the first report on postwar urban planning for the city council, titled "Planned or Unplanned Growth." Moore surmised that industrial growth was very feasible in Austin because of its small size, its relatively small amount of industry, and the absence of legal restrictions in the undeveloped land surrounding the city, which indicated that the city could easily zone any outlying area for industry if it chose to do so. Moore, however, was equally concerned about keeping industry out of the central areas of the city, encouraging the council to develop the downtown and riverfront for recreational and consumer purposes. Austin's primary attractive features, its livability and pristine environment, would not benefit from centrally located industrial production facilities. He advocated removing the railroad from West Austin, essentially discouraging any kind of industry near downtown.[13]

After the war ended, and amid concerns that the depression would reemerge because of the drop-off in production and the return of soldiers, the chamber of commerce hurried to assess the city's economic situation. In 1946 chamber president Walter E. Long received "Planning of Research and Development Work," a pamphlet indicative of the growing importance placed on knowledge. The pamphlet asserted that the need for efficiency during the war effort brought together planning and research for the first time, but the greatest change that the war created in terms of business was that "the key productive unit in Research is the scientist." Scientists' social capital also increased markedly, as technical jobs in the private sector, academia, and directly with the federal government flourished and cities sought to create incentives to attract skilled workers. In the immediate postwar period the dearth of American scientists and engineers was viewed by many as the most critical issue in terms of human capital, and universities across the country, as well as the federal government, rapidly began prioritizing research-based science and engineering.[14]

Richardson Wood confirmed the chamber's beliefs about the centrality of knowledge to Austin's growth ideology, and he likewise linked it to the city's desirable natural qualities and lifestyle. Wood felt that, as a part of a growing region but still a relatively small city that lacked manufacturing and large-scale commercial activity, Austin should take advantage of its existing attributes rather than make a radical shift. This policy imagined the city as a niche market that would not regularly compete with larger Texas cities, such as Dallas or Houston, for business. In 1950 Travis County had the lowest percentage of manufacturing employment of any county with over one hundred thousand residents in the United States. Only six manufacturing outfits had more than 250 employees, and none had more than 500, despite Austin's first major industrial relocation in 1947, by

Jefferson Chemical. Wood encouraged the city council to advertise Austin as a pleasant city with a wealth of natural resources, especially the Highland Lakes and their environs, which differentiated Central Texas from the more arid, and more industrial, metropolitan regions of the Southwest. The state government and the university, with large percentages of professional workers and professional services, could be used to attract more skilled professionals in the private sector. Wood thought promotional literature should also market Austin's cultural institutions and natural landscape, which would appeal to the types of workers that fit into the city's profile. Above all, Austin needed a unified economic policy that marketed economic opportunities to businesses and directed growth from within.[15]

Wood wisely foresaw the university as the primary locus of economic potential, both because of its ability to facilitate business and as a producer of the increasingly sought-after commodity, human capital. The major problem, however, was convincing potential businesses that Austin's cultural and educational assets could be industrial attributes; this meant selling the university as a center of knowledge production and convincing businesses that knowledge workers would like living in Austin more than other places. Business and engineering departments would engender small shops and design laboratories that could then generate other small manufacturing operations; he discouraged large firms.[16] One of the keys to successful growth was to create technical industries that would absorb surplus labor and allow University of Texas graduates to have careers and live in Austin rather than leave the city because there were no jobs for them. This policy aimed to create the infrastructure that would allow Austin to take advantage of its primary resource: a skilled labor pool that would reproduce itself every year with almost no local capital investment. Not all businesses would be suited to Austin, but determining those that would be was paramount. Wood surmised that "Austin should aggressively go after all businesses that seem better suited to what Austin has to offer than they are to any other city in the region." In an era of increasing interurban competition, Wood advocated for cooperation between the city, the region, and the university in an attempt to develop Austin's already strong reputation as "the Friendly City," its longtime moniker. Wood's plan indeed laid the foundation for the urban marketing that Austin began only months after his report was unveiled at the first Austin Area Economic Development Foundation (AAEDF) meeting in April 1948. But in keeping Austin "friendly," Wood also implicitly argued that Austin's destiny was as a white-collar, professional city that did not intend to benefit the working classes, or the racial minorities, that made up a good deal of the population in the late 1940s. *Friendly* was thus imbued with notions of class and race.[17]

The AAEDF, initially made up of some of the most powerful people in Austin, served as the organizing center for Austin's economic growth and marketing. The group followed Wood's recommendations closely. It stressed the need to maintain Austin's way of life as a selling point for businesses in the community and also acted as a body that would coordinate information between city, county, state, and federal agencies for existing and prospective businesses. The AAEDF also functioned as a clearinghouse for all matters of economic development, answering questions from and sending information to businesses looking to relocate from the North and East. Perhaps because of the obvious economic advantage for Austin, the AAEDF also viewed the decentralization of industry as a patriotic duty with respect to the emerging U.S. Cold War defense program and sought to take on industry for the benefit of the country.

While marketing and boosterism were essential to developing the city, creating partnerships across institutions was also paramount. Within a few months the AAEDF had entered into a legal agreement with the University of Texas in which the university's research organizations and development agencies as well as individual professors would provide service to business concerns and individuals looking for data or professional analysis. One of the first joint ventures was the Texas Personnel and Management Conference held in 1948, where the foundation and the University of Texas cohosted over one thousand Texas industrial leaders and brought in prominent business speakers from New York. The conference was designed to market Austin as an emerging center of business for the trade area and the entire state of Texas. At the conference held in 1949 the program included a speaker from the University of Chicago, Laird Bell, whose talk addressed "cooperative planning for education and industry," indicating the growing relationship between technical knowledge and business. This type of relationship between private business and academia may seem self-evident, but the pact between the university and the AAEDF was considered the first of its kind in the United States and was central to the foundation's marketing initiatives.[18]

The AAEDF also began an aggressive, national marketing campaign in late 1948, with the centerpiece being a monthly magazine. Titled *Austin and Industry*, it both promoted the city to potential businesses and informed local businesspeople about what the AAEDF was doing to promote Austin.

Along with helping coordinate Operation Waterlift (discussed in chapter 4), the AAEDF encouraged national and regional business magazines to write stories about Austin and its friendly business climate and relaxed lifestyle. In its first two years of operation the AAEDF had promotional articles published in *Business Week*, *Tide*, *Barron's Weekly*, and *Modern Industry*. The foundation partnered with the Missouri and Kansas City Railroad on a series of promotional articles in the magazine *News Reel*. By 1950 the AAEDF and the University of Texas were also beginning to work together to promote scientific and engineering research as a business, rather than just business services for prospective companies. In that year the groups sponsored an aeroballistics symposium at the university that was jointly attended by academic scientists, businesspeople, and private research scientists. *Austin and Industry* promoted it and reported on it in an article that also discussed the new defense contracts being fulfilled by multiple research groups at the university.[19]

Most importantly, *Texas Parade* wrote a piece titled "Industry without Smokestacks," in November 1950, which became a definitive statement of Austin's dual emphasis on attracting technical skill and clean industry. It also sought to differentiate Austin from regional cities whose economies were increasingly based on more intensive, largely Fordist types of industrial production, such as oil refining, aerospace and defense manufacturing, and petrochemicals. The AAEDF sought planned and diversified economic expansion that was "consistent with the high character of the city," an obvious nod toward Austin's highly educated, white-collar population. "Research and development laboratories," "artistic skill," and "creative talents in technology and design-professional activities and skilled-labor endeavors of all types" were all phrases describing human capital and light industry found in *Austin and Industry* by 1950. Austin would let other cities attract "heavy industries and other large-scale production and distribution facilities," one article claimed, while Austin concentrated on "specialized technological activities" undertaken by smaller, more specialized outfits. Light technological production was natural for Austin, directly in line with the city's "individual magnetic forces." This included fostering startup companies, which could be little more than a graduate student with an idea for a company, and also attempting to find outside capital to fund local startups.[20]

Early postwar growth was robust. Between 1938 and 1949, for example, income for Austin residents increased by 236 percent in real dollars; much of this was caused by a near 400 percent increase in state and federal payrolls, which made up over one-third of Austin's total income in 1950. Between 1940 and 1950, Austin's overall income rose by 22 percent. Roughly 25 percent of all Austin workers were employed by all levels of government. Trade and service income was on the rise as well, accounting for almost exactly one-third of Austin's total income in 1950. By contrast, manufacturing represented only 4.1 percent of total income in Austin (while it was reaching its historical high of over 30 percent in the United States in 1950), a clear indicator of the disinterest in creating a traditional industrial landscape in Austin. These numbers also suggest that Austin was largely benefited by being the "center" of Texas, since the State of Texas payroll, the University of Texas payroll, and university student spending were all main sources of revenue in Austin. The University of Texas payroll doubled between 1944 and 1950, to over $9 million, and student spending in Austin rose to nearly $18 million in 1951. Clearly, Austin was growing rapidly, largely because of significant increases in government and university operations.[21]

Still, the importance of the AAEDF's rhetoric of competitive advantage, based in both human capital and natural resources, and niche place making just five years after the end of World War II cannot be overstated. Simply having investment-guaranteed employers such as the University of Texas and the state government in Austin benefited the city's economy and provided much more stability compared to other cities that lacked such assets. Universities, and especially large, well-endowed institutions such as the University of Texas, mitigated risk by absorbing much of the initial cost of research and infrastructure development. But Austin's growth was largely determined by how the AAEDF conceived of, marketed, and developed those assets. The AAEDF actively created Austin as a technical and knowledge-based center with a unique mode of production and a specialized economy that offered assets no other city in the Southwest could match to specific producers. Austin's urban marketers were among the first to imagine the university as a center of knowledge capital, and they also foresaw lifestyle amenities for potential knowledge workers as an important attractive force from a very early time. In the discourse of capitalism, what the AAEDF envisioned for Austin was a prototype for the radical shifts in the mode of industrial production that became commonplace by the 1980s and gave the city a competitive advantage. At the university, and especially at the growing Off-Campus Research Center, a new approach to the knowledge economy and new ties between academia, business, and profit were being developed.

Decentralized Knowledge Labor and Academic Capitalism at the University of Texas

University of Texas administrators saw the war as an opportunity for entrepreneurial activity related to knowledge production. The university had limited success in earlier efforts to expand via research. Most of its research efforts before World War II were in service to Texas industries and, while valuable, they were not lucrative.[22] One of the first university efforts to profit directly from academic knowledge was the Texas Research Corporation, a private entity founded by the board of regents in 1940. Its sole purpose was to acquire, own, and use patents created by researchers employed by the university and to contract with private businesses and government agencies that wanted to use university patents. Shortly after the corporation was established, the regents applied for patents on air-conditioning improvements and flash-freezing techniques that were developed by various engineering departments. Scott Gaines, a University of Texas lawyer, argued that although the university could not sell patents, it could profit from contracts with private businesses to put the patents to use and to do research. In effect, the corporation was one of the first university-sponsored entities to understand knowledge as a raw material and to actively attempt to commodify scientific knowledge through legal means. It also created the legal status necessary to protect university research copyrights, which had the added benefit of encouraging scientific research by faculty members, who now had a new profit potential as added motivation.[23]

The university's unique ability to centralize and coordinate academic knowledge also brought in some defense contracts during the war through two groups called the Defense Research Laboratory and the Military Physics Research Laboratory,[24] which undertook research and development while fulfilling contracts for the Navy Department's Bureau of Ordnance, working on machine-gun technology. These groups expanded drastically after the war ended, and military research and development became the primary driver of postwar economic growth, particularly in the Sunbelt. The university, though, would need a dramatic expansion of its scientific facilities and an even greater coordination of research-based resources.[25]

After the war, University of Texas administrators and local politicians were emboldened by the success of federally sponsored wartime research and they sought to take advantage of continued investment. Congressman Lyndon Johnson and Austin mayor Tom Miller, both ardent New Dealers with successful experience securing federal funding for infrastructure, also saw growth as paramount. Before the war ended, the chamber of commerce and a group of university professors recognized that future growth could be driven by the magnesium plant. For the university, the existing infrastructure was suitable to at least begin expansion of research facilities that the engineering departments, bureaus, and board of regents desired. Yet spatial constraints and a cramped main campus had long been a problem for the university's science and engineering departments.[26] The plant would likely still have a good deal of technical equipment that could be immediately put to use in laboratories, and the complex would certainly provide much needed space away from the confines of the centrally located main campus. The plant was to be divested at the end of the war by the Reconstruction Finance Corporation in accordance with the Surplus Property Act of 1944, which allowed the chairman of the War Assets Corporation to sell assets at a discount. Production stopped in late 1944, and the plant remained mostly abandoned and unused, far from Austin's center, until after the war ended in August 1945.[27]

While some of Austin's business leaders wanted to keep the facility as a private magnesium plant, by early 1946 University of Texas administrators and a handful of civil engineering professors convinced Miller and Johnson that the plant should be converted into a research facility. Miller sent out two civil engineering professors, C. Read Granberry and J. Neils Thompson, along with University of Texas president T. S. Painter and a representative from the Reconstruction Finance Corporation, to see if the site was suitable for housing in light of the severe housing shortage after the war. Instead, the group found a potential research center. Granberry, a longtime growth advocate for Austin, wrote to Johnson that "the section of the plant requested would be a fine nucleus for a top flight research center for years to come." As Granberry had just reached a blanket agreement with another government agency for a substantial increase in research contracts, the timing for his find could not have been better.[28]

Johnson wrote to the Surplus Property Administration asking that the plant be disposed of for commercial purposes. Johnson and Thompson began negotiations and acquired the plant in 1946. The Off-Campus Research Center quickly became the center of military-sponsored research and development at the University of Texas.[29] In April 1949, the future of the Balcones Research Center (BRC)[30] was solidified when the War Assets Corporation approved the transfer of the property title to the university. Thompson, who quickly assumed the role of lead professor in the negotiations, was able to get the corporation to reclassify the plant from industrial to nonindustrial, which secured a 100 percent price discount rather than the customary 70 percent. Instead of paying for the facility in cash or through a loan, the transfer stipulated that the university would pay for the $8 million facility by undertaking research over a twenty-five-year period that would benefit the public good, meaning that once again the federal government was essentially investing a large amount of unencumbered capital into the university and Austin. It was the U.S. government's largest equipment transfer and the sixth-largest real property dispersal to an educational institution after World War II. The transfer also made the nascent expansion of scientific research into the primary goal for the university, as the space and capital needed for a long-term research agenda were secured. The AAEDF was excited as well; President C. B. Smith, who had already begun preliminary work to assist with the expansion of university research and development, called the facility one of the three foremost scientific research centers in the country. The BRC had already played a significant role in attracting the Jefferson Chemical industrial research lab. It was expected to house numerous labs run by various university groups, many of which were funded by a growing number of government contracts.[31]

Like Frederick Terman at Stanford, Thompson wanted to engender and facilitate growth, publicly for the university and privately through university assets, which included graduates. He was instrumental in defining the BRC's research trajectory and in implementing policy and working directly with private and government concerns to create an economically self-sustaining research facility. Thompson, as vice president of the Austin Chamber of Commerce, also coordinated efforts directing economic growth in the 1950s. As a scientist with a particular skill in management and human relations, Thompson created a system of horizontal integration for science laborers and defined a particular mode of production that came to dominate Austin both within and outside the university.[32]

For Thompson, the major function of the BRC was to facilitate all types of sponsored research in an effort to grow the university, promote Austin as a scientific city, and increase the state of Texas's industrial strength. His commitment to the BRC as an engine of economic growth was a strategy that grew out of the obvious need for new scientists and engineers at the outset of the Cold War and a national defense program that was willing to direct an incredible amount of capital to universities toward that end. Universities would in turn fulfill technical and research contracts for government departments and simultaneously train more undergraduate and graduate engineers than ever before. As early as 1948, the federal government was responsible for 53 percent of the organized research in the United States. Thompson saw that the BRC could provide something that most other universities' facilities could not: research partnerships with private firms that found the space and materials at the BRC attractive. The city's payroll and the university's funding and prestige could grow together if the BRC was properly organized and administered. And the BRC could further the public good in Texas by assisting with the state's industrial expansion as well as providing a surplus of skilled labor that would fill the needs of private industries as Austin's technological and research economy grew. Above all, Thompson clearly understood that technical knowledge was increasingly at the root of competitiveness and productivity, both in the public research domain and in the private industrial arena. Austin, unbound by Fordist-style production facilities and unskilled labor, and boasting a continuous flow of skilled graduates and now the BRC, was in a prime position to take advantage.[33]

Beginning just before the BRC was transferred to the University of Texas in April 1949, Thompson outlined his plan to develop the facility in a paper he delivered to the Southwest Section of the American Society for Engineering Education. The greatest challenge facing the physical sciences at the outset of the Cold War was the lack of established research scientists able to work and teach at the university level. This shortage was, to many commentators, an issue of national security and safety as well as education and was the focus of President Harry Truman's Scientific Research Board studies, which recommended a dramatic increase in federal subsidies for higher education research. Thompson was keenly aware of what was at stake. In 1948 John R. Steelman wrote a report to Truman on manpower and research, which contained a section called "The Crisis of Science in the United States." Steelman found that there were not nearly enough scientists to carry out the necessary research work and to train future scientists. Thompson saw this shortage as an opportunity to both attract federal funding and define a higher education agenda that placed an emphasis on applied research and teaching as a dual function of university professors.[34]

Thompson seized the opportunity by outlining a plan for development that focused on his role as director of the BRC and also as coordinator of research for all the university's engineering departments. He recommended creating an independent university agency to coordinate research among the various departments and the graduate school and to offer assistance with attracting grant money for that research. The agency would plan for the use of equipment and manpower among the various research groups, work with the university administration on contracts and proposals, help to further develop patent policies, acquire facilities and equipment, and report research to the media. Coordination created efficiency, and it also was a potential asset in attracting talent to the university, particularly if the agency was able to develop a patent policy that allowed for individual researchers to benefit financially from their work. To Thompson, creating this type of agency to manage research would give the University of Texas an immediate advantage in securing contracts and assessing its ability to carry out diverse types of research among disparate researchers.[35]

Essentially, what Thompson sought was an agency that would assume the managerial function, and let researchers concentrate on specific scientific research and teaching, while also coordinating and promoting university research and aggressively seeking sponsors. He stressed organizational flexibility and creative autonomy and encouraged small, group-centered work rather than autarkic modes of production. The key to being successful in research, Thompson claimed, was the ability to adjust quickly to the various programs that developed rapidly based on the needs of the military. Unlike private industrial research laboratories or government departments, which had a narrow focus, the eclecticism of university scientists could only be an asset if research groups could be assembled and dissolved to meet the needs of particular programs. When one program was completed, scientists were assigned to different programs by the coordinating organization. By pooling resources, the agency could also facilitate a basic level of vertical integration, where ancillary but necessary services could be centralized (e.g., a machine shop, one of the first shops installed at the BRC) at the university rather than contracted outside of it. Pooling all research department funds, as well as money coming from outside sources, also allowed for the agency to buy necessary equipment and materials in bulk, which cut down on costs.[36]

This very flexible mode of production was developed to be efficient as well as adaptable from a business perspective, but it also suited the project-based research and development model much more than a traditional industrial production model. One reason for this is related to the rise of human capital and the growing need to attract and keep the specific type of skilled laborers who could perform scientific research. Thompson viewed scientists as active, creative workers who needed both constant intellectual stimulation and a business-minded organization to free them from non-research matters. He envisioned flexibility as a tool that would improve the university from a business perspective but also as an attractive asset to scientists who needed creative change. The need to attract talent led directly to the development of innovative and flexible management styles at the BRC and eventually in Austin's research firms. A second reason was the process of scientific work. Unlike most modes of industrial production, the university research and development mode of production was not intended to reproduce specific products; its nature was rather to reorganize itself efficiently and fluidly to meet an endlessly changing variety of potential projects. The production of knowledge, as a process, is of course much less static than manufacturing a good or material, and hence a creative environment was considered more valuable than rational, efficient modes of production or depressing labor costs were. In some ways, then, the more flexible system of production employed in scientific research constitutes a revalorization of the particular, specific skills of the laborer, who was viewed as something like a craftsperson or an artisan by management. Reproduction of this type of labor power requires investing resources into potential laborers, rather than exploiting workers.[37]

By August 1950, seventeen research laboratories had been relocated or established on the campus, and only a small percentage of the building space was being used. Most of the major early laboratories were funded in large part by military-sponsored contracts. The Military Physics Research Laboratory was studying airborne flight control, while the Navy Department was sponsoring supersonic air-flow studies by the Fluid Mechanics Lab. The Nuclear Physics Research Lab had a state-of-the-art Van de Graaff atom smasher paid for and installed through a contract with the Atomic Energy Commission, and the Electrical Engineering Lab was studying radar waves for the Office of Naval Research. The Defense Research Laboratory, the oldest defense-related lab at the University of Texas, was working with one of the first mass spectrometers to be used by a public institution. Donated by Humble Oil in 1948, the spectrometer was available to the labs as well as to private Texas industries. The BRC's early success was demonstrated by the volume of contracts that were almost immediately won by its research groups. By mid-1951 fifteen of the now nineteen labs were performing government-sponsored research related to defense. By 1952 the BRC was financially self-sustaining through research contracts, meaning that its operations required no tax-supported assistance from university coffers.[38]

Thompson spent much of his first few years at the BRC in Washington securing military research contracts and making contacts in the various departments he organized among the research groups. His Washington contacts were essential both in the acquisition of the property and in much of the federal funding that underwrote the center's early operations. By 1953, however, after the BRC was operating exclusively on external contract money, Thompson began implementing long-range plans for the facility that focused more on developing business for Texas and Austin. His function was essentially that of manager: neither Thompson nor any other directors had any say in everyday research operations, and scientists were basically left alone to work with their groups. Thompson assessed spatial and equipment requirements and evaluated the groups' products. The BRC was already training over two hundred graduate students who were employed by the labs, and faculty received what Thompson called "a stimulus, a continued renewed enthusiasm" because their work varied from project to project. The center, Thompson wrote, "does not supervise." Its only function was to optimize resources. None of the labs were completely autonomous, but neither were their day-to-day operations dependent on the central managing body.[39]

While publicly Thompson addressed BRC research and lauded its benefit to the public and to the Cold War effort, by 1953 he also began to outline how the BRC could grow business and industry in Texas. In 1954 he helped create the Institute for Advanced Engineering within the Division of Extension, which offered courses to practicing engineers that helped keep them apace with new theoretical and research work being done at the university and around the country. The program was the first of its kind in the Southwest, and Thompson felt that it would play a large role in keeping University of Texas engineering graduates in Austin to enhance the city's skilled labor pool. When fully developed, the BRC would keep "research sponsored by Texas industry . . . at home, rather than going to Illinois, Michigan, and M.I.T."[40] Beginning in the fall semester of 1962, and owing to the success of the BRC's research program and improved federal funding for research, the university began to separate research from teaching completely by creating three new, nonfaculty research positions: research scientist, research engineer, and systems development specialist. The positions were the first nonacademic, nonadministrative, and nonclassified research positions at the university, and each was funded completely by state or federal grants or private sources rather than by tax dollars.[41]

The BRC provided the central technological space to attract companies looking to profit from proximity to academic knowledge. Private firms, many of which were reorienting themselves toward science and engineering and away from what Thompson called "markets, manpower, and raw materials," could increasingly be courted by cities. For cities, aggressive marketing brought in technological companies, which would then be "self-generating," meaning that they would naturally engender other similar companies due to advantages of propinquity. For Austin, this policy meant marketing the city's natural and social advantages along with its human and scientific resources as something of a work-and-leisure combination that offered constant stimulation to potential laborers as well as an early "sense of place" built on technology and leisure. As Thompson told Governor John Connally, "The most marketable product of the future is improvement of the community, so growth must be focused at the local level." It went without saying that the types of workers the policy sought to attract were those making higher wages; despite an abundance of cheap, southern blue-collar workers in the 1950s and 1960s, Thompson and most of Austin's leaders chose to court higher-wage, higher-skilled personnel who they believed would add more value to the community.[42]

Increasingly, local communities needed to assume risk and take initiative in pooling capital to attract research outfits. Thompson was "convinced that a reconversion period [was] imminent in many areas of industry" and warned that "unless we prepare for it we could experience many heartaches in adjustment and wasted manpower in technology areas." An important function of municipal and state governments was therefore to support an increasingly entrepreneurial policy; that meant not just marketing "centers of excellence" but also providing capital and attractions for potential investors and other people likely to create jobs related to technology. The city and university needed to work together to bring in industries that would keep University of Texas graduates in Austin.[43]

By the mid-1960s Thompson identified that the region's economic health was tied to technological development rather than to the earlier idea, more common in the 1940s, that attracting other kinds of industry could be beneficial to Austin. The issue of quality of life now appeared far more important as a redevelopment strategy than simple economic growth did. Still, aggressive marketing policies were paramount in attracting the capital necessary to grow in this model, and that task could be accomplished best by a cooperative growth effort that comprised the university, the city, and the state. The importance of this strategy certainly lay in the cooperative vision that it evoked, but more so in the idea that the three entities shared a particular business model, one that in later decades came to define regional and urban competitiveness more broadly, yet was clearly evident in Austin during the early 1960s. Austin and the University of Texas were both extremely entrepreneurial in their own right. This model was defined by an aggressive effort aimed primarily at attracting technology businesses and investment, bringing in talented individuals, supporting development locally, and facilitating creative and entrepreneurial enterprises both financially and socially. For both the city and the university, the BRC was the key feature in attracting human and investment capital and would enhance the conditions of technology-based production in Austin; Thompson's role was to help researchers collaborate and build social and academic networks that would, in time, become self-sustaining and eventually a marketable commodity for other creative workers and industries.[44]

Knowledge Labor, Knowledge Space

By 1960 dozens of small research and development and precision manufacturing outfits dotted Austin's economic landscape. White Instrument Laboratories, Lacoste and Romberg, and Texas Nuclear Corporation, all of which had ties to university engineering departments, were some of the largest and most successful.[45] The founding of the research and development corporation Texas Research Associates (Tracor)[46] in 1955, however, best demonstrated the growing links between publicly sponsored research and private development in Austin. By the early 1960s the company's success was also an indication to the city council and the chamber of commerce that research and technological manufacturing could generate high levels of economic growth in Austin. Tracor, which gradually grew into a manufacturer of scientific instruments by the 1960s, was founded and run by University of Texas professors and graduates, many of whom were simultaneously working for both institutions.[47]

Though more successful than most other early technology firms in Austin, Tracor was representative of the city's postwar growth paradigm. The president of the company, Richard N. Lane, was a University of Texas physicist who recruited colleagues from war-related laboratories to found Tracor. In particular, Tracor built its human capital from the highly specialized defense-oriented labs at the university, especially the Defense Research Laboratory and the Military Physics Research Laboratory, which merged in 1964 to become the Applied Research Laboratory. Tracor played a large role in keeping elite, young engineering talent in Austin as the city steadily grew through the 1960s. The flexible mode of production, both in research and development and increasingly in instruments manufacturing, at Tracor also mirrored the style employed by Thompson at the BRC and by many nascent technology firms in Silicon Valley. Tracor's mission statement in its own catalog identified the relationship between the company and the university without overtly portraying the two as overlapping entities: "Although there is no operational connection between Texas Research Associates and the University, close personal liaison is maintained between TRA scientists and their friends on University staff." The list of senior staff at Tracor in the early 1960s was made up almost entirely of researchers and professors who also held positions at the university. Norman Hackerman, who was recruited to the University of Texas to chair the growing Department of Chemistry in 1952, joined Tracor's staff as a senior scientist in 1956 and eventually became president of the university in the late 1960s. Tracor took advantage of a distinct increase in very specialized engineering talent at the university after World War II and simultaneously became an attractive force for more young engineers interested in Tracor's style of knowledge labor and the possibility of holding positions in the company and in academia.[48]

From its outset, Tracor's mode of operation was extremely flexible and group-centered, similar to operations at the BRC. Lane claimed that a "spirit of adventure" led to the company's founding, something that the original group's employment at the Defense Research Laboratory allowed for. Initially, the small company had no office; employees carried "offices" around in their briefcases and conducted planning sessions in a spare bedroom at Lane's home, before opening a small office just two blocks from the university's main campus in a remodeled grocery store in 1960. Lane's relaxed management techniques and the novelty of a scientist being an adept businessman were the subject of a *Texas Business and Industry* article in 1969. The article marveled at Lane, whom it referred to as the "new breed" of businessman, "a man of science and a man of business, a man of action, on the move in a dynamic business world." His transition from university research scientist to businessman was uncomplicated by traditional management training. Instead, Lane adopted management skills that better fit the human-centered scientific mode of production. "There's a real knack to managing the output of scientists without appearing to manage them," Lane claimed, which to him was fundamental to running a successful research and development laboratory because people were the key form of capital. "Science is our business and *people* are our main assets—really our *only* asset, in dealing, as we are, in advanced technologies."[49]

Tracor's emphasis on human capital was certainly an attractive feature for potential high-technology laborers, and its commitment to University of Texas graduates and to expanding Austin's market for those laborers did not go unnoticed by Austin's growth promoters. Promotional articles focused on Tracor's clean industrial manufacturing and research and also on the company's fit for Austin due to the academic, nonindustrial, and nonhierarchical modes of production employed by the company. They also concentrated on Tracor's diversification into myriad projects with dozens of different, constantly changing aims. The curious, eclectic lifestyle afforded to Tracor's laborers was also a selling point for the company and the city. Most of Tracor's researchers had active academic lives while working for the company. In November 1964, for example, one employee was giving a paper in Rotterdam, three physicists were writing a textbook on gas flow, one employee was preparing the keynote address for an American Mathematical Society conference in New York, two employees were contracting out work with an architect on acoustical designs for a recital hall, and dozens of other projects were under way. The focus was on the diversity and variety of work, but articles on Tracor also reflected the image of constant stimulation, of a work environment that encouraged curiosity, direct links with evolving academic knowledge, and, above all, active and sustained creativity made possible by a mode of production that emphasized freedom. These images were interwoven skillfully with general discourse about Austin's work environment, where the workday was portrayed as anything but monotonous for skilled workers.[50]

Lane viewed Tracor as a new paradigm for Austin's urban and business growth. Instead of focusing solely on attracting capital investments from more established industrial regions, Lane encouraged the city, the university, and the state to work together to foster internal growth in the Tracor mold. Lane was well aware of the successful high-technology models already developing near Palo Alto and outside of Boston. These young agglomerations both matriculated out of federally supported university engineering departments and were initially funded largely by defense contracts. Local spin-offs, along with the startup capital generated by federal contracts, are what drove early economic growth and geographic clustering in those regions. From there, growth would be what Lane called "self-generating," and smaller companies could build up rapidly, feed off one another, and keep momentum going by creating an "entrepreneurial atmosphere." The function of universities and cities would be to facilitate this atmosphere. Like good managers, cities would provide the infrastructure and freedom, especially economic freedom, to nurture new businesses. The university would function in its traditional role of educating and communicating with private researchers but also hold symposia, create adjunct positions for technology workers, and generally stimulate creativity in scientists.[51]

Tracor's success through the 1960s was spectacular, largely due to a focus on developing very specific types of knowledge and concentrating attention on niche markets, a tactic developed by Lane and then improved on by Frank McBee, Tracor's second president, who took over in 1970. By 1967 Tracor was easily the largest private employer in Austin, with 1,531 workers.[52] In 1967 Austin received its first major relocation, a branch of the global giant IBM that moved into a new suburban campus two miles south of the BRC campus. IBM chose Austin because of its cheap utilities, university spin-offs, and, owing to the new American emphasis on leisure, "the cultural life of the city." By 1972 Texas Instruments and Motorola had opened branch facilities in Austin as well.[53]

Geographically, Tracor was closely tied to the BRC after 1965 when it opened an 80,000-square-foot facility on Research Road about five miles northeast of downtown Austin. The two scientific complexes formed the basis for Austin's technological agglomeration—one mostly public and the other mostly private—that grew around the BRC and on Austin's eastern border near U.S. 183, which directly connected the two areas. In terms of human capital, the university quite literally engendered the private research and development market, and later semiconductor and other electronics production, in Austin.[54]

By the mid-1960s Austin looked far different from the city that Richardson Wood arrived at in 1947, physically as well as culturally. As in much of the Sunbelt, Austin's economic and demographic growth throughout the 1950s was brisk, consistent with other Texas cities.[55] In the 1960s, however, Austin began to demonstrate growth patterns that far outpaced the region. Austin had the most stable labor market in Texas during the late 1960s; in 1968 and 1969 the unemployment rate stayed below the threshold for full employment, 4 percent, and reached below 2 percent frequently. The manufacturing sector grew by over 20 percent in Austin in 1969 alone, largely due to the growth of electronics research and production outfits. Surplus capital generated by Austin's economic growth fed the building industry, which underwent the most intense boom of any economic sector in Austin in the 1960s; the city had the highest rate of homes built per capita in the United States in 1968. As in most metropolitan areas, residential construction was almost entirely on the urban periphery; the physical size of the city grew by roughly 70 percent in the 1960s, from fifty-one to eighty-six square miles, yet the overall density declined by 17 percent. Growing population and decentralization led to a dramatic increase in automobile ownership and usage; the number of automobiles owned in Travis County doubled between 1960 and 1972.[56]

In 1965 *U.S. News and World Report* listed Austin as one of the fourteen most desirable cities in which to live, based largely on its nonurban qualities such as public parks and schools. The city quickly began incorporating the honor into its marketing discourse. In 1967 the Bureau of Labor Statistics rated Austin as the least expensive metropolitan area in the United States. By 1967 Austin had also moved ahead of the oil refining region Beaumont–Port Arthur as Texas's fifth-largest economy after attracting IBM, its first major high-tech relocation. Texas Instruments came in 1968. The de facto local investment provided by the university and the state government along with the new tech firms kept Austin's unemployment among the lowest in Texas during the 1960s.[57] In Travis County, the population grew by a robust 39 percent (from 212,000 to 296,000 residents, or over 83,000 people) between 1959 and 1969 (the national average was 17 percent); close to 85 percent of Travis County residents lived in Austin proper in 1975. In 1973 over half of Travis County's population had not lived in the area in 1963.[58]

With the population increase came an equally dramatic rise in income and employment: between 1959 and 1969 Austin's per capita income rose by 41 percent and family income rose even further, from $5,795 to $8,459, an increase of 46 percent. The total number of jobs in Austin grew just as sharply, from 79,000 in 1959 to 132,000 in 1969. Vic Mathias, manager of the Greater Austin Chamber of Commerce, wrote in 1969 that the economy was so robust that there were "job opportunities for all who want to work."[59] Throughout 1968 and 1969 Austin's unemployment rate was projected to remain below 2 percent, far lower than the Bureau of Labor Statistics' threshold for full employment.[60] The Federal Reserve Bank's business activity index determined that Austin had the fastest-growing economy in Texas in 1968 and 1969, and between 1968 and 1969 Austin's manufacturing sector grew by nearly 24 percent, largely because of the growth of electronics production firms such as Tracor and IBM. Despite the growth in manufacturing, Travis County still ranked in the lowest 1 percent of manufacturing employment among counties with over one hundred thousand residents, indicating a paucity of blue-collar jobs. In spite of a minor national recession, however, job growth was robust and consistent in the late 1960s as well, averaging a 6 percent annual increase. Between 1968 and 1969, retail sales increased 7 percent in Austin, indicative of major increases throughout the 1960s as well as of a general increase in surplus consumption throughout the decade.[61] At one major bank in Austin, total available capital rose over 300 percent between 1963 and 1968, from roughly $24 million to $73 million, a sign of intense economic activity in the city. Between 1958 and 1968, retail sales in Travis County increased by 84 percent to over $400 million, and bank deposits grew by 177 percent to $675 million. The sharp increase in both population and personal income also had profound effects on the city's tax rolls and budget. In the four years between 1965 and 1969, Austin's city budget grew from $47 million to $75.5 million, an increase of nearly 60 percent.[62] By all statistical accounts, accumulation increased expeditiously in Austin during the 1960s and into the 1970s.[63]

Surplus capital generated by economic growth and relocations was also increasingly reinvested in the secondary circuit of real estate, which underwent the most intense boom of all capital investments in Austin in the 1960s. Although Austin was only the sixty-seventh-largest U.S. city in 1968, it ranked sixteenth in the value of construction permits, with a total value of over $131 million spread throughout forty-six hundred total permits. Both amounts broke building records set in 1967, indicating a sustained output of new construction in Austin during the decade.[64] Employment in the construction sector rose 19 percent in just one year, reaching eight thousand workers in early 1968, or roughly one in twenty adults living in Austin. Much of the building was generated by a large increase in the geography of government institutions surrounding the capitol; between 1959 and 1968 six new buildings were added to the capitol complex just north of downtown, totaling construction costs of $15.7 million.[65] In terms of residential building, Austin saw heavy competition for middle- and upper-income structures, but the market for families with low-to-moderate income had "practically disappeared" according to a 1971 report.[66] For minority residents segregated on Austin's Eastside, however, the city's economic boom only exacerbated the conditions of inequality they had endured for decades.

The city's population increased by over 400 percent from 1940 to 1970, while its physical area grew from about thirteen square miles in 1950 to seventy-two square miles in 1970, largely due to the annexation of undeveloped land and the addition of new subdivisions in the northwestern portion of the metropolitan area near the BRC. The northwestern neighborhoods of Allendale, Northwest Hills, North Shoal Creek, Crestview, and Brentwood were all on rural land annexed after 1946 and developed between the late 1940s and early 1970s. By the 1960s Austin had two large decentralized shopping centers, each with retail sales equal to those of the entire downtown commercial district. Highland Mall, one of the largest modern enclosed malls in Texas, opened in 1971.[67] Most of the research and development facilities were located in the northwestern portion of the city, within a short drive of the BRC. Light manufacturing was clustered to the southeast, closer to the neighborhoods of semi- and lower-skilled workers. The two industrial agglomerations were connected by U.S. 183, later named Research Boulevard, which circumvented the city from northwest to southeast. The annexed land and new subdivisions demanded extensions of services and utilities, which drew municipal funds from the center (often specific areas of the center, as discussed in chapter 6) to the periphery. Austin was one of the southwestern cities most willing to subsidize developers' costs in new subdivisions in the 1950s and 1960s.[68]

In the culture that emerged among the skilled labor force in Austin, research-based work was similarly imagined as part of a lifestyle that emphasized creative autonomy and access to natural space. One excited physicist articulated the relationship clearly in a 1961 interview: "[In Austin] I discovered I could earn a good living for my family while surrounded by trees and lakes instead of dirt and skyscrapers, and now you could not drive us away."[69] The same article lauded what researchers provided for the city: "The personnel [scientific research] attracts is [of] an income level welcomed by merchants, and of an education level likely to fit in beautifully with the city's cultural, philanthropic, and civic projects."[70] Other scientists pointed to the idea that Austin's lifestyle was as important as the University of Texas in attracting and retaining skilled workers in what business professor Stanley A. Arbingast called "footloose industries," because they had myriad locational choices. Austin offered new homes and attractive suburban neighborhoods in close proximity to work, a low cost of living, a moderate climate, and "a refreshing absence of the rat-maze complexities of living commonly bundled together under the heading of 'urban problems.' "[71] Interviews with executives from seven other firms that relocated to or opened in Austin in the 1960s echoed the sentiment that "particularly attractive living conditions" were important in their decisions. Each firm chose to build or lease a new suburban facility on or near the city's northwestern periphery.[72]

The physical and social landscape that emerged after World War II reflected a new emphasis on economic growth fueled by suburbanization, knowledge labor, and the federal largesse that supported both. Austin became one of the nation's fastest-growing cities without developing industries that would pollute its landscape, attract unskilled workers, or undermine quality of life for its Anglo residents. Yet the benefits of growth did not accrue evenly; in fact, decentralization, infrastructural and environmental improvements, and reliance on knowledge labor had deleterious effects on minorities, who were not allowed to participate in the growth. Chapter 6 looks at the effects of postwar development on Austin's minorities and their systematic containment, physically and socially, as the city grew.

6 Building a City of Upper-Middle-Class Citizens

Urban Renewal and the Racial Limits of Liberalism

In May 1967, the newly formed Austin Human Rights Commission presented a text copy of the Austin Fair Housing Ordinance to the Austin City Council for approval. The commission, which the council formed solely to produce the document as a response to federal fair housing legislation, ran the full text of the ordinance in the *Austin Statesman*. The ordinance, which mandated that no person could be discriminated against on the basis of race, color, religion, or national origin when buying, leasing, or financing residential property, was printed on May 24, one week after the deciding council vote.[1] It was enacted on May 27. Immediately, the Austin Board of Realtors, many of whose members supported the ordinance in its early stages, called for a referendum vote against it, which, realtor Nelson Puett later claimed in a letter to Congressman J. J. Pickle, "is not a racial thing . . . not a civil rights thing . . . [but] just another government attempt to restrict and control your individual freedom and to tamper with the most basic human right, private ownership of property."[2]

Publicly, Puett and other powerful Austin realtors such as Hub Bechtol were equally as committed to defending private property rights against fair housing. Locally, the ordinance was denounced as "forced public housing," despite the fact that it did not call for any forced relocation of citizens or forced integration of any kind. Nor did it call for any new public housing. Although referenda were rarely successful in Austin, the Austin Board of Realtors chose to collect the necessary signatures to proceed with a public vote. Within ten days the petition was signed by twenty-seven thousand residents, nearly one-third of Austin's voting-eligible population and roughly 10 percent of Austin's entire population, including minorities and minors, in 1968. It was by far the most-supported petition in the city's history. When put to referendum, the Austin Fair Housing Ordinance was soundly defeated, leaving discriminatory real estate and lending practices difficult to prosecute in Austin into the 1970s and reinforcing the idea that the races should remain strictly separated. In 1969 all three liberal members of Austin's city council, all of whom had backed the ordinance, were defeated, bringing the short-lived period of progressive racial politics in Austin to a close. The overwhelming grassroots support for institutional residential segregation reflected both national and southern trends from the period.[3]

If open housing drew large and vociferous opposition from Austin's whites, urban renewal proceeded with far less consternation. Indeed, in 1967 entire neighborhoods were claimed by the University of Texas and the City of Austin using eminent domain legislation and, over the coming years, were evacuated, demolished, and replaced with different structures meant for different people. Austin's powerful real estate council used the sanctity of private property right to undermine open housing, yet the council fell silent when property-owning citizens were dispossessed, removed, and contained in other places. Contextualizing the defeat of open housing in Austin with urban renewal illustrates the logical inconsistencies of the race-neutral appeal to property rights employed by Austin realtors and supported by many citizens. Even as poor minority residents were dispossessed and neighborhoods evacuated, oftentimes with no plans for relocation, opponents of open housing successfully deified private property rights.

The irony in this logic reflects the fundamental premise that drove accumulation in postwar Austin: improving the garden meant containing the city. Urban renewal must be viewed, however, within the dual framework of historical racial discrimination in Austin and the city's decision to encourage economic growth through nonindustrial development. Austin's rapid economic expansion during the 1960s had very little positive benefit for its minority communities, as city leaders and businesspeople focused on attracting external workers to expand skilled labor markets in the city and especially at the university. Austin capitalists had never concentrated on producing adequate internal labor power because of the nonindustrial quality of its industries. Thus most unskilled laborers were highly expendable because the reproduction of their labor power served little purpose in a local economy with such a paucity of heavy industry.

This facet of production also helps to explain the extreme mental segregation exhibited by Austinites: unlike in areas with more Fordist production, which by the 1950s generally indicated some workplace integration, in Austin minority and white members of the same class rarely worked together, as each group filled specific niches in smaller industries. Not only did the growth of the 1960s remain unfulfilled for most minority Austinites; urban renewal sought to expand accumulation by taking advantage of the surplus created by the boom, which meant profiting from expanding real estate values around but not necessarily in central East Austin as well as on the urban periphery. For minority residents, discrimination actually increased: socially sanctioned residential and public segregation in the 1930s grew into aggressive socioeconomic oppression by the 1960s that included appropriating minority space for profit. Viewed through this lens, it becomes increasingly apparent that, while the end of segregation may have adversely affected the cohesion of previously concentrated minority communities, the concomitant dispossession of thousands of minority residents likely had a similarly deleterious effect on the community.

The simultaneous processes of accumulation via decentralization and containment via concentration were underwritten by the liberal Keynesian state. As discussed in chapters 4 and 5, federal subsidies paid for dams, highway and residential construction, and university upgrades and enrollment and fueled the knowledge economy that drove growth in Austin. From the 1930s to the 1960s, being progressive in Austin usually meant supporting New Deal policies, encouraging strong ties to the federal government as a source of funding, and promoting responsible, nonindustrial growth. Fighting racial inequality or supporting federal legislation aimed at ending discrimination took a decidedly back seat. Federal rules regarding eminent domain and urban renewal served to enhance segregation and widened the chasm between races in the city. To be progressive indicated a desire to be modern, and to be modern meant building a nonindustrial city that catered to university-related growth models, enhancement of natural areas, and suburban-style residential and commercial development.

The city's physical transformation in the postwar decades also illuminates the racial limits of liberalism and the myriad ways that whiteness conferred economic, health, and social benefits. Austin also had a far more liberal political and social culture than most other southern cities.[4] Racial violence rarely occurred in the city, and there were no major riots to speak of in Austin.[5] Owing in part to strong support for New Deal policies among politicians and a tolerant, open-minded university atmosphere, the area gained a reputation as a center of progressive politics in the South by the 1930s. In 1938, at the behest of Lyndon Johnson and Mayor Tom Miller, both ardent New Dealers, Austin was one of the first cities to erect federally sponsored public housing. The city itself received over $1.5 million in New Deal funding, an extremely high figure for a city of its size. In the 1940s the removal of President Homer Rainey from the University of Texas and the Hemann Sweatt integration case at the university's law school both precipitated liberal outbursts against conservative ideologies. Emma Long and Ben White became the first two liberals on the Austin City Council in the early 1950s, long before liberals won seats on most municipal councils in the South; by the mid-1960s three of the five city council members were liberals. African Americans and Latinos had a substantial, though unofficial, presence in the Austin political scene as early as the 1930s. Even conservative elements in Austin's political culture were more like the business elites in other cities and not overt segregationists who employed the language and tactics of white supremacy.

But while there was always a significant liberal presence in Austin politics, racial relations reflected a more conservative tone. Federal policy and subsidies also made reorganizing the city and reinscribing segregation much easier. The same highways that drew people and resources away from the center created a barrier that separated races. Unlike minorities, displaced whites were more able to acquire loans to move to peripheral residential neighborhoods. The geography of open space in Austin reinforced racial barriers, and environmental improvements benefited whites far more than minorities. Federal urban renewal money was allocated locally and eminent domain was administered locally, which allowed for elites to drive spatial transformation in ways that reflected white interests. The outcome was a policy where even residents who owned up-to-standard property in areas targeted for renewal were often dispossessed. Texas urban renewal legislation made it illegal to build public housing with urban renewal funding. Whites fought to maintain racially segregated schools, which limited minorities' ability to compete in the city's increasingly white-collar labor market. The outcome was increased segregation and concentration for Austin's African American community, a policy that amounted to containing racial minorities as a key facet of urban accumulation.

Bifurcated City

Most accounts of African American and Latino life in Austin from the 1930s through the 1950s portray a generally positive period marked by high levels of community cohesion and a relatively vigorous economic life defined by small businesses and networks of familial and neighborhood support. Despite municipal negligence in nearly every aspect of life, segregation brought minority communities together and kept relatively high levels of economic diversity in Eastside neighborhoods.[6] One contemporary book counted fifty-two small businesses in the commercial corridor on East Twelfth Street just east of East Avenue. Eastsider and business owner Ora Lee Nobles recalled "beauty shops, a barber shop, a little grocery store, a

Despite stark differences in opportunity and in infrastructure, Austin's African Americans reported high levels of community cohesion in the 1950s. Here children practice for a Maypole celebration in Rosewood Park, the city's African American–designated public park, in 1959. PICA 24201, Austin History Center, Austin Public Library.

meat market, a little dry cleaner, a doctor's clinic, an ice cream parlor," which had "all been there for years."[7] In hopes of creating additional incentives to force minorities to the Eastside, the city improved segregated facilities during the 1930s, including funding a large public park, building a library, and improving Anderson High School.[8] Although public housing in Austin was strictly segregated—its function was more to demonstrate support for New Deal policies than to assist poor minorities and it was not intended to house the city's poorest residents—it was welcomed by minorities.[9]

Yet major disparities in quality of life still existed between East and West Austin, and Eastside residents were consistently subject to poorer, more dangerous living conditions, had less access to jobs and education, and were generally not considered part of mainstream economic, political, or social discourse in Austin. Aside from parks and beautification projects designed to attract tourists and businesses, Austin was, like many southern cities, reluctant to invest in managerial-type infrastructure. Amazingly, all sidewalks in Austin were privately funded until 1969. As of 1958 only 45 percent of Austin's surface streets were paved; a higher percentage of streets were unpaved in South and East Austin, where concentrations of minorities existed. Eastside residents complained in letters about dangerous conditions in their neighborhoods from dilapidated infrastructure or municipal negligence. In 1955 the sewers became so clogged in one Eastside neighborhood that sewage backed up into the street for days before the city acted. When the University of Texas let out for summer, many of the city's bus routes stopped running, which obviously had deleterious effects on residents without access to automobiles.[10]

African American and Latino residents suffered everyday forms of subtle and overt racial discrimination that ranged from contentious to dangerous. As of 1956, the City of Austin did not employ African American bus drivers, even on routes that went through large sections of East Austin.[11] The municipal government hired few minorities at all, outside of filling janitorial positions. Numerous impoverished African American citizens felt mistreated by welfare agency personnel, some going so far as to forgo their assistance checks rather than dealing with the agency.[12] One domestic worker wrote her congressman for assistance after the Travis County Welfare Agency would not help her when her disabled husband used up an entire social security check, leaving her and three children to fend for themselves.[13] In the transitioning neighborhood of Windsor Park, one man wrote to Senator John G. Tower, "I would greatly appreciate it if you could tell me if there is no way that treason by a nigger can be handled," after an unspecified incident.[14] In some cases, overt forms of racism could have deadly consequences. During the 1950s, at Brackenridge Hospital, the closest hospital to East Austin, white nurses were not required to care for black patients, who could easily be left unattended in any kind of medical condition.[15] Black doctors were not allowed to practice at the hospital either. Discrepancies in access to health care were often simply facts of life, and unfortunately death, for even prominent African Americans in Austin. In the early 1950s, Carl Downs, president of Samuel Huston College, died from appendicitis after not having access to an emergency room at Brackenridge.[16]

Austinites living east of East Avenue, particularly in areas designated for African Americans and Latinos, had a much lower standard of living than Westside residents throughout the 1950s as well.[17] The Mexican American–majority area bounded by East Avenue on the west, First Street on the south, Springdale Avenue on the east, and Seventh Street on the north had the highest percentage of dilapidated housing in the central city, over

East Austin house, circa 1970. Many East Austin neighborhoods lacked paved streets and suffered other infrastructural shortcomings well into the 1970s. PICA 02580, Austin History Center, Austin Public Library.

56 percent. "Dilapidated" was the worst classification of housing available according to Austin's Urban Renewal Agency, indicating that the structure was not habitable and should be torn down. The majority Latino area just to the south and the African American neighborhood to the north did not fare much better; along with the downtown they were the only neighborhoods in the 41–55 percent dilapidated category. No neighborhood in South or West Austin had over 25 percent dilapidation, and a vast majority had fewer than 10 percent. A 1948 report by the Austin Housing Authority found that roughly 75 percent of all dwellings between First and Nineteenth Streets within one mile of East Avenue (roughly 1.5 square miles of virtually all African American and Latino residences) lacked a private bathroom.[18]

The central Eastside reported a far greater percentage of social and health problems than the rest of the city as well. Approximately two-thirds of all juvenile delinquency cases in Austin occurred in the central eastern neighborhoods, despite the fact that the area's residents made up less than 25 percent of the city's population. Upwards of 75 percent of major crimes (aggravated assault, murder, rape, robbery) were reported in central East Austin. Often, clear cases of white-on-black violence were dismissed by police or victims were purposely deemed unreliable.[19] The vast majority of calls to the police also came from the area, indicating a high rate of minor crime and other daily municipal problems. Central eastern neighborhoods also saw a rate of tuberculosis far greater than that of the rest of Austin, perhaps owing to a severe lack of health care professionals on the Eastside and legalized discriminatory practices among physicians in other areas.[20] Finally, slumlords were prevalent in central East Austin. In a 1962 study of 1,057 homes in a heavily dilapidated section of central East Austin, 96 percent of residents were either African American or Latino, while over 55 percent of the real estate was owned by whites. Evidence suggests that the rate of home ownership in much of central East Austin actually declined between 1949 and 1962. Housing conditions were exacerbated by a de facto municipal policy that ignored zoning requirements on much of the Eastside. Poor renting conditions were another fact of life for marginalized East Austin residents.[21]

From Social to Institutional Disparity

Beginning in the late 1950s, the passive disinterest that the city had shown toward the Eastside took on a new, aggressive tone as urban renewal spread across Austin and other cities around the country. From a community perspective, it appears obvious that East Austin was accustomed to institutional and often overt individual discrimination from the city and some white residents and to a lower overall standard of living as well. Despite segregation, minority communities appear to have had a relatively high degree of autonomy and both formal and informal networks of resistance when oppression became intolerable. But, as in many other American cities, urban renewal represented an entirely new and more damaging framework for race and class oppression. Segregation in Austin was for the most part a social phenomenon engrained in the landscape of the city and its collective consciousness. Urban renewal brought an overtly economic aspect to segregation; its policies encouraged politicians, university officials, developers, financiers, and contractors to profit from redeveloping large portions of minority areas to create jobs for white workers and to stimulate the economy by improving infrastructure and the beauty of the central city.

In Austin, the powerful University of Texas was also a major factor in urban renewal. Administrators and regents viewed renewal as an opportunity to expand the cramped main campus in an era of skyrocketing enrollment, to promote a new focus on research and larger research facilities, and encourage rapid urbanization around the campus. They viewed the university as entrepreneurial in the sense that it increasingly functioned as a business—not just in terms of research contracts or allocations for profit-making departments and facilities but also in terms of real estate and physical expansion.[22] In 1965 the state legislature institutionalized this aspect of university entrepreneurialism by granting the board of regents the right of eminent domain. Thus urban renewal represented a regulatory aspect of capitalist development under the welfare state regime of accumulation: already profiting from widespread development on the urban periphery, power brokers in Austin now had carte blanche to imagine an inner-city landscape based on the needs of capital and sanctioned by federal investment. This new landscape, of course, came largely at the expense of the dispossessed and minority residents living in East Austin.

Although, legally, urban renewal in Austin did not begin until 1962, the groundwork was set in motion by Title I of the Federal Housing Act of 1949, which provided for urban slum clearance and redevelopment, funded largely by the federal government. The cleared sites were then given to private developers, who created new housing, often at tremendous profit. In Austin Title I precipitated a study of East Austin by a private consulting firm, which found an alarming amount of substandard housing. A plan was discussed to create additional public housing for African Americans just outside the city limits in northeastern Austin on a plot of land that was being used as the site of an orphanage for black children, but the proposal was struck down because the land could be developed for white use. African American community leader Everett Givens discouraged the public housing project, arguing that it would damage race relations by attempting to integrate too quickly.[23] In 1954 a second housing act was passed that made renewal more enticing to developers by offering FHA-guaranteed mortgages. The 1954 act prompted the city council to continue studying the housing market in East Austin. For this task the council appointed the city's first committee dedicated to the area east of East Avenue, the Greater East Austin Development Committee, which was charged with studying the area almost exactly analogous to the designated African American and Latino neighborhoods on the Eastside. In 1957 the city created its own urban renewal department; it was then adopted under Texas state law by a contentious referendum in 1959, in which few minorities were able to vote, and thus the Austin Urban Renewal Agency (AURA) was formed.[24]

Partly as a reaction to the impending certainty of urban renewal and partly in anticipation of continued urbanization, in 1953 the city council also voted to fund an updated professional master plan for the city. Although the city carried out some smaller studies in the 1940s and 1950s, particularly regarding business development, no comprehensive plan had been developed since 1928. In 1953 the city council revised the city charter, providing for a planning commission that could make and modify master plans, offer zoning recommendations, and establish general parameters for development that could be continually amended. Published in 1958 after five years of research, the plan imagined a geography that facilitated multiple forms of industrial production and knowledge work as well as altering the existing residential and commercial landscape of the city through zoning and urban renewal. The plan recognized that, as national economic surplus increased and the economy diversified drastically during the postwar expansion period, people and businesses would make locational decisions based on "the attractiveness of the community." What this entailed was an "orderly removal and replacement of those areas which have become obsolete and fallen into disrepair." Fulfilling these conditions would not only attract prospective businesses, residents, and tourists; it would also create a landscape "of which the people are proud."[25]

To provide for the attractiveness of the community while simultaneously encouraging economic growth, the master plan proffered industrial development away from areas planned for retail and residential development or from public areas. The first industrial area, which conformed to the rough parameters of the BRC tract and its environs to the northwest of residential and commercial areas, appeared as a research and development facility that would be central to Austin's informational cluster. This area would provide the centerpiece for accumulation via knowledge work while retaining a low profile away from civic and public centers and yet be in proximity to Austin's growing middle- and upper-middle-class subdivisions. The second industrial zone was planned for the area near the Perimeter Loop (now Highway 71) and the East Avenue expressway, far to the south in an undeveloped area of the city. Finally, the third major industrial area was bounded by First Street, East Avenue, Seventh Street, and Loop 183. This area conformed to the majority Latino neighborhood on the Eastside, which contained the highest concentration of Latinos in the city, some poor whites, and African Americans, as well as the highest concentration of dilapidated housing. In the small adjacent area between First Street and the Colorado River, the master plan envisioned a public parkway separated from the industrial zone by a wide boulevard. The parkway would also displace hundreds of residents living on the river's floodplain below First Street. Under the plan's provisions they would not get to reap the benefits of the Longhorn Dam, which when completed in 1960 eradicated the major flooding that often occurred along the river. To facilitate transportation between the research-oriented park to the north and the production facilities to the east, the plan also advocated finishing a major highway, known locally as Research Road, which circumvented the residential areas of the city.[26]

Austin's master plan thus inscribed a geography of power and dominance onto the landscape of the Eastside, with little regard for the community. Despite an almost endless supply of undeveloped land throughout the city and on the periphery, the master plan advocated razing an entire neighborhood to centralize industrial production. The plan's brash statement about the neighborhood envisions it as little more than space to be emptied. Industrial discourse erases people from the landscape: "Austin can take advantage of the urban renewal legislation in the industrial development of the East Area. This primary industrial area is now cut into many parcels. Approximately one-third of its area is covered by housing, much of which is substandard. The area will need to be cleared and the parcels reassembled into sizes and shapes more suitable for industry before this prime location can achieve its potential." Under the urban renewal law passed in Texas, even homeowners whose property was deemed acceptable could not save it from eminent domain if the structure was not "consistent with the plan for the area," indicating that an industrial district would be zoned for single use and all residents would be removed regardless of the quality of their structure. Even residents whose homes were deemed substandard were often perfectly satisfied with them and considered their communities as homes populated by families. One Eastside resident related the widespread notion that "slum clearance" indicated that urban renewal would replace minority neighborhoods with slums, not that the neighborhoods were slums in the first place.[27]

Although the eastern industrial zone did not materialize, it was an ongoing concern for Eastside residents who anticipated removal and dispossession based on the city's racial history and widespread urban renewal projects initiated during the 1960s. The neighborhood did become the center of most industrial-style production in Austin, however. In place of the public park that the master plan envisioned for the north bank of Town Lake east of Interregional Highway 35 (IH-35), in the late 1950s the city began dispossessing the mostly Latino residents on a twenty-two-acre parcel of land adjacent to the reservoir to make way for the Holly Power Plant. Built during the 1960s, the plant at its peak provided 20 percent of Austin's electricity and had a capability of 570 megawatts of power. Dozens of residents were evicted and their homes razed to accommodate the power plant, the city's largest and most centralized power-generating facility at the time. For years the plant was a source of constant irritation for neighborhood residents due to the noise and pollution it created. Numerous times it became a dangerous environmental hazard to the neighborhood; major fuel spills occurred nine times between 1974 and 1993, some of which ignited into fires. Eastside environmental activists have claimed that the plant has caused numerous health problems among residents, including tumors and birth defects.[28]

One major infrastructural project that urban renewal ideology facilitated in Austin was the completion of IH-35, which further institutionalized the symbolic and actual barrier between East and West Austin when it was built directly over East Avenue, for decades the line of racial demarcation between Anglo West Austin and minority East Austin. In many larger cities, federally funded expressways eviscerated existing working- and middle-class neighborhoods and worked in consort with other urban renewal projects to segregate poor minority residents. In Austin, the expressway, completed in 1962, reinscribed a physical and mental landscape of segregation on central Austin in a much more brutal and impassable form. East Avenue was a wide parkway with a naturally landscaped center and multiple cross streets connecting east and west. Residents on either side could enjoy the parkway and also easily view the other side. In its place rose the mammoth structure, twenty feet high in some areas, which created an actual wall between the already disparate communities. Together, IH-35 and the new Research Road around the city's northern and eastern perimeter expedited traffic flow through and around the city and, as in many metropolitan areas, allowed for a significant increase in development around the urban periphery and along the nodes created to the north and south of town. They also inscribed a more rigid form of segregation on the landscape as race and class barriers.

Along with the new power plant, Austin's minority residents also endured other industrial facilities that polluted the air and water and hampered quality of life. In the 1950s a gasoline storage facility, known as a gas tank farm, was located in a heavily Latino residential area far inside East Aus-

Compared to bucolic East Avenue, Interregional Highway 35, part of the Eisenhower Interstate System, created a much larger and more imposing barrier between East and West Austin when it was finished, circa 1975. PICA 11779 and PICA 25021, Austin History Center, Austin Public Library.

tin; it also caused health problems and diminished quality of life for residents well into the 1990s. University of Texas siting choices also promoted less appealing uses for land east of IH-35 that the school had confiscated through urban renewal laws. Instead of building new classrooms, offices, and laboratories in the newly cleared East University renewal area, the university chose to build a new physical plant and baseball stadium, which increased noise pollution and traffic for adjacent neighborhoods. The portion of East University on the west side of the expressway was dedicated for the new Lyndon Baines Johnson Library and School of Public Affairs. Numerous vacant lots on the Eastside were used as casual dumping grounds for people who wished to avoid paying for trash pickup.[29] The city sited a disproportionate amount of industrial facilities in East Austin as well, including its cleaning station for garbage trucks and other municipal vehicles. Private recyclers opened facilities in Eastside locations zoned for heavy industrial use.

Austin's Eastside residents were also more likely than residents on the Westside to face environmental and health hazards. Despite one of the strongest overall housing markets in the United States in the late 1960s, Austin's residential choices for minorities were scarce. Zoning problems, a lack of federal loan guarantees, and a market geared toward larger, more expensive homes meant a paucity of housing choices and often overcrowding and severe dilapidation in Eastside neighborhoods. A 1977 housing study found that as many as 65 percent of houses in one Eastside neighborhood were dilapidated.[30] In 1970, for example, East Austin contained 12 percent of Austin's population, over 50 percent of the minority population (and a much greater percentage of the city's African American population), and 44 percent of the city's substandard housing. While most of Austin became much more affluent during the 1960s, East Austin's housing stock actually deteriorated; the percentage of substandard housing there rose by roughly 8 percent. Concerns about slumlords were valid as well. Between 1960 and 1970, 60 percent of all residential construction in Austin was apartment dwellings. The percentage of apartment buildings constructed between 1971 and 1973 was even greater, in large part due to the federal income tax shelter for development of rental property. By 1973 one-third of all living spaces in Austin were part of apartment complexes. In 1960 only 10 percent of Austin residences were apartment dwellings. While the sharp increase in student population certainly accounts for some of the increase in apartment living, a relatively high percentage of Eastside residents rented rather than owned their homes, especially after urban renewal initiatives gutted hundreds of Eastside houses and found housing for only a fraction of those displaced.

As the city's repository for refuse, pollutants, and other urban waste, the Eastside also suffered from a dramatically different set of environmental problems than the Westside did. Spills at a gas tank farm in East Austin totaled fifty-six between 1950 and 1992 and released numerous toxic substances into the air, soil, and water during that period. In 1960 a large power plant was sited in a Latino residential area because it was close to a hydroelectric dam that could furnish cooling water; in addition to causing daily noise pollution, the plant caught fire dozens of times. Municipal wastewater treatment plants, garbage dumps, and other polluting industrial facilities were located in East Austin, often near schools, parks, or residential areas. Cumulative zoning allowed private industrial firms to locate in residential neighborhoods on the Eastside. While Westside environmental groups focused on preserving open space and quality of life by mitigating development, Eastsiders faced far more extreme environmental threats and suffered a higher percentage of pollution-related illnesses and daily inconveniences.[31]

Containment

Urban renewal projects altered the Eastside landscape dramatically during the 1960s, even as the civil rights movement crested in Austin and elsewhere. Because nearly all the power to determine the quality of structures, neighborhoods, and public facilities was legally given to the city, residents had almost no say in the fate of their property. Language and images distributed by AURA assuaged what little opposition to urban renewal remained among Anglo Austinites. The urban renewal agency simply needed to declare 50 percent of the structures in any given area "dilapidated beyond reasonable rehabilitation" or otherwise blighted in order to condemn the entire area. Because, historically, the municipal government did not consider zoning important in East Austin, and because it was extremely difficult for minorities to acquire loans to buy or improve property in East Austin, a large number of structures on the Eastside were considered substandard.[32] All five major urban renewal projects in Austin affected some areas of the Eastside, and two focused exclusively on the central Eastside neighborhoods of Kealing and Glen Oaks. Large tracts of the central Eastside were razed; the exact number of acres that were redeveloped or residents that were dislocated is unclear, but as of June 1966 nearly 1,000 acres

A map of the Kealing urban renewal district in East Austin in the 1960s showing that many of the houses proposed for demolition were in either "good" or "rehabilitable" condition. Austin History Center, Austin Public Library.

were scheduled for clearance or rehabilitation in East Austin, and at least 250 of those acres were in central East Austin, which was virtually all African American.[33] This statistic did not include the university's proposed eastward expansion, which targeted the northwest portion of the traditional African American area.

The greatest conundrum for policy makers and planners was relocating displaced residents, particularly the elderly and the impoverished. The

Texas Urban Renewal Law specifically forbade that any property acquired through urban renewal be used for public housing, which meant that private low-cost housing would be necessary for thousands of displaced residents.[34] The urban renewal law did not, however, give any indication of how dispossessed citizens without accommodations would be handled, other than giving cities the power to "plan and assist in relocation." While Austin was one of the first cities to construct public housing specifically for African Americans and Latinos in the late 1930s, by the 1960s there were long waiting lists. Since public housing was still technically segregated until 1968 in Austin, early victims of dispossession could not apply for relocation to white-designated public housing units. An AURA newsletter confirmed the dire situation that African Americans in need of public housing faced in 1964: the entire city had only 429 African American–designated public housing units (or less than one per thirty African American residents), and only 32 more units were planned, despite the obvious dislocation of many disadvantaged citizens that urban renewal would create.[35] A 1966 issue of *Austin in Action* claimed that most residents displaced from Kealing "had been moved to better living facilities through their own initiative," but this assertion appears spurious. Even those African American residents who found their way into segregated public housing often faced unhealthy and sometimes dangerous conditions. At the Booker T. Washington homes in the eastern part of central East Austin, which contained roughly three hundred low-rise units built in 1953, over one hundred fires were reported in the site's first twenty years of existence. In 1960 rat infestation became so overwhelming that two local companies donated over one thousand pounds of poison to residents. By 1984 the entire complex was deemed unsafe for habitation and was evacuated for five years.[36]

Although data regarding displacements are lacking, the *Austin American-Statesman* reported that over four hundred acres were cleared and more than one thousand residents displaced.[37] Numerous vociferous members of the Eastside community voiced their displeasure with urban renewal in letters to community leaders and politicians. In response to a letter from an evicted woman whose new apartment would not be ready for months, provided that one was available, Congressman J. J. Pickle admitted to a severe shortage of low-income housing in Austin, which made relocating disadvantaged citizens more difficult than anticipated.[38] Though Pickle demonstrated empathy in his letter, he voted against open housing while serving in Washington, which would have dramatically increased the available housing stock for Austin's African American citizens.[39]

Others wrote to alert Pickle that urban renewal had negatively affected many citizens in Eastside neighborhoods. Perhaps most poignantly, Frederick B. Scott noted that many African Americans, some living on a pension, were unable to purchase a home or afford a suitable apartment with the money that the government paid them for their property. To Scott, the ubiquity of African Americans who could not afford property after they were forced to sell indicated that urban renewal officials and real estate agents lied to community members about the availability of affordable housing in Austin. Efforts to redevelop sections of the Eastside were viewed as bald attempts on the part of developers and politicians to increase accumulation by further dispossessing marginalized citizens.[40] The Blackshear Residents Organization (BRO), representing one of the poorest, most dilapidated Eastside districts, fought against the $1.8 million urban renewal plan for its community. At a 1969 meeting between the city and Blackshear residents, J. E. Mosely, president of BRO, claimed that urban renewal would result in "bulldozers cleaning us all out" and that redevelopment would not benefit citizens in the neighborhood. BRO was able to postpone urban renewal in Blackshear until federal funding was not renewed in 1973, preserving almost the entire neighborhood.[41]

Compounding the issue of real estate dispossession was the virtual absence of new housing starts for low- and moderate-income families in Austin during the late 1960s and early 1970s, despite the vigorous real estate market in Austin described in chapter 5.[42] In 1978, for example, the average building permit value of a new single-family house in Austin was $50,545, a figure that was affordable to only 7.7 percent of non-Anglo households in the city.[43] The paucity of new low-income housing in Austin was due to a number of factors, including rapid growth in more lucrative middle- and upper-class markets. But again, institutional racism under the guise of private property rights appears to have played a significant role in home building and lending practices. Unlike other southern and southwestern markets, in Austin loans through the FHA's Section 235 program were "practically an unknown quantity" according to a 1971 report. Created under the Federal Housing Act of 1968, FHA 235 loans were designed to assist low-income citizens by allowing them to pay a set percentage of their monthly income toward their mortgage. The government could pay a significant percentage of the mortgage interest directly as well, subsidizing monthly cost for the recipient. In Austin, however, the strong resistance to racial mobility, an emphasis on personal freedoms, and the institutional framework of real estate and banking networks made securing FHA 235 loans very difficult, especially for minorities. By 1969 even the executive director of AURA was forced to vaguely admit that "it is presently impossible to help everyone [who was displaced] because of the many problems and situations that exist."[44]

In the private market, mortgages and home investment loans were literally nonexistent in central East Austin. In 1976 and 1977, just 0.2 percent of real estate investment from banks and savings and loan associations went to central Eastside properties. The difficulty in securing loans led the Austin Human Rights Commission to charge banks and savings and loans with redlining in 1979.[45] Haphazard land uses caused by lax zoning laws and municipal disregard for minority spaces also deterred investment. The small, almost entirely Latino neighborhood called Barrio Unido, just east of IH-35 between Fifth and Seventh Streets, for example, had a land use pattern that was "totally mixed and conflicting," with 27 percent residential use, 39 percent commercial or retail use, and 34 percent heavy industrial use. As a result, by 1984 zero new homes had been constructed in Barrio Unido since the 1940s.[46]

Similarly, the apartment market maintained virtual segregation through advertising practices and racially based social networks among real estate agents and their clientele. The same 1971 report baldly stated, "The Austin apartment dwellers (non-students) have a society that is pretty much their own. Because of this, the most effective form of advertising is word of mouth. Although they do watch newspaper ads, approximately eighty percent of people interviewed indicated that they heard about the apartment they rented from a friend."[47] In terms of residential building, Austin saw heavy competition for middle- and upper-income structures, but the market for families with low-to-moderate income had "practically disappeared" according to the same report. Developers had little incentive to build new subdivisions in minority areas; only one new African American–majority subdivision was built before 1970. From 1970 to 1976, only 8.9 percent of the ten thousand new single-family homes built in Austin were located on the Eastside. Over 93 percent of the applications for new subdivision construction on the Eastside filed between 1972 and 1977 were outside central East Austin in areas that contained few African Americans; only one cluster of subdivisions in northeastern Austin contained more than 10 percent African American residents. Owing to the lack of available property, residents in segregated minority areas paid a far higher percentage of their income on housing than whites did in the 1970s. Profound racial homogeneity in new subdivisions prompted the Austin

Human Rights Commission to conclude that "the building industry, while intentional or not, is pursuing an overwhelmingly Anglo market."[48] Existing residential and social segregation thus largely determined networks for finding accommodations in Austin and contributed to ongoing physical and mental segregation for both white and minority residents.[49]

In Austin, urban renewal policies were obvious indicators of the status of African Americans and Latinos in the larger community, as well as examples of what David Harvey calls "accumulation by dispossession," or directly assuming control of another's resources for a nominal price and turning that property into profit.[50] Almost completely devoid of agency or political voice in West Austin, poor minority residents were often dispossessed at the whims of politicians, developers, and University of Texas administrators. Discrimination was sanctioned largely by a discourse that, as espoused by much of the business and political community in Austin, considered urban renewal a key facet to modernizing Austin—to make it more attractive for investment and continued urbanization in a specifically non-urban mode. Ironically, real estate developers and growth-minded politicians who claimed that private property rights were sacrosanct had no problem developing land that had been acquired by the state under eminent domain and against the wishes of property owners. Despite civic leaders' claims to the contrary, in the decades after urban renewal the central Eastside endured a sharp rise in poverty and crime, as residents of means moved farther east and northeast and poverty became concentrated. Although the neighborhood's central location gave residents access to many other areas after segregation ended, the area actually became more economically and socially segregated from the rest of the city. Overall lack of income was endemic to historically minority neighborhoods in central East Austin. In 1977 87 percent of central East Austin was deemed "low-income" by the Community Development Block Grant application for that year.[51] In 1970 the central Eastside had a poverty rate of 37.5 percent, 70 percent greater than the African American poverty rate and about 350 percent greater than the poverty rate for Anglo Austinites. By 1990 the rate of poverty in central East Austin had risen to a staggering 52 percent, a total that, while almost unbelievable for a city with one of the highest rates of economic growth and employment in the United States during the 1980s, was over 500 percent greater than the total percentage for the city.[52]

The housing stock in Austin's central Eastside locations did not show marked improvements in the years following urban renewal either. Unlike the areas that the university annexed, parts of the Brackenridge Tract and the University East area, which demonstrated a significant decline in substandard housing by 1977, the concentrated African American and Latino neighborhoods farther into the Eastside had similar levels of dilapidation as they did when surveyed in the early 1960s. The historically African American community area bounded by IH-35, Martin Luther King Jr. Boulevard, Airport Boulevard, and Seventh Street had 51 percent substandard housing stock according to the city's Community Development Block Grant application for 1977. The two historically Latino neighborhoods to the south (together bounded by IH-35, Seventh Street, Springdale Road, and the Colorado River) separated by First Street had 53.4 percent and 65 percent substandard housing stock, respectively. Clearly, urban renewal projects did not target the most disadvantaged minority citizens in Austin; they were rather processes of accumulation that favored Westside business and political interests in remaking selected portions of the landscape for profit and growth.[53] Perhaps because of continued discrepancies in availability of decent housing and a possessive investment in whiteness, a 1976 survey found that only 8 percent of African Americans and 18 percent of Latinos did not see a need for more integrated housing in Austin. In stark contrast, 68 percent of white respondents answered in the negative. In central East Austin 90 percent of respondents indicated that the city needed more integrated housing.[54]

Access to Knowledge

Similar to their influence on residential neighborhoods, Austin whites worked to keep schools segregated and collective civic and social experiences among races at a minimum, while also ensuring white students an array of financial, social, and educational advantages over minorities. As in most of the South, African American students in Austin attended rigidly segregated public schools before *Brown v. Board of Education* engendered desegregation in 1954. In the 1940s the school board explicitly stated that improvement to schools in East Austin should be undertaken solely to keep minority residents from leaving the Eastside. School site selection was dictated by segregation.[55] Almost no minority teachers taught in majority white schools. Because of an obscure law dating from the founding of Texas, Austin's Latinos were not legally segregated in schools, but de facto segregation remained very strong, even in relatively integrated neighborhoods.[56] After 1954, the newly formed Austin Independent School District (AISD) implemented a variety of plans to circumvent new desegregation laws or to slow down the process of integration. One of AISD's methods to keep white students segregated while legally integrating schools was to simply draw boundaries that integrated existing African American and Latino neighborhoods. Because, legally, Mexican Americans were considered white, AISD hoped that integration between Eastside African Americans and Mexican Americans would be enough to keep federal courts out of Austin without integrating Anglos with minorities whatsoever. AISD also adopted a "freedom of choice" plan in the late 1950s that allowed students to attend whatever school they lived closest to; because residential segregation was still so ubiquitous, the plan had the desired effect of stalling integration. In 1959 less than 1 percent of the 5,512 African American students enrolled in AISD attended a majority white school.[57]

Desegregation in Austin stalled throughout the 1960s despite unwanted external attention from the U.S. Department of Health, Education, and Welfare (HEW) as well as the Dallas Education Branch of the Office of Civil Rights. Both groups investigated AISD closely and attempted to reach a solution that both the city and the U.S. government could find acceptable. AISD's new version of the freedom of choice plan, basically the same one it was operating under at the time, was rejected by HEW. AISD countered by offering a redistricting plan and new buildings program, which was also rejected by HEW. In 1970, after years of stalemate, the Federal Office of Civil Rights filed suit against AISD for failing to comply with desegregation guidelines. As part of the first federal case in 1972, the court collected data on the racial makeup of AISD schools. Eighteen schools in East Austin enrolled student populations that were more than 90 percent minority; in two-thirds of those schools over 90 percent of students were African American or Latino. Only one school in East Austin was not more than 90 percent minority as of 1972. AISD's final attempt to assuage the Office of Civil Rights was to integrate just one level across the city, but this plan was also rejected. Additionally, of twenty-nine AISD schools opened since 1954, nineteen were over 90 percent white, while twenty-one had zero black students. Although Mexican American students were more evenly distributed, they often acted as buffers between the other two races or had a very small presence in heavily Anglo schools.[58]

In Austin, as in many southern cities and some northern urban areas, court-ordered busing was implemented as a last effort to integrate schools after AISD proved unwilling to move desegregation along.[59] Federal circuit courts heard a series of three cases against AISD throughout the 1970s, beginning with the initial 1972 case, which provided statistical information to demonstrate AISD's negligence and also began the process of setting guidelines for integration. Before busing was even formally mentioned in the courts, however, Austin attorney Bill Lynch formed the Austin Anti-Busing League in 1970. For Lynch, busing represented not only the manifestation of socialism and communism in the government; it was also an impingement on individual rights.[60] Even though the liberal AISD council might not want busing, Lynch claimed that members had "no quarrel with the social mixing aspect. That's where we differ. It's wrong to impose a socialistic state."[61] Although Lynch thought that the antibusing league was unfairly characterized as racist, he did admit that many members were racists. He thought that the Left was characterized by Communists and that busing represented unwarranted federal growth in one of its more "repugnant forms."[62] To Lynch, if citizens were to demonstrate their true feelings about busing, "they'd hang Will Davis."[63] He added, "They'd go to the Supreme Court and say 'you fellas better leave the country.' "[64] While Lynch's virulent rhetoric may have been more hyperbolic than many Anglo Austinites could tolerate, the league's stance was popular among them. Numerous smaller groups formed in defense of the neighborhood school system and the AISD's freedom of choice plans in the early 1970s. In a show of grassroots opposition that mirrored the open housing issue three years earlier, over twenty thousand citizens signed an antibusing petition that was delivered to HEW in 1971.

The circuit court, which rejected each plan proposed by AISD, found no other way to create a unitary school district. Despite pleas against busing from AISD and complaints of white flight and other social maladies that the district related to busing, the second federal desegregation case against AISD, heard by a circuit court judge in 1976, deemed busing a suitable method to integrate AISD schools. Busing was ordered by circuit court judge Jack Roberts in 1979 after AISD proved unable to create a unitary school system. Beginning in 1980, AISD complied with the order, which specified that the district would be declared unitary and the court case dropped if results were satisfactory after three years of extensive busing. Elementary schools in East Austin and West Austin were paired; first through third graders were bused one way and fourth through sixth graders were bused the other way. The Austin Schools Project, a 1998 study conducted by law professor Elvia R. Arriola, found that by 1983 busing created a unitary system in most Eastside schools, reversing the wholly segregated demographic of those schools in 1971.[65] The case closed in 1983 after the district was perfunctorily decreed unitary.

Busing ended in Austin almost directly after the enforced period of desegregation concluded in 1986, and the former neighborhood school format became the system of choice for AISD. In 1987 AISD replaced busing with a system that allocated more funds to schools with large minority populations, in an attempt to address past injustices without continuing busing. Almost immediately, levels of integration dropped precipitously. Although they did not reach the segregated levels of the 1960s or early 1970s, by the mid-1990s 75 percent of Austin's elementary schools were distinctive majority-minority schools. Another perfunctory change made to encourage continued desegregation without busing was the magnet school program, which AISD adopted in the late 1980s specifically to improve integration while creating elite programs for the city's most talented students. All three magnet schools were placed inside Eastside facilities, which meant that Anglo students would usually need to be bused to schools. The busing, however, did not exactly imply integration. Arriola's study found that magnet schools generally functioned as "schools within schools," where the heavily Anglo magnet school operated almost entirely independent from the regular minority-dominated school. Arriola also found that instances of minority attrition were much higher than Anglo attrition at the magnet schools as well.[66] Owing to racial discrepancies, the failure of busing, and the historical lack of investment in education, minorities in Austin lagged far behind whites and rarely competed for white-collar jobs. In one predominantly Latino neighborhood, in 1979, only 24.1 percent of adult residents had graduated from high school. In that same neighborhood 36 percent of adults reported zero to four years of formal education.[67]

The Increasing Significance of Race

While most narratives portray a slow improvement of race relations after the contentious battles fought over civil rights in the late 1950s and early 1960s, in Austin racial tension appears to have increased throughout this period. Even in the 1970s Whites, African Americans, and Latinos remained in almost entirely distinct communities with little shared history or space outside of Austin's emerging nightclub scene. School desegregation in Austin was as contentious as anywhere in the South; among other tactics, the Anglo school board attempted to appease federal judges in the 1960s by integrating African Americans with Latinos, who were considered white under Texas's antiquated constitution. A thirteen-year federal school desegregation suit that began in 1970 followed. Tension in the schools and the incredible grassroots opposition to the Fair Housing Ordinance demonstrated that a wide variety of white Austinites, not just politicians and business interests, were hostile toward integration. Despite a number of liberals on the city council and a growing progressive movement emanating from the university in the 1960s, Austin's racial tensions mirrored those in much of the South.[68]

In the three decades after World War II segregation became more rigid because it became more linked to economic growth. While African Americans did begin to move out of the central Eastside in the 1960s and 1970s, their dispersal reflected an extension of segregation patterns rather than an erosion of them. Most African Americans moved farther east and northeast into new housing developments that were highly segregated. West of IH-35, only one neighborhood, Clarksville, which was unincorporated, historically African American, and without municipal services, was less than 85 percent white. Outside of Clarksville, in 1970, only one neighborhood on the Westside was more than 5 percent African American, and only one neighborhood in the growing upper-middle-class area west of Lamar was more than 1 percent African American. That tract was less than 5 percent African American.[69] Minority concentration in East Austin was even more acute. In 1977 the historic African American neighborhood bounded by IH-35, Martin Luther King Jr. Boulevard, Airport Boulevard, and Seventh Street was 99 percent minority. The historically Latino neighborhood bounded by IH-35, Seventh Street, Springdale Road, and First Street was 92 percent minority. That percentage increased as most of the small remaining Anglo community in the East Cesar Chavez neighborhood left during the 1970s.[70]

Economically, Austin was an increasingly bifurcated city by 1970 as well. The poverty rate in central East Austin, for example, was 37.5 percent, which was nearly double that of the rest of the city and during a period of less than 2 percent unemployment in Travis County.[71] Wealth in Austin in 1970 was highly concentrated. Although per household income was $10,529, 33 percent of Austin households had an income of less than $5,000 and 68.5 percent had incomes less than $10,000. These data suggest that only 44 percent of Austin residents could afford to buy houses in 1971 because of the paucity of low-to-moderately priced residential buildings. The data also suggest that urban economies based on high-tech production in the 1960s demonstrated a high concentration at the top of the socioeconomic spectrum and a high level of economic marginalization. Many minorities were cut off from not only home ownership but also from any kind of mainstream economic life in Austin, regardless of their location. Economic data by race clearly demonstrated a highly bifurcated city in 1970 as well: per capita income for African Americans was 52.3 percent of the average, while per capita income for Latinos was 59.3 percent of the average. Because minority statistics were included in the average, minority income was almost certainly an even lower percentage of the average income of Anglos in Austin. Clearly, Austin had a significant economic gap during the 1960s, and that gap was highly characterized by race.[72]

Grassroots, race-neutral appeals to private property rights in the fight over open housing in 1968, then, rang hollow despite their enormous popularity, because they rested on the racist application of eminent domain policies and reflected the irony inherent in Austin's growth. African American residents, dispossessed and uprooted by federally sanctioned, locally sponsored urban renewal, did not enjoy those same private property rights or share in the city's overall economic improvements in the postwar period. In fact, they suffered the abuses of government-sponsored renewal disproportionately and saw little benefit from the accumulation generated by the improvement of their property. Ideologically, a wide chasm existed between Keynesian attitudes about the ability of the liberal state to create a better way of life and liberal social politics that sought to confer that way of life to all citizens. Rigidly segregated residential and social geography was thus also an important component of the city's identity as nonindustrial. The focus on skilled labor markets, along with strict residential segregation, allowed *Austin in Action*, the chamber of commerce magazine, to boldly claim that Austin "is primarily a city of upper middle-class citizens" in an article that documented the city's attractive features. The piece went on to quote an article from *Industrial Development* at length: "Austin can concentrate on offering the specialized facilities of an intellectual center without having to duplicate within its own limits the service functions of a large city or the large scale facilities of a heavy fabrication center. . . . This gives the city the opportunity to build up its amenities and retain the small-city convenience and flexibility that might well be lost if it were trying to push its way to the fore as an all-purpose metropolitan center in its own right."[73]

The intended message that Austin was not really a city, without the major problems manifesting themselves in many American cities throughout the 1960s, was a very attractive feature for Austin's growth advocates and presumably by the "laboring class of cultured intellectuals" they wished to attract. To create this image, however, minorities not only needed to be kept out of white areas; some of their neighborhoods needed to be refurbished and improved to facilitate nonindustrial accumulation. Segregation thus helped to maintain a nonindustrial image that city leaders used to market Austin as a pleasant place for knowledge workers to live and do business. Urban renewal allowed Austin's leaders to reimagine and reinvent the city's physical landscape, yet their claims that urban renewal benefited people rang hollow. By the 1970s, Austin's African Americans were more segregated and isolated than they were in 1950. African Americans and Latinos lagged behind whites economically just as much as they had in 1950, and they failed to share in the dramatic economic growth the city experienced in the postwar era. They continued to live in substandard housing at the same rate and were also subjected to more environmental hazards; as the city grew, its necessary functions did as well. A power plant, a gas tank farm, garbage dumps, and other facilities that attended growth were disproportionately sited in minority areas. For most minorities, improved standards of living and economic opportunities remained elusive in the midst of urban growth.

7 More and More Enlightened Citizens

Environmental Progressivism and Austin's Emergent Identity

In October 1969, University of Texas officials, determined to expand the main campus because of increased enrollment, sent bulldozers to Waller Creek to remove trees to provide room to augment the football stadium. A group of students met the bulldozers at the creek on the morning of October 20, carrying environmental protest signs and occupying trees slated for removal. On Wednesday, Board of regents chairman Frank Erwin made his way down to the creek and instructed the workers to begin removing trees regardless of who was in them. He summoned dozens of police vehicles and implored officers to arrest anyone who did not comply. By the end of the day twenty-seven people had been arrested, and the trees, considered part of the university's sense of place and a natural oasis by many students and residents, were destroyed. Environmentally conscious students and residents alike coalesced into an environmental movement aligned against Erwin over the next year, calling for his removal and dubbing him "Axe Erwin." Longtime activist Roberta Crenshaw founded the Austin Environmental Council in the wake of the Waller Creek incident, claiming that "more and more enlightened citizens are going to become restless as they witness further environmental abuse" from university expansion and unchecked urban growth. Two weeks later the Young Democrats and Young Republicans jointly sponsored protests of university development policies. Over the next two decades popular environmentalism became the primary focus for Austin progressives, with urban growth at the center of the debate. Land development policy became the core of what many commentators saw as a battle between real estate development and the forces of capital, on the one hand, and slow growth environmentalists who valued sense of place, on the other hand.[1]

After World War II American environmentalism grew from a peripheral movement that focused on pristine natural areas to a mainstream movement that saw nature being threatened everywhere.[2] Millions of acres of land were rapidly developed as metropolitan areas expanded with tract housing developments and suburban shopping malls. Suburbanites and other residents around the country began to notice that the natural landscape was being transformed to make space for development and houses and that problems like flooding, soil erosion, and water pollution were becoming more common. People also began to take note of the waste generated by the most affluent consumer society in human history and the chemicals used to grow its food and produce its consumer goods. A number of popular works documenting environmental degradation and the dangers of unmitigated growth emerged in the early 1960s, and by the middle of that decade hundreds of suburban-focused grassroots environmental groups appeared.[3]

Like many social movements, mainstream environmentalism in Austin coalesced around a specific and intense incident that clearly demonstrated how vulnerable the natural landscape could be in a growing city. Yet various Austinites had voiced concerns over environmental issues prior to the Waller Creek incident, and some had begun to collectively challenge growth-oriented development policies and the pro-growth business community. The economic gains and massive new construction in the 1960s bolstered the growth community's confidence, but also helped to set the resolve of environmentalists. The Waller Creek incident gained notoriety in the local press and galvanized Austin's environmentally minded citizens on overdevelopment as the most significant problem facing the city. It also merged formerly separate interested groups into a coalition and gave a diverse set of environmentalists, from wealthy housewives to radical university students, a common goal. By the 1970s and especially in the 1980s, thousands of Austinites believed that the city's natural environment was its most important characteristic and that real estate development, driven by a profit-oriented growth machine, was the most significant threat to Austin's sense of place.[4] Conversely, the higher density, mixed uses, and industrial character of the Eastside, as well as its location away from Austin's environmentally significant spaces, made it easy for environmentalists to disregard or, in the case of the waterfront, to view it as a place to challenge the growth coalition without changing social relations.

This chapter traces the emergence and growth of mainstream environmental groups in Austin from the late 1950s through the 1970s, demonstrating how Austin's progressive identity became defined by protecting and enhancing the city's natural spaces and quality of life against overdevelopment and the business community that profited from it. In the postwar era, growth became a type of psychology for urban boosters, particularly in the

South and West, and they often assumed political power to facilitate it.[5] Chamber leaders often promoted environmental thinking, but they imagined it as aesthetically improving the landscape to enhance accumulation; early grassroots environmentalists challenged this definition by emphasizing the environment as a collective sense of place that benefited all citizens, not just the business community. Early environmentalists used the city's planning commission to preserve public open space and improve natural areas for communal meaning. They engaged in letter-writing campaigns designed to slow development and privatization of public open space and often viewed open space and nature as civic assets, similar to Austin's boosters during the early twentieth century. By the 1970s, liberals assumed power in the city council as well as in the department of planning. The department of planning created a public participation initiative called Austin Tomorrow, which sought to curb and direct urban growth. The legacy of early environmental progressivism was a strong cohort of environmentally focused neighborhood associations that grew in the 1970s and 1980s and eventually won two battles over developers that set the course for Austin as an environmentally responsible city.[6]

Yet the mainstream environmental movement's victories also point to the deep racial and geographic chasm dividing the city and provide valuable evidence about the illusive and malleable nature of the possessive investment in whiteness. While these groups often fought local development battles, their local focus emphasized neighborhood environment and quality-of-life concerns rather than social justice or cross-class concerns. Despite a generally liberal attitude, a lack of overt racial discrimination, and in some cases a willingness to incorporate minorities, the ways they imagined the environment, and some of the plans they implemented, subordinated minority spaces and communities. Narratives of popular environmentalism usually portray grassroots environmental groups defending natural space and promoting a social sense of place against capitalist developers, yet often there are more interests at play. Rarely are oppositional groups seen as fractured and fluid like the growth machines, and often fractures come from within environmental groups (such as divisions between mainstreamers interested in quality of life and property values and radical groups interested in stopping development entirely) rather than as separate groups with largely separate interests.[7] By attempting to co-opt Eastside space for environmental purposes, using existing segregated districts in public planning, and imagining different uses for public water space, mainstream environmentalists demonstrated a subtle yet important willingness to protect and enhance nature at the expense of minority community interests.

While the historical geography of racism certainly perpetuated the chasm between environmentalists and minorities in Austin, the different ways that the two groups understood what constituted "environment" was equally important. Latinos and African Americans responded to environmental planning initiatives in ways that reflected a more expansive vision of what part of the environment needed to be improved. They argued that the environmental improvements embedded in Austin Tomorrow came at the expense of socioeconomic and health issues that they faced. In the wake of comprehensive planning that sought to destroy their neighborhoods and in the midst of urban renewal, they pointed to an unresponsive and uncaring municipal government as reason to doubt the intentions of planners. They recognized that their spaces were increasingly used as sites to facilitate growth, and they defended them against unwanted new land uses and recreation activities that were not intended for their community. Defending their environment, which they conceived of as both the built and natural landscape, helped to form a sense of identity that linked space and race into a social consciousness. "Defending the barrio," as activist Paul Hernandez wrote, was an act of defending a unique sense of place against outside interference but also an act of community and racial empowerment.

Chants Environmental: Resisting Growth and Improving Green Space

Some of the first articulations of modern metropolitan environmentalism in the postwar era were attempts to curtail development on rapidly expanding urban fringes. In response to a severe housing shortage after World War II, the federal government rapidly expanded New Deal–era programs that subsidized housing construction, guaranteed mortgages, and offered economic incentives to home buyers. The lack of existing housing stock and new mass production technologies meant that peripheral urban areas became ideal sites for large tract developments where homes could be built quickly and affordably. By the early 1950s over 90 percent of new residential construction was built on formerly undeveloped land, and the suburbs came to dominate the American landscape. New earth-moving technology also meant that developers could now clear and develop large tracts on hills and in wetlands that previously would have been impossible to transform. Additionally, hastily planned developments often lacked planned communal open space and eradicated any vegetation present in the area. Concerned suburbanites organized to protest environmental degradation, pointing to the loss of agricultural land, less undeveloped land to mitigate flooding and runoff, and, not least, the negative aesthetic impact of tract housing and the lack of access to nature, outdoor recreation, and open space.[8]

While demographic and physical growth in Austin was robust in the 1950s, it was nowhere near the level of expansion in major metropolitan areas such as New York, Los Angeles, or even Atlanta, so tract housing was not seen as a primary concern until the late 1960s. Austin also had almost no suburbs in the 1950s; nearly everyone lived under the same municipal jurisdiction. In Austin, grassroots environmentalism initially took hold not on the suburban periphery but around the city's water spaces, mostly near downtown and in older, established middle-class neighborhoods near downtown. Austin's proto-environmentalists targeted the city's growth coalition, a mix of growth-minded political and business leaders, in an effort to preserve public open space near the city's major sources of water. In doing so, they institutionalized an antigrowth narrative that would define the environmental movement for the next four decades and also established access to water amenities and public outdoor space as vital components of Austin's high quality of life.

Austin's nascent environmentalism emerged in the late 1950s in response to a plan to privatize green space around the new Town Lake, a reservoir that impounds the portion of the Colorado River adjacent to downtown Austin. By late 1957, with the completion of the Longhorn Dam scheduled for less than two years away and the upper lakes already creating millions of dollars in annual revenue, Austinites began planning for a downtown lake. The small body of water, which was intended both as a recreational space and as a cooling lake for the new Holly Power Plant located in East Austin, extended into the Eastside of the city under the IH-35 bridge. Although the Colorado River had always provided water at that spot, the dam's completion ensured that the reservoir would always be filled to the same level and would never again destroy property in the central part of the city. The last infrastructural water improvements on the lakes, right in the heart of Austin, had dramatic effects on the city and demonstrate the extent to which water dominated Austin's image, both as a public good and as a marketing and economic force for the city.

In November 1957, the Austin Chamber of Commerce announced plans to move its headquarters to a public park on the shores of Town Lake, near

Congress Avenue, the city's main north-south commercial thoroughfare. Although the chamber did not announce it, many residents understood the plan as the first step in a process where lakefront property would continue to be privatized by the chamber and other development-oriented outfits. Growth-related discourse rose precipitously beginning during World War II, and by 1957 Austin was in the midst of its most intense population and economic growth to date, largely because of the lakes. The revised Austin Plan, which was created between 1955 and 1961 and sought to update the master plan of 1928, emphasized using zoning to better regulate land use and to improve downtown for commercialization: business, conferences, and tourism. Immediately, letters from concerned citizens protesting the chamber of commerce's plan flooded Mayor Tom Miller's office. While the letters demonstrate a collective fear of rapid development and a sense that Austin's unchecked, capitalist growth would have deleterious effects on citizens' quality of life, they also reveal the extent to which water was ensconced as the central motif of the city's social and cultural identity.

Most letters focused on Austin's unique natural landscape, which centered on the pastoral nature of the downtown area, and linked it with a sense of civic pride that the area was public. To many writers, putting the interests of private business over public good was not only wrong but also damaging. Lillian Peek objected to the development because she believed that Austin's most attractive feature was its inimitable landscape; Austin, because of its relaxing atmosphere, was not "like any other commercial city." Fred S. Webster viewed the downtown waterfront as both relaxing and offering a potential economic benefit, but only if it remained open public space. "How many cities in Texas can match the relaxing vistas of Austin's downtown waterfront with the green expanse of its public park?" he asked. "Surely the Austin Chamber of Commerce . . . must realize that a major selling point to outsiders is Austin's natural beauty and uncrowded physical layout." To Webster, capitalist development of public space along the river was shortsighted. In the long run Austin could prosper more by maintaining its natural spaces, while other cities continued to destroy theirs. Elizabeth F. Gardner made a similar point, identifying the natural landscape and the downtown river as central to Austin's identity. "Where will the continued encroachment upon and destruction of the natural landscape and beauty of Austin lead us?" she asked state senator Charles F. Herring. Gardner continued by emphasizing the importance of natural beauty to Austin's identity, writing that "the great charm and attraction of Austin lies in what Nature has given us, not what men have created." Here Austin's sense of place is again defined by the natural or pastoral. The antiurban logic of open space advocates was taken to its conclusion by Ruth Isley, who derided the chamber of commerce plan because "next time someone may want to build a factory—once we make an exception."[9]

Citizens were likewise concerned that private business development would come at the expense of Austin's natural landscape, and many voiced displeasure with the potential increase in business activity symbolized by the chamber plan. Two concerned women argued that the city's novel natural characteristics had long been part of the city's draw, writing, "The people of Austin have been very far seeing in maintaining our heritage of beauty and uniqueness." Business and industrial development threatened to disrupt the very history of the city. Webster noted that the plan would "set a precedent which would certainly encourage those who seek to exploit the recreational and aesthetic values of public park land for their private gains," indicating that recreation and beauty were civic characteristics of Austin in danger from waterfront development. Mrs. R. Q. Underwood wrote the most aggressively antibusiness letter to the city council: "The citizens of Austin . . . react with cynical thoughts about these 'servants of the people' who are so much businessmen first that they sacrifice this important and long-dedicated open space to the swollen desires of their fellow businessmen for a fancy meeting place." The 140 members of the Travis Audubon Society also voiced their displeasure with the chamber of commerce, claiming that the plan would set a dangerous precedent and undermine the history of Austin as the capital of Texas and the bastion of public culture for the state.[10]

The dozens of letters and multiple newspaper articles condemning the chamber of commerce plan had their desired effect. Less than three weeks after the letter-writing campaign began, the chamber accepted an offer to lease a space in southwestern Austin, far from the waterfront. This small victory for nature enthusiasts in Austin was decidedly antigrowth, portending many civic battles between economic growth advocates and environmentalists in the coming decades.[11] Ironically, though, it was the nature enthusiasts that understood Town Lake would be a better asset to Austin's economy if it remained public and peaceful rather than commercialized. One citizen wrote, "If . . . the council keeps it free of commercialization and the attendant objectionable features, the lake may well become something of which Austin citizens may be proud, something which they will be happy to point out to their visitors, and which in the long run will bring more business to Austin than will the commercialization of the lake." Others echoed this sentiment as well.[12] But it also demonstrated the increasingly pastoral identity, associated with the river, that citizens considered the defining characteristic of the city and in some cases a tool to develop the city responsibly. They considered the water and its shoreline public property, something bestowed upon Austin by nature and guaranteed by the founders of Texas, an essential component to Austin's lifestyle and culture.

Questions of how to use the new reservoir took the same tone as the privatization debate after Longhorn Dam opened in 1960. Longhorn Dam completed the Highland Lakes chain and meant that the entire length of the Colorado River from Lake Buchanan to the eastern reaches of Austin was impounded by dams. The chamber and the city council made the Austin Aqua Festival (AAF) the premiere event to showcase regional water resources, but they also attempted to make Town Lake a centerpiece for real estate development and commercialization in Austin. Environmentalists, excited by their easy victory over the chamber in 1957, also sought to develop lands around Town Lake. But their plans emphasized picturesque public open space that was devoid of commercialization and would attract tourists and revenue while also enhancing downtown Austin's beauty. Their visions of the city were not for no growth but rather for growth based on peaceful recreation in a beautiful, serene setting that showcased the capital and cultural center of Texas for all residents. Done properly, one prominent open space advocate claimed, the area around the lake and the new auditorium on the south bank could become "Austin's front yard," a social and cultural center that reflected Austinites' interest in arts, recreation, and aesthetics.[13]

The city council and the chamber of commerce usually viewed waterfront real estate both as an economic engine for the city's business elite and as way to increase tax revenue, an important function in a city where hundreds of acres of prime land were owned by the state and university and thus not subject to property tax. Rising municipal taxes and higher utility rates were the subject of citizen protest in the late 1950s, and the council created a committee to initiate business and marketing campaigns to generate revenue from Lake Austin and Town Lake.[14] In the 1960s the city council established another committee to study the physical aspects of Town Lake and its environs.[15] Through the mid-1950s, Austin's western hinterland communities were the primary beneficiaries of the reservoirs, though visitors often came to Lake Austin in the western part of the city. The completion of Longhorn Dam gave the chamber and the city council impetus to showcase the downtown and potentially transform it into a more lucrative business and convention center.

Environmentalists imagined waterfront space as a long-term asset that would showcase the city as one of Texas's most beautiful, sharply contrasting it with larger cities that many Austinites considered unattractive, such as Houston and Dallas. The idea that access to open space near water was an important component of Austin's quality of life drove environmentalists. Roberta Crenshaw, who headed the Parks and Recreation Board (PRB), repeatedly argued that commercialization of the waterfront was a short-term plan that would leave Austin's downtown less attractive and less profitable than if the waterfront were to be maintained as open space. Her letters to newspaper editors and to council members articulated the importance of the waterfront to Austin's long-term economic growth and to the revitalization of a downtown deteriorating because of decentralization. She wrote that poor choices in redeveloping Town Lake could "destroy one of Austin's greatest economic and aesthetic values" and that "Town Lake must be integrated into downtown if downtown is ever to come back," since outlying subdivisions and malls were drawing people and commerce away from the core. She grew frustrated with the "downtown business interests'" lack of foresight because they "have not seen their own welfare" in maintaining open space.[16]

The completion of the dam and the adoption of the new Austin Plan in 1961 initiated a decadelong battle between Crenshaw and the city council that would help to define the parameters of progressive environmentalism in Austin. Crenshaw and her allies on the PRB and in the Sierra Club and the Audubon Society began to organize support for an antidevelopment movement that intended to use municipal law to fight privatization and business-oriented planning. While the PRB understood that events like the AAF could promote the region and create economic activity, members feared that the waterfront would turn into a permanent site for the chamber's "carnival dreams" if left unchecked.[17] They began by enlisting architects from the University of Texas to articulate the importance that academics were placing on environmental studies and the need for municipal oversight of parks and open space. In 1963 the PRB surveyed twenty-eight other midsize cities and found that Austin was the only one that left the maintenance of parks to the Department of Public Works, which had little time or budget for park maintenance.[18] The director of the school of architecture wrote an editorial in the *Austin Statesman* claiming that the city needed to create an effective parks and recreation department, staffed with "philosophically-oriented personnel" and professional landscapers, to preserve and maximize the city's beauty.[19] The PRB also sent letters to the council arguing that clean, well-maintained parks actually pay for themselves because they augment the city's ability to attract businesses and "substantial citizens as residents."[20] The campaign convinced the city to establish and fund the Parks and Recreation Department to oversee park maintenance and create recreation programs, which began operating in 1965. Following University of Texas architect Alan Y. Taniguchi's call to make the water a peaceful retreat in the bustling city, the PRB in 1962 recommended that the city set aside the shoreline for public use, outlaw motorized crafts on Town Lake, not build an expressway on the north bank of the lake (as the 1961 plan encouraged), and set aside funds for landscaping and natural vegetation.[21]

By 1965 Crenshaw was bolstering the small Austin cohort by linking it to the emerging national environmental movement and taking advantage of federal programs aimed at redeveloping urban open space. By the mid-1960s urban planners were beginning to take the importance of open space seriously, and a number of national organizations dedicated to more responsible development emerged.[22] Crenshaw served on the board of trustees for the National Recreation and Parks Association as well as on the Council for Outdoor Recreation. In early 1965 first lady and Austin naturalist Lady Bird Johnson informed Crenshaw about the White House Conference on Natural Beauty, a reaction to the growing interest in urban environmentalism and conservation around the nation.[23] Crenshaw attended the conference in May, met other activists from around the country, and learned about new federal programs aimed at improving urban open spaces. The conference emphasized the plight of upscale suburban areas that ostensibly had the most to lose from overdevelopment.[24] Later that summer she cited a Conservation Foundation report on the conference to inform the city council about potential federal funding for beautification and improvement programs. Housing legislation that followed was pivotal. Title VII of the 1961 Housing Act authorized grants to state and local governments for undeveloped land for open spaces in urban areas. Title IX of the 1965 Housing Act expanded Title VII to include improvement of public spaces such as malls and waterfront areas, lighting and benches in urban parks, tree planting, and cultural activities. It promised that the federal government would provide mortgage insurance on loans to cities for open space development.[25] Crenshaw's awareness and documentation of environmental legislation helped to validate the movement locally, and the city won a federal development grant to improve parkland around Town Lake in 1968.[26]

The White House Conference on Natural Beauty, which included symposia on urban greenbelts, also helped to expand the PRB's focus to include other areas around the city that were deemed particularly beautiful or important. The floodplain surrounding Shoal Creek was turned into a trail by a private citizen and dedicated as parkland in the early 1950s, but other major waterways, such as Barton Creek and Waller Creek, as well as other public outdoor spaces remained essentially unregulated.[27] Crenshaw wrote to Mayor Lester Palmer warning of the dangers of developing Mount Bonnell, a popular public bluff overlooking Lake Austin, and urged him to listen to the multiple citizens concerned about development there and to think about the "nationwide emphasis on preserving natural beauty" when considering how to zone adjacent land.[28] By 1965 she and parks director Beverly Sheffield began to advocate for easements to create public trails along Barton Creek, a tributary to the Colorado in South Austin, and by 1967 liberal Austin mayor Harry Akin was speaking publicly about the need to preserve greenbelts along urban creeks throughout the city.[29] In cases where existing zoning made it impossible to protect creeks from development, environmentalists organized to minimize the impact. The Sierra Club petitioned the Texas Highway Department to consider alternative routes for Highway 360 across Bull Creek in northwestern Austin; the number of bridges crossing the creek was reduced from six to three.[30]

Crenshaw and the PRB used the momentum provided by their connection to other movements to fight the development coalition and to create their own plan for the redevelopment of Town Lake in the late 1960s. After the initial success of the AAF, the chamber of commerce sought more permanent sources of revenue from waterfront areas, and members considered a number of proposals. One of the largest was an amusement park called Little Texas, which Mayor Palmer began to promote in 1966. According to Palmer, the park, which was scheduled to be built on the south bank of Town Lake next to the IH-35 bridge on land leased from the city, would draw over one million visitors in the first three years and become "one of the top tourists attractions of all of Texas and the Southwest."[31]

Board members also developed their own master plan for the waterscape. Their environmental vision for Austin was grounded in creating a sense of place based not just on preserving the environment but also on augmenting and enhancing the landscape for middle-class recreation. Town Lake, as well as the entire chain of Highland Lakes extending over sixty miles northwest from downtown Austin, was the product of infrastructural improvements, an envirotechnical system that used technology to improve nature. Crenshaw and Sheffield saw improving Town Lake as an extension of other technological improvements. After successfully blocking construction of Little Texas, they urged the city council to preserve the entire frontage of Town Lake as public open space.[32] Crenshaw began planning for improvements to Town Lake that went far beyond protecting and parking open space. In 1967 she publicly unveiled the Lake of Lights proposal to link pedestrian pathways, rail bridges, and auto bridges with a system of lights. The plan sought to create a small island in Town Lake, with a foot bridge and water taxis, as well as a man-made waterfall and a system of underwater lights.[33] A park on the lake's south bank would include a series of fountains and an open-air venue for concerts and other cultural activities. Some of the recommendations came to fruition in the 1970s when Lady Bird Johnson adopted Town Lake as a project and led some highly publicized cleanup events that established environmental concerns as primary for Austin's progressives.

While the focus was on downtown Austin, proto-environmentalists, including the growth coalition, sought to promote their visions for growth by using open space on the Eastside, demonstrating in subtle but nonetheless important ways that minority space was thought of as an end in the development debate rather than as an environment where people lived. Crenshaw's proposed park system extended well past the IH-35 bridge along the north bank of Town Lake to a floodplain inhabited mostly by Mexican Americans. Longhorn Dam made the floodplain safer (although it was occupied by poor whites and Latinos for decades), but the area contained a park that, in 1967, was mostly undeveloped and unimproved open space and was used almost exclusively by Latinos during most of the year.[34] Crenshaw and the board presented a plan to develop the park into Laguna Gardens, a show garden for Texas foliage, and Festival Beach, a proposed site for large public festivals in Austin. Where the growth coalition had proposed a highway along First Street, the PRB proposed a parkway separating the open space from Latino residential areas to the north and connecting the parks to IH-35 as well as a bridge linking the complex to the south bank of Town Lake. They also wanted to use roads to separate the park from the residential portion of the neighborhood, rendering it more difficult to access for neighbors but easier to find for people arriving in automobiles via IH-35.[35]

The way that environmentalists bargained with power brokers about the use of Town Lake also demonstrates how they imagined non-water minority space as being outside their definition of environment. One of the longest-standing and most charged debates about the proper use of Town Lake was over the type of watercrafts that would be legal on it. The discussion began as soon as plans for Longhorn Dam were announced, when the city held a public meeting to consider options. Crenshaw and others argued that Town Lake should not be open to motorized crafts or waterskiing because Lake Austin, which was almost entirely privatized, already served as the urban waterway dedicated to sports and motorboats. Town Lake, according to one person, should be "a place for people to just be," with nature trails, paths, and a quiet calm atmosphere.[36] Throughout the early 1960s, the PRB recommended that the city ban motorized crafts from Town Lake. Despite attempts from the chamber to "commercialize Town Lake," the water remained free of motorized crafts.

Yet while the board members fought the use of motorized watercrafts on central Town Lake relentlessly, they put up little opposition to the use of motorized crafts east of the IH-35 bridge. Beginning in 1962, that section of the lake, adjacent to Latino neighborhoods, was used for motorboat races during the AAF, one of the most lucrative, but also most crowded and noisy, events. The barren park normally used by Latinos was transformed into a stadium with bleachers and grandstands for the boat races and waterskiing competitions. In the early 1970s Austin's Chicano movement ironically coalesced around the same issue: noise pollution and traffic from Anglos coming east to watch speedboat races engendered the first organized Latino protest group, the Brown Berets.[37]

By the 1970s, the AAF events on Town Lake in East Austin became symbols of racial discrimination for some Latino residents in a social landscape increasingly marked by minority protests aimed at uneven community development, urban renewal, and economic stagnation in East Austin. The festive atmosphere of white privilege in the area during the AAF proved more than many Latinos could bear. During the late 1960s and throughout the 1970s, a number of Latino-based social and economic organizations emerged to defend their neighborhoods against what they saw as real estate development that benefited Austin's business interests at the expense of minority spaces. The East Town Lake Citizens Neighborhood Association, El Concilio, the Austin branch of the Brown Berets, and the East Austin Chicano Economic Development Corporation focused on defending the Eastside against outside development. These groups were also part of larger Latino movements for civil and economic rights and cultural autonomy and pride that flourished just after the Civil Rights Acts of 1964 and 1965.[38]

To many residents, the AAF's use of space in East Austin was a clear demonstration of the city's disregard for the neighborhood and its largely minority population. During most of the year, very few of Austin's Westside residents came to the Eastside. But during the AAF the portion of the Holly Street neighborhood on the north bank of Town Lake accommodated spectators for boat races. Paul Hernandez, one of the founders of Austin's Brown Berets and a longtime community activist in East Austin, related the tensions created for residents by the motorboat races. Hernandez discussed how the neighborhood was disrespected by festival-going whites: "It bothered the old folks. And it bothered folks who lived in the immediate area. These people with the flashy boats and the flashy litter didn't have any respect. They littered the neighborhood. Those upstanding citizens should have been charged with indecent exposure because they were urinating all over the goddamn place. They would piss right by the car in somebody's yard, and it was that kind of disrespect that got people angry." Others related stories where drunken viewers would park their cars on private property, destroy property, and leave trash on lawns. Police presence was minimal, and, according to Hernandez, the police were not interested in the welfare of residents. The boats were noisy and fumes lingered in residential areas long after festivalgoers departed. As they would do in later years, early Latino environmental activists looked at their neighborhoods and homes as inextricably linked to the environment. To them, health and quality of life were linked with community autonomy and cultural history; defending their neighborhood against encroachment and environmental degradation became the lynchpin for Latino activism in East Austin. The AAF motorboat races also produced East Austin as a space that was simply a good place to watch a spectacle designed to be consumed by "Town Lake Cowboys."[39] Hernandez argued, "It was called the 'boat race issue,' but that's really a misnomer. It was really a land development issue. It was an issue of community rights and an issue of how the poor and the people of color and elderly people are treated vs. pleasure, luxury, and profit."[40] Hernandez was keenly aware that the lakes were sites of white pleasure and profit that undermined minority autonomy. The boat races were just one example of spatial domination; for Austin's Anglos the Eastside was a place to store things that did not fit with the city's clean, natural, and bucolic image. Despite a petition signed by over fifteen hundred residents in 1972 and protests at city hall for the next six years, the boat races continued on east Town Lake.[41]

Latinos march in protest of the Aqua Festival speedboat races in East Austin. The protest helped to solidify a Chicano neighborhood defense movement in the 1970s. PICA 11698, Austin History Center, Austin Public Library.

When activists protested the boat races in 1978, police quickly intervened. Land use was a central concern for East Austin residents during the AAF. Festival Beach, a large park on the north bank of Town Lake, for most of the year served as a primary public space for the Mexican American community of Austin. Like all of Austin's metropolitan parks, it was used by residents free of charge for most of the year; however, during the AAF the city fenced off the park and charged a fee to get in to watch the boat races. The 1978 protests used the entrance fee at Festival Beach as a rallying point, but they were designed to focus attention on the unfair treatment of minority residents and on the misuse of private property, more so than on the actual park space. During the protests, Hernandez and other Brown Berets were forcibly removed from the area by police. Hernandez was photographed being "manhandled" by police officers, and charges of police brutality surfaced from all over the East Austin community.[42] The publicity did force the city to relocate the boat races for 1979, constituting a grassroots victory for Latino political activists. The city council also gave the neighborhood $60,000 for improvements along Town Lake as a result of the protests. So while the boat races demonstrated that many Anglos saw the Eastside as a place to put undesirable functions, they also engendered activism and protest in the Latino community.[43]

"We Done Told You What We Needed": Austin Tomorrow and the Failure of Participatory Planning

By the early 1970s, the environmental movement in Austin and in the United States had become mainstream. The first Earth Day was celebrated in April 1970, and later that year the EPA was created. Environmental writers, such as Barry Commoner, achieved fame, and most every major city had a growing environmental movement. Throughout the 1960s and early 1970s, environmentalism grew in popularity in Austin as the city's economic and demographic growth hastened with a dramatic increase in the size of the state government, the university, and the research and development sector. The Waller Creek incident became a focal point for seasoned environmentalists and brought numerous college students and younger people into the environmental orbit. By the early 1970s progressives had gained control of the PRB and had begun changing the ethos in the city's department of planning. As in many other cities, they began to view land use and the mechanism by which use was determined, that is, zoning, as the primary arena where battles over the course of development would unfold.[44] Mainstream environmentalists, most of whom lived in more suburban West Austin and were middle- or upper-middle-class, educated, and liberal, sought to change the course of developmental practices in Austin by changing the law. And they sought to change the law by changing the process by which laws were made.[45]

The rise of urban environmentalism coincided with radical changes in urban planning ideology, which also emerged from broad progressive and populist movements of the 1960s. In Austin and elsewhere, urban planning was historically used by business and political elites to create comprehensive plans that ordered cities: roads, parks, business and residential districts, industry. While these plans were often well conceived and did bring order to cities, they also reflected the very narrow interests of urban growth coalitions and rarely considered the needs of ordinary citizens. Thus for most of its history urban planning has been used as a means for the powerful to reorganize space to fit their interests.[46] As part of the 1960s turn toward more grassroots, participatory forms of local governance and rejection of autocratic decision making, community-based planning emerged as a viable alternative to professional planning that operated closely with urban business and political elites.[47] The new federal Model Cities Program, adopted in 1966, provided federal subsidies for alternative planning initiatives.

Similar to their attempts to develop waterfront open space, political elites, supported by the powerful Real Estate Council of Austin and the Builders Association, rarely considered environmental regulations on development and attempted to develop much of Austin's public open space as the population grew rapidly. By the mid-1960s city leaders, such as automobile dealer and mayor Roy Butler and furniture dealer and mayor *pro tem* Louis Shanks, stood to profit heavily from increased development. Comprehensive planning reflected their interests. The new comprehensive plan, prepared in stages from 1955 to 1961, sought to strictly separate land uses to make new peripheral developments as lucrative as possible. Planners used zoning to encourage middle-class real estate and commercial development on the city's periphery, particularly north and west of town. In 1968 builders put up more new houses per capita in Austin than in any other large city in the United States. Growing population and decentralization led to a dramatic increase in automobile ownership and usage; the number of automobiles owned in Travis County doubled between 1960 and 1972.[48]

Austin's demographic, economic, and spatial growth during the 1960s was dramatic. Due largely to Sunbelt migration, increased university enrollment, and the growth of Texas's economy, Austin's population grew by a robust 39 percent in the 1960s. The per capita income increased by 41 percent. Austin had the most stable labor market in Texas during the late 1960s; in 1968 and 1969 the unemployment rate stayed below the threshold for full employment, 4 percent. The manufacturing sector in Austin grew by over 20 percent in 1969 alone, largely due to the growth of electronics research and production outfits. As in most metropolitan areas, residential construction was almost entirely on the urban periphery; the physical size of the city grew by roughly 70 percent in the 1960s, from fifty-one to eighty-six square miles, yet the overall density declined by 17 percent. For many environmentally conscious residents, the rapid expansion of the city posed a severe threat to quality of life and even the natural environment itself.[49]

Austin planners, led by environmentalist and progressive Dick Lillie, applied for and won a grant to create a participatory planning program that would invite all Austinites to attend planning meetings. Called Austin Tomorrow, the idea won praise from local environmental leaders as well as the business and political community, and a new planning document would then be produced based on citizens' input. Lillie, a seasoned urban planner and liberal with roots in the democratic movements of the 1960s, was versed in community organizing practices and engaged with the emergent participatory planning literature. He was also a member of the grassroots environmental movement in Austin and a close associate of Roberta Crenshaw, founder of the Austin Environmental Council and the most vociferous opponent of the city's growth coalition. He helped to create the city's Office of Environmental Resources Management in the early 1970s, which oversaw the building of public utilities, including power lines and sewers, and made sure that development was not harmful to the environment.

The extensive planning needed for a project as large as Austin Tomorrow began in 1972. It was divided into three stages. In phase I, a preliminary stage, planners gathered and assessed data from the 1970 census. After initial research was completed, Austin Tomorrow's mission statement indicated the importance of community participation and choice: "The research is aimed at identifying the realistic alternative for the future, and citizen participation allows people to select and refine the most suitable of these alternatives."[50] Phases II and III began to incorporate citizens into the process. The department of planning organized a 250-member Goals Assembly, made up of citizens who were identified as socially active. That cohort was trained to lead a larger group of fifteen hundred neighborhood volunteers, who would then organize and lead meetings open to all Austin residents. Fliers announcing all Austin Tomorrow events were placed in the *Austin American-Statesman* and, cleverly, mailed out with the utility bills the city sent in February 1974. The plan and growth program, Lillie guaranteed in an address to the Concerned Citizens for the Development of West Austin, "will go beyond the physical needs to the social needs of communities."[51] National commentators agreed and lauded the democratic nature of the process as "an example of government asking people directly," which undermined the stranglehold that developers and real estate powers had on urban planning.[52]

Yet, in practice, Austin Tomorrow often failed to live up to its democratic promises. Many public critiques of Austin Tomorrow emphasized the homogeneity of the most active participants, who were largely selected from the ranks of proactive environmentalists. *Statesman* editor Sam Wood, initially an enthusiastic supporter of Austin Tomorrow, lamented that while most Goals Assembly members were "a pretty well organized group of people along the same line of thought, the general public did not participate."[53] And many working citizens could not take part even if they wanted to. Meetings were scheduled during normal working hours, which precluded large segments of the population from participating. Many residents felt that their voices would not be heard over the bifurcated leaders of the growth proponents on one side and the environmentalists on the other side. And many thought that the program was nothing more than a ruse used by Austin's business leaders to propagate the illusion of democracy. The first *American Statesman*'s article on Austin Tomorrow reported that three neighborhood groups planned to withdraw from the Austin Tomorrow program because NLM, the consulting group the city hired to oversee the process, wrote a letter to the planning commission in which it indicated the public hearings would be a rubber stamp for elected officials.[54] NLM left the Austin Tomorrow process in early 1974, but attendance at phase II meetings continued to decline amid accusations of "stacking the deck" by both environmentalists and businesspeople. One of NLM's last press releases accused the planning commission of stacking the Austin Tomorrow Goals Assembly with no-growth advocates and students.[55] Citizens were also skeptical because nearly 15 percent of the Goals Assembly consisted of University of Texas students whose interests in the program were short-term and decidedly no-growth. By the end of the program there were only 176 out of 250 projected Goals Assembly members, meaning that students likely made up an even higher percentage of the cohort. No-growth advocates were very active on campus, and young, progressive politicians recognized the potential of engaging students in environmentally friendly urban planning. The rapidly expanding university population provided a large, fairly organized segment overwhelmingly supportive of the liberal agenda in Austin city politics.[56]

Phase III, which sought to incorporate regular community members from all Austin neighborhoods, also began inauspiciously. When informed that Austin Tomorrow was a long-range program designed to direct land use, a large portion of the one hundred people at the meeting left; their only concern for attending the meeting was the desire to install new traffic lights in their neighborhood.[57] Attendance remained sparse throughout the ten weeks of phase III meetings that were held around the city, although most citizens who attended understood that the program was intended to address long-term planning issues. A variety of issues were addressed at each meeting, from the problems of suburban sprawl to utilities planning, public transportation, and billboards.

By far, however, environmental concerns were the primary topic covered throughout the phase III meetings. The Austin Tomorrow interim report concluded that the foremost concerns of phase II and early phase III meetings were creek development, solid waste disposal, uncontrolled development, and public space.[58] Occasionally, developers, builders, and trade unions sent representatives to meetings to advocate for growth. Austin Tomorrow planners set aside one particular meeting in affluent West Austin where the Austin Association of Builders met with planners and citizens. One hundred members of the Austin Builders attended and debated with 120 suburban Westlake residents, almost all of whom were against continued urbanization.[59] Organized labor and business interests became increasingly skeptical of Austin Tomorrow as phase III progressed and increasingly demonstrated what they considered to be a no-growth or slow growth approach. A local American Federation of Labor and Congress of Industrial Organizations (AFL-CIO) president, initially an active member of the Goals Assembly, dropped out after becoming disenchanted with the Goals Assembly's increasingly militant antigrowth stance. Labor and business leaders as well as newspaper editors worried that the program was increasingly characterized by militant antigrowth supporters and did not represent a legitimate cross-section of Austin residents.[60]

Despite the progressive environmental outlook, Austin Tomorrow also demonstrated the way that environmentalism and racial politics existed in tension and approached ideas of urban improvements from vastly different perspectives. No group, publication, or association, not even the department of planning itself, failed to make note of the paucity of participation among minorities, and the planning commission made numerous attempts to increase involvement among Eastside residents. Publications aimed at the African American community, such as the *Austin Tribune*, also supported Austin Tomorrow as a means for East Austin voices to be heard by other politically active residents.[61] Although as the program moved forward more Eastside residents did participate, largely because of special measures taken by the planning committee, they did so less than any of the other ten districts in the city, and the concerns they voiced often fell on deaf ears. Their experience with the Austin Tomorrow program clearly demonstrates the geographic and social fractures that separated city from garden in the minds of environmentalists and minorities.

Owing to the segregation and economic bifurcation discussed in chapter 6, Austin's Eastside residents were much more likely to face environmental and health hazards as well as limited access to housing and jobs compared to residents on the Westside. Given these disparities, Eastsiders shared more pressing concerns about jobs, municipal services, environmental health, and public safety, while other zones concentrated on environmental and land use issues. Their conception of "environment" was much broader, with labor, public health, housing, and pollution being issues of concern. One of the few Eastside meetings during phase III made this abundantly clear. In what newspapers characterized as a complaint session, twenty-five residents addressed a wide range of problems specific to the Eastside meetings. They asked for the city to give attention to the absentee landlord problem and to enforce housing codes and zoning regulations in their neighborhoods.[62] Residents at the meeting also focused on basic issues of public health, safety, recreation, and jobs. One of the most alarming issues was the need for a citywide health education program along with free clinics and improved ambulance service. East Austin did not have a hospital, and ambulances took longer to get to East Austin residents. The Eastside had far fewer doctors per capita than the Westside, and many of the early migrants out of East Austin in the late 1960s and early 1970s were middle-class professionals. Although the Eastside was geographically smaller and denser compared to other areas in the city, it lacked the open municipal park space that the rest of the city enjoyed. The Colorado River and reservoirs, along with Barton Springs and its creek, so crucial to the recreation of residents downtown and on Austin's west and northwest sides, did not benefit Eastside residents nearly as much, and they suffered hot summers with few public spaces near water. Eastside residents also felt the lack of public transportation much more acutely than did other Austin residents, who were much more likely to own a car. Finally, every meeting on the Eastside had a focus on the economic instability there and the lack of available jobs and industry, the type of development that most Westside participants actively sought to curtail. While Austin's economy as a whole grew profoundly in the 1960s based on university and government growth and an emergent research sector, minorities found little improvement in their economic opportunities or their quality of life during that decade.[63]

An even greater issue was that so few of the city's minority residents chose to participate, even though Austin Tomorrow leaders made minority participation a main goal. Each of the minority-run newspapers and publications in Austin provided information about Austin Tomorrow and usually encouraged readers to take part. Participation, however, was lacking during the phase II and III portions of the program but also well beyond that, in the period between when Austin Tomorrow's recommendations were published and when the new version of the Austin City Plan, based in part on the recommendations, was adopted. In 1978 an Austin Tomorrow Ongoing Committee sent out questionnaires to residents and neighborhood associations about the city's progress concerning major issues detailed in the final Austin Tomorrow report. The planning commission received hundreds of completed questionnaires from all over the city, except from the zones representing areas on the Eastside that were majority African American and Latino. One resident partially completed a questionnaire, and all of the nine neighborhood groups declined to participate in the study.[64] The lack of support among minorities, while probably a bit stunning to the committee, was not out of line with the consistent frustration minority residents felt with both the Austin Tomorrow program and the city's specious efforts to provide equal municipal services or its attempts to incorporate the Eastside in earlier planning processes.

The frustrations inherent to East Austin politics were reiterated by independent newspaper editors in Austin's African American and Latino communities. *Villager* editor T. L. Wyatt, an active member of the Austin Tomorrow Goals Assembly, was nonetheless very skeptical of both the program and its effects on his largely African American readership. Wyatt claimed that the completed report would be "of very little or no value" largely because it was redundant and would follow the data recommendations regardless of what the population desired. His comments on his readers' participation in Austin Tomorrow demonstrate an even greater apathy: "As far as Austin Tomorrow goes, my readers have had no interest in it from the beginning, because they've always seen it as a do nothing program. . . . Every time the time comes for them [the city] to make some improvement in this part of town, the money always runs out. . . . The only way to get them involved again is to take action on some of the things the city already has to do."[65] Wyatt was also initially disheartened because the original Austin Tomorrow staff lacked any black or Chicano representation throughout 1973, until Sharon Fisher was hired to be the planning department's minority participation coordinator.

Marcelo Tafoya, editor of the short-lived Latino publication *Echo*, was even more acerbic in his condemnation of the program and of the city's treatment of minority residents: "You have to remember that for countless number of years the Mexicano has never gotten his fair share. . . . So, here they come with this Tomorrow program and a big bang—'we need you, we need this, we need to know this . . .' And all the Mexicanos say 'we done told you what we needed, why go through the hassle all over again.' So, to the Chicano, all they thought is another cover-up by the city to get federal funds. . . . 'We'll [the city administration] get our money and then use it somewhere else.' "[66] Tafoya, well aware of how urban renewal operated in Austin and of the city's industrial vision for his community, was even more aggressive in emphasizing how little Austin's Latino community cared about Austin Tomorrow's findings; when asked how much the community cared he replied "zero." He continued, "We found out that in the master plan East Austin doesn't exist. In the master plan, East Austin is going to be the industrial area of Austin and everybody is to be moved out into the different suburbs. . . . The city has always said 'well, why go in there and rebuild it, when it's not going to be there anyway twenty or thirty years from now.' "[67] Tafoya also explained that he went downtown to meet with Austin Tomorrow planners on several occasions, but rather than listen to his opinions on how to incorporate East Austin into the program, they seemed more interested in putting him on radio or television to endorse Austin Tomorrow.

Wyatt and Tafoya voiced issues that appear to be emblematic of the frustration and exasperation felt throughout African American and Latino communities, especially on Austin's Eastside. To them and other residents, the self-serving, discriminatory interests of the city and most Austin Tomorrow participants were certainly axiomatic. It is clear from data analysis that Austin's booming period during the 1960s largely skipped over minority residents. As in many urban centers, civil rights gains in Austin during the 1960s did not come with attendant socioeconomic benefits for most minorities; in fact, the promise of civil rights victories without economic gains became a source of tension in minority communities. Although minority Austinites did not riot on the scale seen in many larger cities in the 1960s and 1970s, discrimination in housing, employment, and social services was acute.[68]

The way planners imagined geography in Austin Tomorrow also reflects a divide between city and garden. Planners never understood that history and geography played key roles in East Austin's frustrations; to them a lack of communication was the key difficulty to overcome, despite the fact that communication about Austin Tomorrow was ubiquitous. Rather than acknowledge and try to understand the historical and geographical reasons for the city's race and class fractures, Austin Tomorrow planners invoked the language of populist democracy and egalitarianism when conceiving and promoting the program. They viewed Austin as largely static and as uniform in matters of spatial and socioeconomic ideology. A focus on the neighborhood scale sought to empower a specific segment of the community, not to give equal voice to all participants. Years of racial segregation allowed for easily definable discourses and mental maps of the city that re-institutionalized patterns of segregation and disadvantage long after municipally sponsored segregation ended.

Institutional and assumed segregation was obvious in the way that Austin Tomorrow's planners conceived of the city's urban geography and how segregated maps implicitly informed the Goals Assembly's visions of future spatial production in Austin. The Goals Assembly's acceptance of predetermined geographical boundaries for the program demonstrates an institutionally segregated city. Rather than reframe Austin based on a racially integrated mental map, the earliest zone maps produced by the planning commission rigidly conformed to antiquated notions of space in Austin and reinforced segregation. Simply, Austin Tomorrow's zones were based on census tracts that were created during segregation and reflected segregated practices and ideologies. Two tracts covered the African American sections of the Eastside (one of them also incorporated an increasingly dilapidated, low-residence central business district), and one zone covered the Latino neighborhoods to the south. Unsurprisingly, these three zones showed some of the lowest participation among all the zones citywide, and the primary concerns they voiced in phase III were not reflected in Austin Tomorrow's citywide recommendations.[69]

This segregated mental geography and fractured zone map were sutured over, however, by a discourse of democratic, egalitarian participation consistently employed by city planners. Instead of encouraging the different zones to each accurately reflect the city's demographic profile or including zones that allowed for more cross-neighborhood dialogue, Austin Tomorrow insisted that the main goal was to attain zone meetings that accurately reflected each zone's population and generic demographic profile. While in theory representative of the entire population, the outcome was largely homogenous zones based on largely homogenous neighborhoods that very clearly reflected the segregation endemic to the city. The majority of participants, most of whom supported low- or no-growth platforms characteristic of burgeoning environmental movements at the time, had their views reflected in the final document's citywide goals. The most positive effect of the neighborhood zones was a great increase in the number of neighborhood associations and in awareness of development taking place in Austin's neighborhoods. Unsurprisingly, new Eastside neighborhood associations in the 1970s focused on defending neighborhood integrity, economic stimulus, and infrastructural improvements; Westside groups tended to focus on traffic containment, open space, and low growth mixed with strong environmental regulation.[70]

Environment and Equity

Members of the environmental movement that emerged in Austin in the 1960s and coalesced in the 1970s saw preserving open space and slowing real estate development as their primary agenda. Led by Roberta Crenshaw and other conservationists, they identified the city's long-standing business community and political elites as a growth coalition that threatened Austin's sense of place and quality of life, defined predominantly by access to water and open space. They linked their local concerns with the growing national environmental movement to validate it and to take advantage of federal programs designed to make cities more ecologically sustainable and livable. In the 1960s their work helped to preserve much of Town Lake and its shoreline as public open space and they increased public consciousness about the role that irresponsible development played in the deterioration of other urban waterways.[71] By the early 1970s, older conservationists and open space advocates joined forces with a more diverse, yet equally single-minded, group of university students, planners, and others who yoked antidevelopment initiatives to a liberal agenda in Austin. They founded an umbrella group, the Austin Environmental Council, in the wake of the much-publicized Waller Creek incident, which brought together disparate groups worried about the negative impact that growth could have on Austin's natural environment; the council counted twenty-four civic groups among its members and sent its newsletter to over thirty thousand people by 1970.[72] Austin Tomorrow provided a chance for environmentalists to institutionalize their ideologies into laws about land use, and the *Austin Tomorrow Comprehensive Plan*, published in 1980 was an environmentally progressive and thoughtful document. Yet many characterized it as a failure because it lacked authority and was disregarded as radical environmentalism by political and business elites.[73] While Austin Tomorrow's recommendations were never accepted or turned into law, the process organized people into neighborhood groups, which numbered nearly two hundred in 2000. These neighborhood groups became the foundation for more successful environmental initiatives, as discussed in chapter 9.

Unfortunately, however, environmentalists and planners were unable to overcome existing geography and racial divisions when forging their own plans for developing Town Lake and when attempting to institute an environmental framework in Austin Tomorrow. Their plans for the public development of Festival Beach and Laguna Gardens sought to usurp minority space and transform it into a show garden for middle-class tourists, while cutting it off from minority areas using a feeder road. They allowed AAF promoters to move motorboat races from the Anglo section of Town Lake across the IH-35 bridge to the de facto Latino section, giving rise to a Latino protest movement. Latinos and African Americans, acutely aware of subtle institutional discrimination, used land use policy and public planning initiatives as ways to simultaneously point out social and economic disparities and increase community cohesion. Mainstream environmentalists, however, failed to take the city's long-standing geography of racial exclusion into account when developing the Austin Tomorrow program, which, despite their diligent efforts, further alienated minorities, who were accustomed to viewing urban planning as a tool with which to constrain, segregate, or dispossess them. The power to plan Austin's future was distributed far more evenly after Austin Tomorrow than before it. Yet its ability to deliver on its democratic promises was limited by a history and geography that denied equal participatory rights to all groups. In this way, despite its robust efforts to include marginalized populations, the program could not heal the scars discriminatory planning had left on Austin's minorities.

Explaining this blind spot among mainstream environmentalists is no simple task, especially considering that most mainstream environmentalists were liberals who shared at least some concern for civil rights and racial justice. They worked to incorporate minorities into planning, yet for minorities deteriorating conditions and loss of space in the midst of robust growth demonstrated how race was essential in determining who benefited and who did not. In Austin, years of seeing the city as socially and geographically divided contextualized mainstream environmentalism. Environmentalists used national programs and discourse to bolster their arguments, but they usually saw the environment from a local perspective, so much so that the neighborhood became the primary scale used to imagine environmental progressivism. Similarly, as Samuel P. Hays writes, "the phrase 'environmental quality' would have considerable personal meaning" for most middle-class adults in the postwar decades.[74] They had little social contact with minorities and shared few collective experiences because schools, parks, and residential areas were so heavily segregated. In some sense the social distinctions inherent in the landscape are reproduced in collective thinking. The maps people had of what the city looked like stopped and started at very different places. In the 1980s, however, everyone's map of Austin would change drastically as the technological sector engendered more intense spatial and social transformation than ever before.

8 Technopolis

The Machine Threatens the Garden

In 1983 Austin, Texas, surprised the American high-tech world when it won a competition against fifty-six other cities to host the Microelectronics and Computer Technology Corporation (MCC), a federally sponsored consortium of elite computing and semiconductor companies. MCC was the first consortium of its kind in the United States and a highly visible experiment designed to reassert America's global technological dominance. It was also a paradigm for future cooperative research endeavors, so its decision gave Austin immediate notoriety. Narratives of Austin's success focused on the competitive economic advantages that the city and the state of Texas offered: free space in a state-of-the-art complex built by the University of Texas, immense state and local subsidies including use of University of Texas funds, cheap skilled labor, and low housing costs and attractive mortgage programs provided by the city's business community.[1] MCC's decision generated economic activity in real estate and investment sectors almost immediately, and numerous other high-tech companies, including another, larger federally sponsored consortium called SEMATECH, located in Austin during the next decade. Since the 1980s Austin has become one of the fastest-growing cities in the United States, driven largely by a robust agglomeration of high-tech firms.

While, initially, discourse surrounding Austin's success focused on the economic incentives offered to MCC and other corporations, the benefits provided by Austin's natural environment and quality of life proved equally important to the high-tech workers and managers who actually relocated to Austin. Austin's natural landscape contained abundant water resources, which differentiated it from other, more arid regions in Texas and the Southwest. The wealth of natural amenities did not go unnoticed. As MCC site-selection committee member Robert Rutishauser commented: "Most of us who had not been to Texas had the image that most people had. You go to Dallas or to Houston, and you watch 'Giant,' and you think that's the state. In the brochure that Austin's Chamber of Commerce had prepared to sell the city, there is water in every scene. It's a picture of Town Lake, or it's a picture of Lake Travis. . . . When I took the helicopter ride over Austin, I looked down at Lake Travis, and I thought: 'Boy, that would be a beautiful place to have my boat.' "[2] Austin's natural environment and access to water-based leisure were great assets in attracting high-tech firms and skilled workers to the city and engendering robust demographic and economic growth during the previous four decades. In the 1990s the City of Austin made the area's natural amenities a centerpiece of its Smart Growth Initiative, which blended economic growth and environmental protectionism to encourage a high quality of life. In an era when skilled laborers had more mobility than ever before, their locational choices were increasingly determined by perceived quality of life in places; for many, access to environmental amenities played a major role in their decision to migrate.[3]

But the irony of the demographic and physical growth was that it undermined the access to nature, outdoor lifestyle, and quality of life that made Austin attractive in the first place. Over the decade following 1983, MCC proved to be less of an independent economic force for Austin than most prognosticators expected when the consortium chose to locate there.[4] But without question, MCC brought a significant social and symbolic benefit to Austin and the University of Texas on a national level, which in turn engendered rapid growth; it was also the catalyst for another major round of urban and economic restructuring in the city based on a rapidly growing technological agglomeration.

This chapter charts Austin's explosive economic, demographic, and spatial growth in the 1980s. It locates growth in the entrepreneurial approach to technological development that was adopted by university leaders, state and local politicians, and local and state businesspeople as a way to reinvigorate and diversify Texas's economy by taking advantage of aggressive federal spending on defense-related technology. Their success in making Austin into a hub of U.S. technological production and high-level research dramatically transformed urban space as the city expanded physically to the north and west. Yet development affected different parts of the city in ways that reflected the possessive investment in whiteness and shaped how whites and minorities defined environment in radically different ways. While the vast increase in peripheral residential and commercial development threatened Austin's pristine natural landscape and quality of life for Westside environmentalists, Eastsiders saw new production facilities as threats to health and community cohesion.

Building "Technopolis": Public Capital and the Knowledge City

One major reason for Austin's agglomerative strength in the 1980s and 1990s was the investment it received as a "technopolis" from the state and other private benefactors that wished to see Austin flourish as a benefit to the state and regional economy. The concept of technopolis indicates something more particular than simply a public realm or city built on technology. Robert W. Preer defines *technopolis* as "a region that generates sustained and propulsive economic activity through the creation and commercialization of new knowledge. . . . A technopolis is not merely a concentration of high-technology firms or research and development organizations. At the center of the technopolis is the creative process of developing new technologies and translating them into commercial products or processes."[5] In an important recent work on Austin, Eliot M. Tretter adds to this definition by arguing that state governments play an important role in this process by channeling resources to cities to make them more competitive, but also by using the quasi-public universities to offer additional incentives to firms relocating.[6] The University of Texas's ability to provide economic investment, capital, industrial/research space, labor, discursive support for entrepreneurialism, and knowledge acquisition was important to MCC and other private companies that either relocated to Austin or began there in the 1980s and 1990s. Austin's lack of prior industrial development, which can inhibit the growth of newer industrial formations and research and development initiatives, was an agglomerative force in terms of quality of life, space, and investment potential.[7]

State governments as well as private businesspeople often view growth based on high technology favorably for a number of reasons: it draws an educated workforce, it is generally cleaner than heavy industry, and it has a high capacity to generate new and potentially lucrative businesses known as spin-offs. A spin-off firm is generally originated by small groups of laborers at a larger firm who find a market niche in their work and seek to exploit it with their expertise. Because they are usually tied into local social relations and to local supply markets, spin-off firms also add to high-tech agglomerations. In high-tech industries, profits are maximized by getting products to market efficiently, because products tend to be unique, instead of by depressing labor costs, so high-tech firms are less likely to seek geographic relocation in areas with lower labor costs. They rather seek regions with highly skilled labor markets. The obvious outcome of this is a revalorization of highly skilled labor as the primary component of success in the high-tech market. Finally, technological agglomerations have demonstrated a strong tendency to perpetuate themselves, meaning that the more robust a cluster becomes, the better the chance it will continue to grow.[8] Universities are key in this regard because they often have the skilled workers and technological resources that form the foundation of technology-based growth.

The creation of successful technological agglomerations, though, is a social process that requires economic and ideological capital as well as an institution that can coordinate and commodify high-level technological research. The Balcones Research Center (BRC) was the primary site of research for the University of Texas, but the development of the Institute for Creative Capitalism (ICC) by George Kozmetsky, dean of the Graduate School of Business, provided the coordination that drove growth for the university and for Austin in the 1980s. For Kozmetsky, the transfer of publicly funded research generated by university engineering departments to engender private accumulation was the lifeblood of technopolis and the key to long-term regional growth. The university, supported by tax dollars and federal research grants, could sustain the risk involved with technological research much better than most private companies.[9] Kozmetsky understood this and saw the university as a profit-making entity that could attract surplus research capital if coordinated properly. Kozmetsky created the ICC to reorganize the university as a technologically driven management center that could initiate and sustain large levels of accumulation for Austin. The institute grew the university as a business by providing access to capital networks, public research space, and programs that coordinated research and business activities.

After arriving in 1966, Kozmetsky transformed the Graduate School of Business from what one writer called "a regional accounting school" into an "internationally recognized training ground for managers of 21st century industries" by 1975.[10] The School of Business Administration was the largest in the university, and graduate enrollment at the Business School had tripled in nine years.[11] Kozmetsky also founded the ICC in 1975 as a free market–oriented think tank that could articulate his vision for technopolis. Understanding capitalism, particularly the possibilities of academic capitalism, in a fluid and ever-changing postindustrial landscape, was the official charge of the ICC. One contemporary commentator referred to the ICC's mission as "directly or indirectly support[ing] private enterprise through research and the distribution of educational materials." The center, he claimed, hoped to become the public version of the Brookings

Institution.[12] For Kozmetsky and his collaborators, the triad of technology, the free market, and the scientific and organizational creativity possible at the University of Texas provided the basis for "technopolis," the utopic spatial and ideological manifestation of the new business-driven social order Kozmetsky envisioned. Though initially characterized socially, technopolis was in reality an economic project, however, designed to reinvigorate capital accumulation through entrepreneurship and the commercialization of federally funded technology research.

If the ICC gave discursive and institutional support to the notion of Austin as a technopolis generated by free market practice and entrepreneurialism within the university, the effort to attract MCC to the city demonstrated how forceful state government, university, and private business cooperation could be as a magnet for capital in a competitive environment. Austin had the benefit of being a center of public investment in Texas because of the university and the state government; the two institutions ensured that the city would always have a large base of public employment as well as cheap labor and spending money provided by students. As pointed out earlier, it also had the benefit of being relatively nonindustrial, nonurban, and without the widespread socioeconomic problems endemic to many U.S. cities as deindustrialization intensified through the 1970s.[13] But, again, the major locational advantage the city possessed was the university and its human and institutional resources, particularly regarding science, engineering, and business. Owing to a decline in the oil and gas industry throughout the 1970s, Governor Mark White as well as a number of prominent Texas businesspeople began formulating plans to attract high-tech business to Texas in an effort to diversify the state's economy. While Dallas was already growing as a midsize center for electronics development and applied research, the university made Austin the logical choice as a basic research and development center that could possibly form a technological development corridor with San Antonio. MCC, a unique private research consortium developed in 1982 to study microelectronics, particularly semiconductors, was the initial proving ground for Texas's ability to attract a national research and development outfit to Austin.[14]

While the role of the state was central in Austin's growth, the city also benefited from new global perspectives and federal projects. MCC itself was the product of a new emphasis on global competition in high technology and the willingness of Ronald Reagan's administration to invest massive amounts of money into defense-related high-technology research. In late 1981, Japan announced a joint government–computer industry program called the Fifth Generation Computer Systems Project, aimed at developing a new supercomputer that would establish its dominance in computer technology. The heavily subsidized Japanese group would have the luxury of looking into the future—projecting out ten years or more—and experimenting more so than a private company, whose profit motive would force a much more pragmatic, market-centered approach to development.[15] The United States was also in the midst of a prolonged industrial slowdown that, according to a growing number of economists, threatened the country's position in the global economic order. Unemployment reached 10.8 percent in late 1982, and other growth numbers, including in technology as demonstrated by patents filed and growth of the technological labor force, stagnated or slowed down precipitously after 1973.[16]

In response, the Reagan administration began adjusting patent laws to encourage technological entrepreneurship and research, especially among university researchers. As the first American research consortium, MCC was instrumental in the landmark National Cooperative Research Act of 1984, which generated a series of research consortia in myriad fields during the late 1980s and early 1990s. As the first, MCC's trajectory was closely followed by the media and the industry. Privately, leading members of the semiconductor and computer industry met in February 1982, at the behest of Control Data Corporation chairman William C. Norris, who proposed that the industry form a research consortium to better compete with the Japanese. To Norris, a longtime advocate of research consortia, "it wasn't until a lot of these companies got the hell scared out of them by the Japanese that they were willing to give [a consortium] a try."[17] The original consortium formed in August 1982, with fifteen member companies and an initial capital investment of roughly $50 million. In December 1982, the Justice Department granted conditional approval to the joint venture, beginning the process of overturning nearly a century of antitrust laws and basically giving MCC the right to organize itself.[18]

As the central research facility for a number of the biggest computer companies in the United States, MCC provided a great locational advantage for a city, as a source of high-tech, well-paid employment, but more so in terms of prestige. MCC promised to enhance a city's chances to grow technological agglomerations, as the winning city would demonstrate its attractive features to potential companies while also offering MCC and its employees as assets. Locational agglomerations, while manifestations of a competitive environment and economies of scale, are also cooperative ventures that reinforce and reestablish their social and economic capital every time they attract new forms of investment. Once a firm locates in a place, it becomes part of the already-existing community and benefits from other similar organizations that fill a role for that community. A national research consortium moving to a particular city is thus much more than the jobs created by the consortium; it is also a symbol of enhanced technological and economic prowess.

For Texas, and increasingly for Austin, MCC was much more than just a corporation that would bring jobs, ancillary industries, economic growth, and prestige, although these factors were vital. It was also a conduit to a new world order that was no longer based on industrial modes of production in established urban centers. Two factors played important roles in Austin's efforts to bring in MCC. The first was the city's emphasis on the future and on the lower cost of living and higher quality of life that Austin afforded residents; both of these characteristics were results of Austin's historical trajectory as a nonindustrial, relatively affluent city in an economically robust, low-tax, business-friendly region. Austin, at the time not considered a national technological agglomeration despite a number of high-tech companies and a fourteenth-ranked electrical engineering department at the university, focused its marketing efforts on the future. The plan to attract MCC promised a diverse culture of improvement where resources would be funneled into education, infrastructure, and equipment that would allow the university and MCC to flourish together.[19] Governor White and lawyer Pike Powers, who headed the task force, knew that MCC wanted to locate to a region where it would be one of the most visible outfits; MCC was, after all, designed as a kind of national industrial cooperative experiment whose success could provide a paradigm for other similar outfits. Hence a region willing to work with the corporation and showcase it was essential. An ideology that revered technological growth as well as a competitive industrial ethos was thus paramount for MCC. For Texas, looking to the future was paramount as well; one document argued that MCC would "[catapult] Texas into the world limelight as the State of the Future."[20]

The second factor was that Austin was deemed Texas's choice for MCC and the state's wealth of resources, both public and private, was invested into Austin's bid to attract MCC. An essential characteristic of the investment was the participation from all pertinent groups in Texas: the private sector, academia, and state and local governments. Again, what MCC desired, other than the obvious capital investments and dedicated resources, was a full commitment from a diverse array of actors. Bobby Inman, president of MCC, sought a community environment that would focus its collective energies on MCC. In early April 1983, after fifty-seven presentations, Austin was named as one of the four finalists for MCC along with San Diego, Atlanta, and Raleigh-Durham. Almost immediately, White and Powers began assembling a marketing committee of civic, business, and academic leaders and devised a slogan for Austin's MCC bid aptly titled "The Texas Incentive for Austin." It was clear that the State of Texas was investing in Austin's future as technopolis.[21]

The committee rapidly put together an incentive package that included an almost incomprehensible array of economic and social benefits for MCC that was very consistent with the practices of subsidizing potential business growth. What is most remarkable about "The Texas Incentive for Austin" proposal is the diversity of the incentives and of the institutions that promised to provide them to MCC. Possibly the package's most attractive feature was the potential long-term capital investment that the university could make in MCC and its own computing and engineering program using the Permanent University Fund. The university system's $2 billion fund, by far the largest public university endowment in the United States, gave the state more economic leverage than any other attribute did. Two-thirds of the assets in the fund were possessed by the University of Texas at Austin and the other one-third by Texas A&M University, which was also a part of "The Texas Incentive for Austin" proposal. While the fund was originally intended to support infrastructure and building on campus, since the late 1950s the university had put it to other uses that bolstered targeted departments and programs.[22] MCC was thus guaranteed that the fund would support both capital needs and academic excellence in prioritized fields. A second vital incentive that Texas offered was development and use of twenty acres of land on the BRC tract, which grew consistently through the 1960s and 1970s, for a cost of $2 per year to MCC.[23] During the late 1970s and early 1980s, the university began amassing world-renowned scientists, most notably physicist John A. Wheeler from Princeton in 1976 and Nobel Prize–winning physicist Steven Weinberg in 1982. In 1981 the regents authorized over $52 million for upgrades to the existing BRC facility, and three large university outfits announced that they were relocating from the main campus to the BRC. The tenants in the new facilities on the east tract were the Bureau of Economic Geology; the Center for Electromechanics, directed by Ben Streetman;[24] and the Center for Energy Studies, directed by Kozmetsky.[25] Along with the low-cost facilities, the most important incentive offered by the university was the establishment of a $15 million endowment that would help to attract and retain top electrical engineering and computer science faculty.[26]

The city and the state combined to offer MCC a host of other benefits that would smooth the transition for MCC employees relocating to Austin. Before the laboratory complex was built at the BRC site, the City of Austin agreed to house MCC at existing office buildings free of charge, many of which were donated by real estate developers who stood to benefit greatly if MCC located to Austin. More important, however, were the economic and social incentives to employees. Ben Head, chairman of Republic Bank of Austin, along with other local lending agencies pledged $20 million in single-family mortgage loans at 2 percentage points below the current interest rates to MCC employees. The city subsidized MCC's travel costs by underwriting relocation expenses for the company at $500,000. MCC personnel were also eligible for a total of $3 million in low-interest gap loans to "facilitate smooth transition financing." The Austin Women's Center and the Office of Relocation Assistance pledged to help MCC wives acclimate to Austin and to provide them with employment assistance. The center hired a full-time coordinator specifically to cater to MCC spouses as well as to provide information about day-to-day social activities, schools, child care, and clubs.[27]

MCC's decision to locate in Austin was seen as a huge political and economic victory by growth-oriented Texans, whose collaborative effort to win the MCC bid demonstrated the competitiveness of Texas and Austin in the emerging global economy. MCC was the lynchpin of Austin's efforts to grow as a technopolis. For Kozmetsky, the ICC, and the board of regents, the MCC decision was more of a sign that enhancing the commercial capacity of the university should be the primary aim of their future policy. Kozmetsky understood the value potential that MCC represented for small companies, spin-offs and otherwise, to a region. Simultaneously, having a large amount of skilled labor was the key to sustaining growth. Skilled workers also tend to be highly mobile, so creating industries, as well as other social conditions, for attracting them was paramount. Essentially, the university's general policy amounted to striking while the iron was hot, and this meant a high level of capitalization for research-based facilities and the concomitant growth of extra-academic functions designed to commercialize knowledge work. For the University of Texas, MCC was the harbinger of a technopolis that would provide profit for the university, savvy investors, and potentially the city and region more broadly.[28]

Along with MCC, a renewed federal investment in defense-related technologies and research and development provoked a new round of commercialization theories and applications among techno-capitalists in the business and engineering schools. Two important federal measures contextualize the increased potential for research-related profit among universities in the 1980s. The first was the expiration of the Mansfield Amendment in 1977. The amendment, passed in 1971, stipulated that Department of Defense research applications must have an applied military function. It severely curtailed federal funding for basic research at universities, which was easily the primary source of funding for university researchers since the beginning of World War II. Universities, cut off from their primary source of research funding, turned to private corporations to assist with basic research outlays.[29] The amendment's expiration, made politically viable by the end of the Vietnam War, opened up federal coffers to university researchers, who now had funding connections with the federal government as well as private sources. The second measure was the Patent and Trademark Act of 1980, which, as discussed in chapter 5, gave universities and their researchers intellectual property rights over federally funded inventions or processes that they patented and heightened incentive to pursue research that could potentially be profitable as well as increasing competition for skilled researchers among universities.

Another large impetus to university research was Reagan's Strategic Defense Initiative, commonly known as "Star Wars," which he announced in 1983 and institutionalized as the Strategic Defense Initiative Organization in 1984 within the Department of Defense. During the 1970s, federal defense policies focused on détente, an effort on the part of both the United States and the Soviet Union to limit the number of weapons that each side produced and accumulated. In contrast, the Strategic Defense Initiative was an ambitious if not impossible project that sought to invest major federal resources in new defense technologies, which necessitated a major investment in university-based research. Reagan also made it clear that he was willing to permit deficit spending for the nearly singular purpose of enhancing defense capabilities; despite cutting funding for the EPA, Medicaid, food stamps, federal education programs, public housing, and federal assistance to local governments, the national debt increased by close to $2 trillion during his presidency. After a short period of stagnant federal support for basic research during the 1970s, by 1983 federal money was available for university scientists and engineers who could support the Department of Defense's agenda. Whereas in 1979 federal research and development expenditures for defense were lower than all other research and development expenditures, by 1986 they were nearly three times greater. In real dollars, research and development spending for defense grew from $13.6 billion in

1979 to over $40 billion by 1986. Defense funding for basic research, primarily done at universities, doubled during that period. In 1985 federal research and development obligations for defense made up over 60 percent of all federal research and development obligations.[30]

The State of Texas followed the university's lead in facilitating entrepreneurship among research scientists. In 1985 the Texas legislature amended the Texas Education Code to require all universities in the Texas system to establish intellectual property laws that would grant ownership of all inventions and patents generated on campus to the university. As the legislature prepared the university system to profit from technological innovation, it also began the process of privatizing university-generated scientific information by revising the Texas Open Records Act to allow the university to keep information with commercial potential outside of the public record, no matter who funded it. These changes amounted to a declaration of open competition for commercialization; changes were made based on the university's ability to reap profits from inventions and to keep information away from competitors, as well as to allow scientists and engineers to research outside the view of media. Finally, the passage of the Ownership Equity Bill in 1987 allowed the university to own spin-off companies that were generated from university departments and other research units. In 1989 the legislature further augmented conflict of interest restrictions regarding university-owned businesses by allowing University of Texas regents and other administrators to invest in companies that had licensing agreements or contracts with the university.[31]

Owing to numerous changes in the federal government's ideology and the possibilities for technological development during the 1980s and 1990s, the University of Texas charted a course that made for-profit research its central function. By 1986 the university held over $137 million in Department of Defense contracts, making defense research one of the most attractive profit-making activities on campus and giving the University of Texas the fifth-highest total among U.S. universities.[32] Although these relationships had existed since World War II, in the 1980s the university sacrificed its traditional role as a seat of higher education in Texas to profit from the increased emphasis on defense technology under the administrations of Reagan and George H. W. Bush.[33] Whereas former technopolis-generating universities worked to facilitate technology business, in Austin the University of Texas became a business. MCC's locational decision had the dual effect of funneling large amounts of public and private capital directly into the university and convincing university officials that for-profit, high-tech partnerships with private business were sound investments.

Thus Austin's business and political leaders, working closely with State of Texas and university officials, took advantage of national and global changes that prioritized defense-related research and favored universities because of their ability to generate valuable high tech research. Federal military spending is also what engendered Austin's intensified growth in the early 1980s and brought the city national attention as a potential technopolis and heir to Silicon Valley. The University of Texas consistently invested in its laboratories that were heavily sponsored by defense contracts, sometimes to the detriment of other programs on campus. This uneven investment in defense-related technology represents the subordination of knowledge, and especially scientific knowledge, to capital. MCC and later SEMATECH, which were both made up almost entirely of defense contractors, were drawn to Austin by public capital that centralized itself at the university. At the center of Austin's agglomeration were the ICC and George Kozmetsky, along with other defense-affiliated administrators including Hans Mark, University of Texas chancellor and former head of the Strategic Defense Initiative. Many universities drew heavy federal funding for defense-related projects in engineering departments and for special university-affiliated research teams. What made Austin and the university unique was their ability to successfully harness that investment, induce a variety of business and political actors to support it, and then generate private development and urban growth from it. Yet for many Austin residents, the growth generated by technopolis came at a steep social price.

Changes in the Landscape

By 1990 Austin had become the defense-related technopolis that Kozmetsky, Governor White, and others had envisioned a decade before. Much of the growth was driven by large firms relocating to Austin. Before MCC, large companies including Lockheed, Abbott Labs, and the ROLM Corporation added thousands of jobs to the local economy. Later in the decade, 3M and Applied Materials Inc. moved large production and research facilities to Austin, highlighting another round of economic growth for the city. From 1982 to 1985, Austin added ten thousand manufacturing jobs, two-thirds of which were in the high-tech category. The university became one of the top five public universities in outside research money awarded and spun off dozens of small startups including Dell Computers, which would help redefine the economy again in the next decade.

Rapid economic growth, though, transformed the landscape in ways that threatened environmentalists and, by the late 1980s, minority environmental justice groups. While the 1980 Austin Tomorrow plan made recommendations about where future development would be prudent, it was never codified into law. The massive demographic and physical growth of the city in the 1980s followed patterns established beginning in the 1950s; office parks, shopping centers, and residential subdivisions were all built on the city's periphery with the intent of providing upper-middle-class white workers easy access to and from their sites of labor, recreation, consumption, and domesticity. Most developers chose the northwestern and southwestern periphery, long considered the most valuable, picturesque, and desirable land in the city. Yet the new developments would penetrate deeper into sensitive ecological areas that necessitated extensive physical augmentation, covered large aquifers, and contained unique wildlife. Almost overnight, massive developments were built, often along watersheds and on rocky hills where runoff was more likely to affect groundwater.

The BRC and the MCC facility remained the geographic hub of Austin's knowledge economy, but by the 1980s a much larger incentive to develop peripheral land for multiple uses emerged. One large driver of this process was globalization, where production shifted from the United States to the developing world; this meant that investors needed to find new profitable outlets for capital. The deregulation of the savings and loan industry in 1980 as well as a shift away from fixed production capital to secondary circuits of investment led to a widespread increase in real estate development in the early 1980s.[34] Another was MCC, which made Austin's real estate sector particularly dynamic. Along with other established private firms, many of which had ties to the defense industry, and the arrival of 3M in 1984, MCC's decision generated growth and an inflated real estate market almost immediately. While the boom was concentrated in the northwestern hills, it affected the entire metropolitan region and rapidly drew national developers to Austin. In the late 1960s, Austin was one of the most affordable metropolitan housing markets in the United States, with an equally low cost of living. By the mid-1980s, explosive appreciation fueled by unregulated speculation turned the Austin area into a volatile market where money was made and lost quickly, environmentally friendly recommendations were ignored, and office space development was ubiquitous.

Texas legislators, enthusiastic about the state's growth potential, facilitated the rapid physical expansion of its cities beginning in 1963 with the Municipal Annexation Act, which allowed cities with over five thousand residents to annex adjacent land.[35] In the 1950s and 1960s, the city facilitated rapid expansion by accelerating the policy of annexation that it began in the 1940s. From 1950 to 1970, the physical size of Austin grew from 37.5 square miles to 80.1 square miles, an increase of just over 200 percent. The heaviest annexation took place in the northwestern corner of the city, where new residential subdivisions and strip malls were annexed along with high-tech facilities in 1975. In other areas annexation specifically targeted high-tech facilities that would generate tax revenue for the city. Motorola, IBM, Glastron, and Tracor were among the annexed sites that added $900,000 to municipal tax rolls in the 1970s.[36]

But annexation laws also required that cities extend services and provide infrastructure to newly annexed areas, which could slow down the process if cities could not raise capital for improvements. In the 1970s, however, Texas augmented its annexation laws to accommodate rapid urban expansion by creating municipal utility districts (MUDs), which were essentially independent planned communities that could sell bonds to fund wastewater plants and other infrastructure. MUDs allowed developers to build sewer systems and water delivery systems into their subdivisions, which negated the need for septic tanks and increased property values. Texas also had relatively unrestrictive laws regarding extraterritorial jurisdiction, where a home rule city could apply zoning laws, provide services, and govern land use within five miles of the municipal boundary. Taken together, MUDs and extraterritorial jurisdiction allowed for developers to finance infrastructure development, rather than the city, and for the city to make sure zoning and building requirements were met, leading to fast annexation. And because in the 1970s Austin lacked strong water-quality ordinances, developers could build subdivisions with little regard for environmental quality; the city's policy of annexing MUDs only after their property taxes could pay off the developer's bond debt actually incentivized denser buildouts.[37] As a result, the city's physical size grew by almost 300 percent between 1970 and 1990. In the 1980s alone, the city added 102 square miles of territory, a growth of 82 percent.[38]

Population growth in Austin during the mid-1980s was among the most intense in the United States, and real estate values skyrocketed across the region in a matter of years. Between 1981 and the end of 1984, the population of Austin's standard metropolitan area grew from 537,000 to 671,000, an increase of nearly 25 percent, making Austin the fastest-growing major metropolitan area in the United States during that period. From 1982 to 1983, population growth was 9.5 percent, almost three times the 3.5 percent growth rate of the previous decade. The value of real estate across the city underwent a boom that was even more intense. Between 1983 and 1984, Austin's total appraised real estate value rose by 52.8 percent, with 94 percent of that increase attributed to revaluations rather than just to growth. Average home value increased by over 50 percent between 1982 and 1983, from $60,000 to $92,000.[39] By October 1984, Coldwell Banker reported that Austin had the most expensive residential real estate in Texas, including the upper-class suburbs of Dallas, and was far more expensive than any of the other thirty-seven Sunbelt cities where Coldwell Banker had an office. An average 2,000-square-foot house in Austin cost $140,000, $50,000 more than in Houston and $5,000 more than in Dallas. In some exclusive neighborhoods in Northwest Austin, land values doubled from 1983 to 1984.[40]

The growing high-tech agglomeration had an equally strong impact on office space and retail development throughout the city, especially in the northwestern portion of the city in proximity to the concentration of high-tech development around the BRC. Even before MCC arrived, Austin's high-tech industry was growing rapidly. The Austin metropolitan region added over 150,000 jobs from 1971 to 1983. From 1981 to 1983, ten high-tech or defense-related companies relocated to Austin, bringing closely to 12,000 jobs.

In 1984 the City of Austin annexed fifteen square miles of land where 3M and Schlumberger built facilities. The 3M Austin Center, located in picturesque hills far from Austin Tomorrow's "desired development zone," was a typical research campus for high-tech workers, "carved into a cedar brake" over 162 acres.[41] Architectural critic Michael McCullar compared the facility to a "shopping mall, providing far more for employees within than for motorists passing by."[42] Phase I contained nine interconnected buildings, taking up 1.2 million square feet of labs and offices. In total, the center took up over 4 million square feet. The complex, according to the architect, was designed to promote comfort and to inspire workers. Yet "the central feature of the site is a wooded valley" near the head of Bull Creek, one of Austin's largest. Schlumberger's much more modest 175,000-square-foot complex was designed to be integrated with Hill Country ecology. The buildings "are pushed to the edge of a bluff that overlooks a wooded valley. Oak trees embrace conference rooms, and buildings are connected by wide wooden walkways."[43] Combined, the two facilities included parking for almost two thousand cars, yet parking lots were designed specially to hide cars away from the buildings. In both cases, architects considered access to natural beauty and parklike campuses as key components in their designs. After annexing the land, the City of Austin promised to provide water and wastewater services to the two facilities. By 1985 the BRC was joined by IBM, Schlumberger, 3M, and MCC in the northwestern part of the city, employing well over ten thousand knowledge workers among them.

These facilities were often accompanied by large residential subdivisions, office parks, and regional malls. By 1985 some commercially zoned spaces fronting the major highways that crisscrossed the area had doubled in value since 1983, indicating the generative effects of MCC on the local real estate market. The following year, ancillary office and retail building exploded in the corridor along the MoPac Expressway (named after the Missouri Pacific Railroad) and U.S. 183, in close proximity to the research campuses. Philadelphia-based Landmark Associates built the aptly named Great Hills Corporate Center, which features native fauna and flora and a pool and waterfall carved out of native limestone to provide an "Austin personality."[44] In late 1985, Northwest Austin had over 1.5 million square feet of office space under construction, second only to the central business district among Austin's commercial areas. Many large-scale developers built high-end shopping malls alongside office parks as work-play areas for white-collar laborers. In the span of just three years, national developers, led by global giant Trammell Crow, which was the largest developer in Austin by 1982, built the Arboretum, Northpoint, Prominent Point, Stratum, and many smaller office and retail facilities, within a few minutes' drive of one another, along corridors in Northwest Austin. Office and retail space averaged nearly the same price per square foot as in the central business district in 1985, indicating a high level of occupancy in the new office parks.[45]

Owing to the extension of the controversial MoPac Expressway, finished between 1976 and 1982 (discussed in chapter 9), Southwest Austin saw significant growth as well.[46] Advanced Micro Devices (AMD) and Motorola were the first high-tech companies to open facilities south of Town Lake on the Edwards Aquifer in the 1970s. By 1988 they were the second- and seventh-largest private firms in Austin, employing a total of close to seven thousand people across five facilities, three in Southwest Austin. In 1981 Barton Creek Square Mall opened near the intersection of the MoPac Expressway and Loop 360, the two largest north-south thoroughfares through West

Austin. One year later, Austin real estate analysts predicted that "the southwest area will attract the most attention from developers in the next several years because of its proximity to downtown, spectacular views of the city and environmental beauty."[47] Owing to the hilly terrain and thick cedar forests, views in Southwest Austin were considered the most beautiful. Real estate developer George Nalle III boasted that his subdivision, Treemont, would have a panoramic view of the city and be the most "environmentally sound" in the city.[48] Here, too, developers understood that customers considered access to nature vital to their developments. Builder Joe Bienvenu argued that Austin had more discriminating customers than Houston did because "buildings that will survive in this competitive market are those designed to fit Austin's lifestyle." He included "beautiful landscaping" and "preservation of the natural environment" as part of the necessary features.[49] These new subdivisions proved attractive to high-tech employees who worked in Southwest Austin. By the mid-1990s over half of Motorola's employees lived in the Edwards Aquifer contributing zone in Southwest Austin.[50] Unsurprisingly, office space also expanded rapidly in Southwest Austin, growing by over 300 percent from 1982 to 1983 alone and briskly for the next five years.[51]

While the western periphery rapidly filled with research campuses, offices, malls, and single-family homes, the Eastside remained devoid of nonindustrial planned developments; the *Austin Business Executive* pointed out that the Eastside was the only section of the city not undergoing a residential and commercial boom in 1984.[52] Statistics gathered by environmentalists supported the magazine's assertion. In the first quarter of 1984, at the height of the boom, 626 new acres of land were zoned for residential development in central and western Austin; in East Austin two new acres were zoned residential. In outlying areas west of IH-35, fifty-three new plats were proposed, forty-three of which were in West Austin. In East Austin, four plats were proposed.[53] As late as 1984, an African American real estate broker claimed, "I couldn't name you 10 new houses that have been built in East Austin except for the ones the city is building."[54] Another developer interested in the Eastside claimed that older sewerage and wastewater systems were too dilapidated to handle larger commercial projects.[55]

Industrial parks, however, were built on the Eastside due to both restrictive zoning across the western periphery, existing production facilities made easily accessible by highways, and cheap land. Tracor moved to a location on East Research Boulevard in the 1970s and Motorola and Lockheed followed in the 1980s. In the early 1980s the former industrial hub in central

East Austin also grew, but more modestly. Bridgeview Industrial Park was developed on 70 acres at the eastern terminus of the Cesar Chavez neighborhood, near the Colorado River, and a 130-acre plot was developed on Research Boulevard near the Schlumberger facility. Even on the Eastside, natural amenities were considered important. The land at these sites, according to one real estate consultant, "is preserved in its natural state with trees and rolling terrain," providing "a secluded environment so desired by research and development type users."[56]

In 1988 the arrival of SEMATECH, a second federally sponsored research consortium of fourteen corporations specializing in semiconductors, however, had the most dramatic effects for another East Austin community. The package that Austin put together to attract SEMATECH was similar to the MCC package. The University of Texas guaranteed roughly $38 million in bonds to build a "superclean" laboratory for SEMATECH, which would rent it for $1 a year. The university also drew $15 million directly out of the Permanent University Fund and made it immediately available to SEMATECH on its arrival. In all, the package, funded mostly by the university but also by the city, the state, and private businesses, totaled nearly $70 million. Like MCC, SEMATECH was a safe bet to bring economic gains to Austin regardless of its success. The Department of Defense, an active participant in SEMATECH, guaranteed $500 million of funding for the new consortium over five years, and the thirteen private firms invested in SEMATECH already produced well over half of the United States' semiconductors.[57]

The university also spent $12.3 million to purchase a vacant plant in the former Data General site in the far southeastern neighborhood of Montopolis, one of Austin's least developed and most environmentally hazardous, poor, and Latino neighborhoods, to turn into the new facility.[58] The area, far removed from Austin's most famous natural sites and wealthiest neighborhoods, already contained the city's two largest industrial parks, Southpark, developed by Trammell Crow, and the 100-acre Missouri Pacific Industrial Park. The SEMATECH facility would be part of a new agglomeration, also anchored by a new AMD plant, much closer to the residential neighborhoods of Montopolis and Pleasant Valley.

From Technopolis to Creative City

By 1996, when global multinational giant Samsung expanded into the Austin market and in the midst of Dell's unparalleled rise to computer industry leader, Austin's technopolis looked less dependent on defense than ever before. The Greater Austin Chamber of Commerce report *Next Century Economy* found that Austin created more jobs in the semiconductor industry between 1990 and 1996 than any other high-tech city benchmarked in the document. Another study found that Austin also led the nation in patent production growth over the previous decade, and the region was second to Silicon Valley in patents per resident. While not exactly an indicator of economic vitality, the number and variety of patents produced in Austin suggest a robust and diverse research climate in the city.[59] What is striking about Austin's industrial growth is the diversity of firms the city produced and attracted within the high-tech framework. While large company growth and relocation, along with the consortia, provided economic stimulus and national prestige, small, indigenous companies in the high-tech industry provided the continuous, consistent growth that sustained the city and region. In 2003, for example, a mayor's task force found that small businesses (those with fewer than fifty employees) constituted 94 percent of all Austin businesses, over 7 percent higher than the nationwide average. In all, over two thousand high-tech firms opened for business in Austin during the 1990s, over 90 percent of which had fewer than fifty employees. At all levels of the high-tech industry, techno-capitalism in Austin continued on its robust path throughout the 1990s.[60]

But judging the success of techno-utopia, that is, the effort to invest in technology-based growth for the benefit of all citizens, is more difficult. In an economic and social landscape increasingly defined by regional and metropolitan competition, benefits did not accrue evenly. Residents from a variety of backgrounds asked what price this phenomenal growth would have on the city—in a sense, the negative reaction to high tech held the promise of the first large-scale grassroots movement that traversed the city's longstanding racial bifurcation. Yet, sadly, after a brief coalition was formed, that large temporal and spatial gap ensured that the social fractures of the past would again create separate entities.

Since the early 1970s, Austin's mainstream environmentalists had grown in rank, organization, and scope. By the early 1980s over one hundred neighborhood groups claimed spots at Austin neighborhood association meetings. A new crucial waterway emerged at the forefront of their campaigns against irresponsible growth. The most attractive areas of Austin's hinterland were unfortunately also its most vulnerable. The Edwards Aquifer and Barton Springs, both part of the same sensitive ecology that made Austin unique and, for many, special, were becoming severely polluted by all the construction. Rainwater that had to go through concrete rather than dirt picked up dangerous toxins on its way to the aquifer. Construction left debris. Thousands of acres of forest were destroyed, sending animals to new environs. And socially, the city's growth promised to attract many people who did not share in some longtime residents' sense of what made the city so special.

On the Eastside, by the late 1980s, long-lingering issues with the environment but also with socioeconomic issues, such as poverty, discrimination, and lack of opportunity, boiled over. Residents looked at their physical landscape and saw not only an architecture that excluded them from social and economic opportunity, but one that actively harmed them. SEMATECH's siting in Montopolis was a lightning rod, but it was also the straw that broke the camel's back, one in a string of many environmental abuses that reflected the city's indifference to its minority neighborhoods. Activists there organized as well, drawing on the growing transnational networks that sought to expose environmental racism, as well as on the particular history of environmental discrimination in East Austin: gasoline tank farms, garbage dumps, power plants, cumulative zoning, and now high technology. While both groups coalesced in response to overdevelopment, each also responded to a particular history and sense of place and had a particular framework for what they considered the "environment." Their quests to improve their city reveal how the environment served a social cause. Their stories are told in chapter 9.

9 Of Toxic Tours and What Makes Austin, Austin

Battles for the Garden, Battles for the City

Almost everyone in Austin remembers when it used to be a better place to live. The pangs of loss are not, as in Dallas and Houston, for the days when the skies were full of construction cranes . . . but for things that are irretrievably gone: the view West from Mount Bonnell of pristine hills, now carved into subdivisions; the low water crossing on a country lane, now turned into an outer loop; . . . the steep hill on the edge of town, now bulldozed into a stubby mesa now topped by a graceless shopping mall.

—Paul Burka, "The Battle for Barton Springs,"
Texas Monthly, August 1990

PODER's Woes Bigger than Springs, Birds

—Eunice Moscoco, "PODER's Woes Bigger than Springs, Birds,"
Austin American-Statesman, July 21, 1997.

On February 10, 1992, new environmental activist group People in Defense of Earth and Her Resources (PODER) organized a "toxic tour" of East Austin to alert locals to hazards and demonstrate the environmental inequity that plagued East Austin residents to politicians, neighborhood leaders, and school board members. The tour, which focused on a gasoline tank farm that was sited in the Govalle neighborhood in East Austin in the 1950s, was not widely attended by the general public but did draw area residents and the invited officials. PODER leaders emphasized the many chemicals that the facility emitted, including gasoline, diesel, benzene, and carbon monoxide, as well as the pollution of the groundwater and soil, and the fires caused by the tanks. The farm had deleterious effects on local residents, who were almost all minorities and suffered ailments, such as bronchitis, headaches, throat and eye problems, and vomiting, at a much greater rate than Austinites as a whole. Residents were unable to grow gardens in their yards and endured greasy, odiferous tap water and petroleum runoff in local creeks and backyards. Astonished school board members and politicians asked how something like this could happen, especially in the wake

A man and his children point to the gasoline tanks near their backyard as part of the PODER protest against the tank farms, 1991. Photograph by Joe Bienvenu. Austin History Center, Austin Public Library.

of widespread environmental victories in West Austin in 1991. In East Austin, PODER representatives responded, industrial facilities that existed before air quality laws were passed in 1972 were not subject to regulations. And numerous industrial facilities existed in East Austin.[1]

Over the next month newspapers reported widely on the tank farm. One article showed an ominous tank, standing nearly fifty feet high, casting a shadow on a small home less than one hundred feet away. Tests of water and soil were administered. Some registered chemical levels seven hundred times greater than federal safety limits allowed. Homes near the farms saw the largest across-the-board devaluation of any Austin neighborhood since the 1980s. Journalists learned that the Texas Water Commission confirmed groundwater contamination four years before the toxic tour and perhaps broke state and federal law by not revealing its findings.[2] They also found that residents had been filing complaints about the farms for decades, and the city had done nothing. Twenty separate Latino neighborhood and activist groups immediately called for the farm to be closed.[3]

The scandal and newspaper coverage embarrassed the Texas Water Commission and some Austin politicians, but it also bridged the decades-long gap between East and West Austin, between the long-standing mainstream environmentalists and the newer proponents of environmental justice. For some, the tank farms became a cause célèbre. Leading environmentalist Brigid Shea said she was unaware of environmental hazards in East Austin until the toxic tour, but pledged her support for their removal.[4] The Lone Star Sierra Club publicly supported regulations on tank farm emissions and demanded that the area be cleaned up to mitigate health problems among residents. The Audubon Society, the Save Our Springs (SOS) Alliance, Green Peace, and the Environmental Defense Fund all volunteered to assist in East Austin antipollution campaigns.[5] State representative Glen Maxey led the charge to create the Tank Farm Citizen Monitoring Committee and publicly chastised the Texas Water Commission for its dubious actions.[6] The goodwill ran from east to west as well. Eastside activists took up the cause of Barton Springs and discussed the environment as something that could bind east and west together. After decades of separation, the tank farms brought concerned Austinites from both sides of the city together. Their combined effort was quick and effective; the tank farm was shuttered for good for in 1993, though its environmental effects lingered for another decade.[7]

Yet, in 1997, Eastside residents' frustrations with mainstream environmentalism boiled over at two meetings discussing land use in East Austin. In the first meeting a group of mostly African Americans argued that efforts to protect Barton Springs, the centerpiece of Westside environmentalism, were shortsighted given the declining quality of life among Eastside residents due to the zoning that allowed industry interspersed with residences. Two weeks later, a mostly Latino audience accused environmentalists of caring more about the Barton Creek salamander and the golden-cheeked warbler, two endangered species unique to the Barton Creek watershed, than they did about minorities, who endured a recent fire at a recycling plant, along with its waste and noise. Speakers were quick to point out the lack of Anglo environmentalists at the meetings despite PODER's support for SOS. "Where are my environmentalist friends who said they would come and support me?" queried one speaker. "Maybe I'll come support them the next time they want to save a salamander or some bugs and bees." Susana Almanza succinctly articulated the Eastside's sense of alienation from Westside environmentalists: "We, too, are an endangered species."[8]

The fleeting sense of camaraderie and geographic reconciliation generated by the tank farm protests perhaps came as little surprise considering

Zilker Park, 1938, with a relatively undeveloped Barton Springs in the foreground and the Colorado River in the background. Many improvements were made in the 1920s and 1930s, yet the landscape still appears sparse. PICA 17205, Austin History Center, Austin Public Library.

Young people enjoy a much-improved Barton Springs pool, 1960. PICA 17741, Austin History Center, Austin Public Library.

the long history of race and environment in Austin, despite the fact that often the groups responded to similar concerns: health and happiness, the imposition of technology, land use, and overdevelopment. The wide cultural and socioeconomic chasm, produced by decades of policy choices that divided the city racially as well as environmentally, was far too great to overcome. Yet the divide allows a view into the way that groups construct ideas and arguments about what constitutes "environment," a concept that can refer to almost any space, and the relationship that idea has to the similarly amorphous concept of "sense of place." In the most fundamental rendering, mainstream environmentalists sought to protect important natural spaces from being ruined by development; Barton Springs, to thousands "the soul of the city" and the symbolic center of the movement, proved pivotal. They argued that communal natural spaces created a sense of meaning and community that all residents shared, and they feared that their destruction would also erase the city's sense of collective consciousness. Alternatively, defending those spaces against greedy developers reflected not just their love of those places but also their sense of Austin's environmentalist community as a progressive, collectivist group that cared about its city and did not want to see it changed drastically. In this way, the mainstream conception of environment indicated Austin's natural spaces, but it also defended the image of Austin as a small progressive town, different from sprawling, capitalist Houston and Dallas, and unique in its natural gifts but also in its social consciousness.[9]

Yet not all groups shared the same depth of feeling for Barton Springs or other cherished natural spaces because those spaces were not equally accessible to everyone. Nor did those spaces share the same meaning for different people. For Eastside environmentalists, the idea of defending their communities was also central to their understanding of the relationship between environment and sense of place, but their definition of environment was much more expansive. Laura Pulido has convincingly argued that minorities tend to conceptualize "subaltern environmentalism," which is different from mainstream environmental thought. For Pulido, minorities, because they face multiple levels of oppression, enmesh environmentalism in a broad framework of subordination and alienation that links environmental injustice with economic, social, and other forms of inequity.[10] For PODER and other vocal Eastsiders, subaltern environmentalism also meant that they imagined the environment as a place where people played a profound role and were, as Susana Almanza said, potentially endangered. PODER emerged in opposition to SEMATECH's choice to site its facility in

Montopolis near impoverished minority neighborhoods, and the group addressed other forms of pollution that threatened their community: power plants, industrial facilities and dumps, the tank farm, and other high-tech facilities. Yet their most sustained battle was against gentrification, something that they were able to categorize as environmental under their anthropocentric definition of environment. In all their writing and activism, community health (economic, social, and physical) took precedent over saving the environment to create a sense of place.[11] They often perceived their goals as markedly different from, and sometimes in opposition to, the goals of mainstream environmentalists. Their battle was likewise against development, but what they defended was a community, which included natural areas but was fundamentally composed of people. More than nature, community and neighborhood, oftentimes places where relatives had been forced to live for decades, defined sense of place on the Eastside.[12]

This chapter follows both environmental movements in the 1980s and 1990s, focusing on how different ideas of what constitutes the environment emerge and how they generate symbols that tie activists together. The particular, place-specific history and geography of Austin play a large role in understanding these different ideas and creating systems of meaning that underpin arguments for what matters when defining environment. Water, for example, has long been a symbol of Austin's quality of life as well as a potentially scarce resource threatened not only by development-generated pollution but also by demographic and economic growth itself. Barton Springs symbolized Austin's soul for environmentalists, but it also symbolized the Edwards Aquifer and the Highland Lakes, the region's principal sources of drinking water. The Barton Springs salamander and the golden-cheeked warbler, two endangered species indigenous to the watershed, symbolized the region's fragile, defenseless ecology in the face of development.

For PODER and other Eastsiders, it was the lack of access to nature that defined symbolic meaning. Because of decades of discriminatory zoning, uneven dispersal of improvements, and municipal indifference, pollution already threatened their water and the trees and gardens in their yards. The built landscape was a constant reminder of their lack of nature, rather than a threat, and it provided the symbols of domination that they wished to deconstruct. Tank farms, power plants, and industrial facilities were environmental hazards that necessitated removal, but they were also symbols of the city's dubious historical treatment of minorities and their spaces. Gentrification provided the most forceful symbol of domination; as soon as

TOWN LAKE

Town Lake is the heart and soul of the open space concept for Austin. It has become the most important public space in Austin and with the completion of work on the landscape development, it should become the center of outdoor activity, sporting events, cultural events, festivals, boat races, art shows, trade shows, etc. The major events celebrating the American Bicentennial in Austin will take place on Town Lake, a fitting tribute to the completion of one of the most unique and beautiful linear parks anywhere. As each creek greenbelt is completed to Town Lake more and more people will have easier access to Town Lake by walking or by bicycling.

BULL CREEK

The 17,000 acre watershed of Bull Creek contains the most beautiful of all the landscapes around Austin. With the completion of the crosstown sewer and Loop 360 the area will be subject to very rapid development. 25% of the entire area is made up of slopes of 15% or greater, many of which are considered to be unstable. Problems with natural factors are compounded by the prospects of increased erosion, and flooding that will inevitably follow development. Acquisition of a greenbelt along Bull Creek has begun with the creation of Bull Creek Park, just off Spicewood Springs Road. Hopefully the Greenbelt will be extended to include the most beautiful portions of the creek.

TAYLOR'S SLOUGH

West Austin neighborhood groups are anxious to link Taylor's Slough, an old lime quarry just off Lake Austin, to Reed Park and to Casis School to the East. Presently the creek that follows this path is undeveloped but in private ownership. Hopefully a way will be found to create the greenbelt even though some form of development of the land seems inevitable. A mutually satisfactory solution that would benefit the development as well as the entire neighborhood could serve as a positive example for future growth along creeks.

BEE CREEK & LITTLE BEE CREEK

These relatively wild creeks run through undeveloped areas of Southwest Austin, primarily through West Lake Hills. People in these areas have proposed a major wilderness preserve of about 400 acres to be known as "Wild Basin Park." The proposed park would include the very beautiful areas of Bee Creek as well as most of the land on both sides that is too steep for building. A great deal of care must be exercised by the Highway Department where Loop 360 is scheduled to cross Bee Creek. A less than sensitive bridge could destroy one of the most unique natural areas in Travis County.

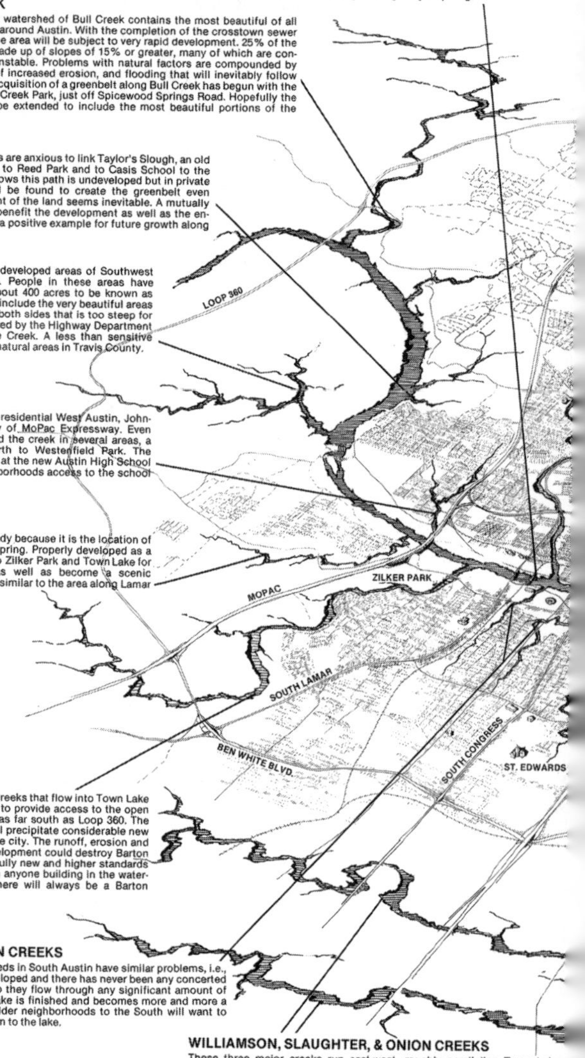

JOHNSON CREEK

A small, completely developed watershed in residential West Austin, Johnson Creek roughly parallels the right-of-way of MoPac Expressway. Even though the road construction has obliterated the creek in several areas, a greenbelt is planned from Town Lake north to Westenfield Park. The proposed trail will emerge from under MoPac at the new Austin High School on Town Lake, giving the West Austin neighborhoods access to the school and the lake.

DRY CREEK

Many people are familiar with Dry Creek already because it is the location of a nature trail and the site for "Safari" each spring. Properly developed as a greenbelt, Dry Creek could provide access to Zilker Park and Town Lake for the rapidly developing southwest areas as well as become a scenic easement along portions of Bee Caves Road, similar to the area along Lamar Blvd. at Pease Park.

BARTON CREEK

Barton Creek is not only the largest of all the creeks that flow into Town Lake but it is the creek with the greatest potential to provide access to the open countryside. A greenbelt has been acquired as far south as Loop 360. The construction of both Loop 360 and MoPac will precipitate considerable new development in the southwest quadrant of the city. The runoff, erosion and pollution that result from poorly planned development could destroy Barton Creek and with it Barton Springs Pool. Hopefully new and higher standards of planning and design will be expected from anyone building in the watershed, and hopefully that will insure that there will always be a Barton Springs.

EAST & WEST BOULDIN CREEKS

These two very small watersheds in South Austin have similar problems, i.e., they are both completely developed and there has never been any concerted effort to care for them, nor do they flow through any significant amount of open public land. As Town Lake is finished and becomes more and more a major focus of activity, the older neighborhoods to the South will want to develop greenbelts to link them to the lake.

WILLIAMSON, SLAUGHTER, & ONION CREEKS

These three major creeks run east-west, roughly paralleling Town Lake, through a relatively undeveloped portion of South Austin. Some of the newer subdivisions in sharp contrast to most of the older ones, have respected the creeks. Several have even created hike and bike trails. All three creeks lead to McKinney Falls State Park several miles east of IH 35. If trails can be developed along these three creeks, tremendous numbers of people would have access to the park.

Onion Creek with a 60 mile long watershed is easily the largest creek in Travis County. Fortunately for Austin the huge floods that can come down Onion Creek reach the Colorado River downstream from Austin.

AUSTIN'S BICENTENNIAL PROJECT

A bold plan to preserve, restore, and enhance the creeks and waterways of Austin.

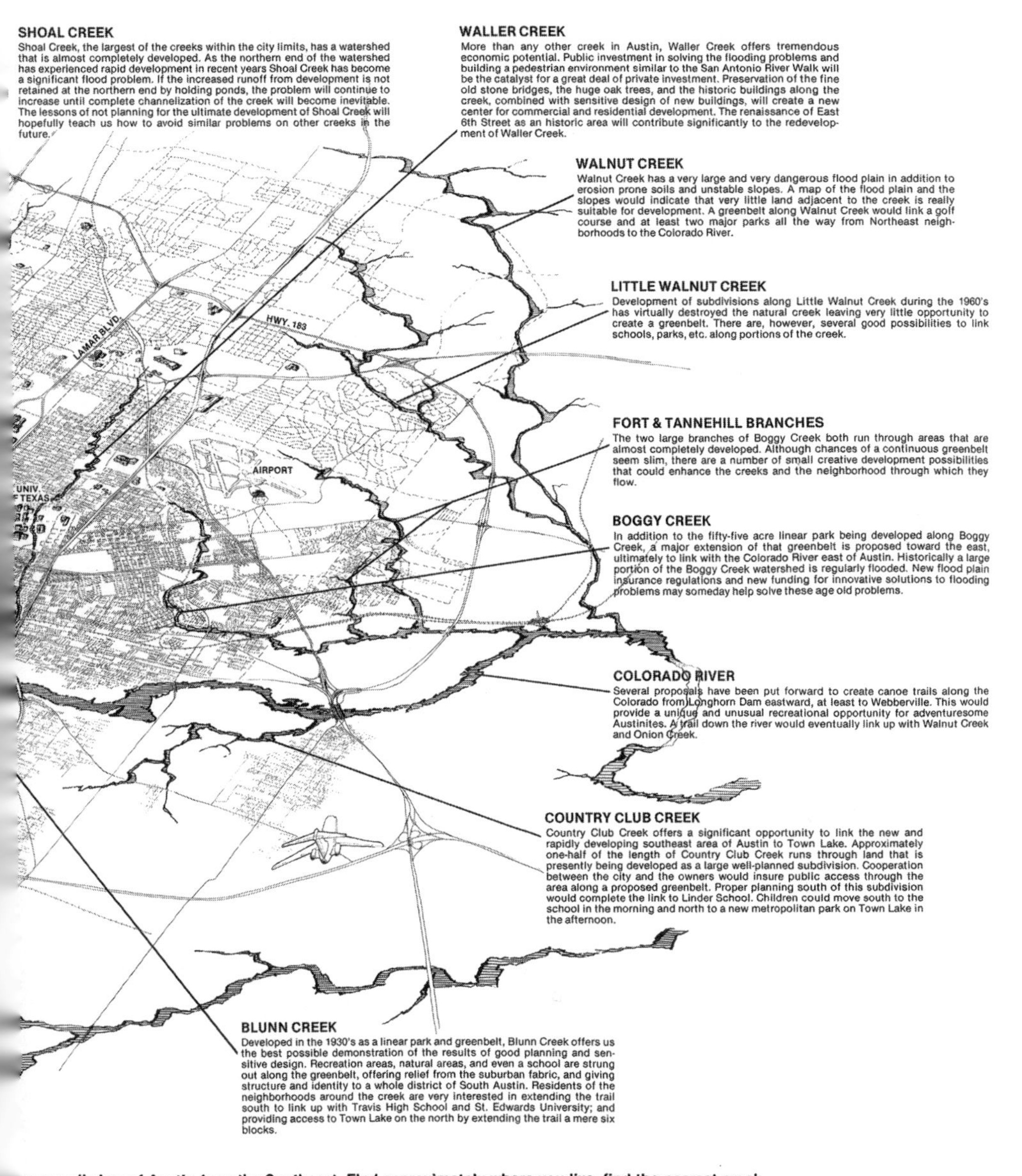

n overall view of Austin from the Southeast. Find approximately where you live, find the nearest creek, ow see where and how far you could travel along that creek if the greenbelt system were complete.

A map of Austin's major urban watersheds, produced as part of a campaign to improve Austin's greenbelts in the 1970s and 1980s. Austin History Center, Austin Public Library.

industrial hazards were cleaned from minority areas, real estate costs increased. The built landscape thus came to symbolize the city's new investment in urban lifestyles by the late 1990s, and they perceived "sustainability" as something that negatively affected their community. To Eastsiders, the powers that be wanted to preserve houses and buildings, not people, which was yet another example of how the built landscape loomed as the primary symbol of oppression in East Austin.

"Barton Springs Eternal": Natural and Cultural Ecologies

From the earliest recorded history, water has been central to the way Austinites have imagined and discussed their city. Earlier chapters that discuss the meaning of the Colorado River for Austin's growth (chapter 1), infrastructural improvements (chapter 3), the commodification of water (chapter 4), and the defense of waterfront public space (chapter 7) all indicate that water has been vital to Austin's existence and growth but also to its identity. While the Colorado River and reservoir system dominated policy discussions and discourse of Austin's capitalist advantages well into the 1960s, no body of water is as invested with symbolism and local meaning as Barton Springs is. The springs and environs are imbued with symbolic meaning that reflects the city's unique culture to many Austinites. They are used to differentiate Austin from larger, more conservative, and more sprawling Texas cities, such as Houston and Dallas. They often symbolize a simpler, more authentic Austin in antigrowth discourses that lament growth and urbanization and tie Austinites to natural and human history.[13]

The springs, which emanate from a small section of Edwards Aquifer just south of Town Lake, were used for water and bathing by Native Americans long before Spanish explorers, who built a mission near the springs, arrived in the eighteenth century. Private owners used the springs to mill grains and as a tourist attraction in the city's earliest decades. During bad drought years the city attempted to purchase municipal water from its private owners. In 1917, during one of the worst droughts in recorded history, owner and businessman Andrew J. Zilker agreed to sell the springs and surrounding property to the city. Today this tract makes up the core of Zilker Park.[14]

In the 1920s the springs became a central motif in Austin's growth aspirations and a center of civic pride. Bonds generated after the Austin City Plan of 1928 was adopted paid to dam the springs to create a permanent pool, sidewalks, and landscaping. A bathhouse was constructed using private funds in 1922; that same year the chamber of commerce and Lions Club began promoting the springs as a potential destination for automobile tourists in Central Texas. In 1931 and again in 1934, Zilker sold large tracts of his adjacent property to the city.[15] In the 1930s Tom Miller's friendly relationship with New Dealers brought federal funds for landscape planning, roads, trails, and ponds throughout Zilker Park. The Civil Works Administration spent more money on Zilker Park than on any other park in Texas. The Hillside Amphitheater was built in 1937, and a new bathhouse was erected in 1947. From the 1930s to the 1950s, the springs and surrounding parkland became a cultural center for Austinites, hosting a weekly Sunday night gospel sing-along, a weekly recreation program, swim lessons, and, by the 1960s, the Yule Fest Trail of Lights at Christmastime.[16] In a more risqué endeavor, in 1937 an estimated five thousand spectators turned out for a swimsuit show. The springs were long considered a special, even sacred, aspect of Austin's landscape as well as a civic center. In his speech accepting Austin's Most Worthy Citizen Award in 1928, Zilker called Barton Springs a "sacred spot" that should be owned and enjoyed collectively, not by one individual. Longtime Department of Parks and Recreation director Beverly Sheffield wrote that "Barton Springs is like heaven, more a state [of mind] than a place."[17] Texas literary and folklore figures Roy Bedichek, Walter Prescott Webb, and J. Frank Dobie famously conducted barefoot impromptu colloquia at the springs, and a statue memorializes their mystical presence there. Yet, like all Austin parks through the early 1960s, Zilker Park and Barton Springs were strictly segregated by race. Indeed, Zilker called the park sacred because it was dedicated to the memories of Confederates Robert E. Lee and Albert Sidney Johnston, not because of its particular natural qualities or availability to all Austinites.

So when Austin environmentalists arrived in the 1960s or 1970s, or even when younger Austin natives imagined Barton Springs, they were immediately dealing with an entity long the construct of human thought, capital, labor, and prejudice. As early as the 1930s, Barton Springs was decidedly a garden-like creation, the blending of natural environment with human technology and political capital. Hundreds of thousands of dollars were spent to purchase the springs and parks and hundreds of thousands more to improve them with an array of dams, roads, and architectural structures to make enjoying the space easier. The springs had already been conceived as a tourist attraction, cultural center, and recreation area and even a place to memorialize Civil War veterans who fought for the South in the nineteenth century.[18] Above all, it was a space that was consistently meant to be enjoyed. By the 1970s and 1980s, environmentalists imagined defending an unchanging natural place untouched by human development, "Barton Springs eternal, the soul of a city" as author James Michener called it. Yet it was specifically human intervention that finished Barton Springs and also made its meaning malleable.[19]

For the mainstream environmentalists in the 1970s, Barton Springs was still more a place to enjoy than a place to defend. The Austin Tomorrow developmental recommendations (discussed in chapter 7) had no legal force behind them, which many environmentalists saw as a defeat. But from a procedural standpoint, the Austin Tomorrow process organized the city at the neighborhood level, and the number of neighborhood-based groups more than doubled from twenty-nine to sixty-six between 1973 and 1980; the Austin Neighborhoods Council provided an umbrella group.[20] Through the 1970s and early 1980s neighborhood groups overwhelmingly addressed what they considered neighborhood issues. Foremost among these issues was the desire to keep apartment complexes and retail stores out of single-family residential neighborhoods. Owing to increasing university enrollment, dislocations from urban renewal combined with a lack of low-cost housing starts, and the appeal of Austin's growing cultural scene to younger adults, the building of apartments was intense during the 1970s.[21] The University of Texas, perpetually looking to expand, induced developers to build apartments to rent to students along the city's northwestern periphery.

Neighborhood groups and vocal citizens almost universally feared the deleterious social effects that apartment complexes and higher-density development would have on residential neighborhoods and the families they housed. One of the South River City Citizens' four goals, for example, was to "restore and preserve the single family character of neighborhoods" by fighting zoning changes that would allow apartments or commercial development. United South Austin members likewise complained about "overzoning" that could lead to higher-density development and environmental degradation. Residents of the upscale Great Hills Trail neighborhood in Northwest Austin engaged in a large letter-writing campaign seeking to keep the area residential amid office park construction.[22] Worries about the physical effects of apartments and stores were often articulated along with fears of the social impact of apartment dwellers. A resident of Allandale in Northwest Austin summarized his community's fears by arguing that apartments "perpetuate transient occupancy," detract from "neighborhood atmosphere," and bring in "trash and vermin." Homeowners likewise were economically "penalized" by apartments. Another resident of the Oak Ridge subdivision in far northwestern Austin was equally as vitriolic; she condemned apartments for bringing in crime and congestion and equated apartment complexes with Austin becoming "another Houston." Others were more direct in arguing that apartment complexes would threaten real estate values. Residents in far southeastern Austin argued that apartment houses would tip the racial balance in their neighborhood and should instead be placed in areas that were not already relatively heterogeneous. Some even doubted the wisdom of the Austin Tomorrow development plan, arguing that more intense uses of older central areas would lead to less desirable conditions and the flight of residents.[23]

While individuals often voiced concerns that reflected a not-in-my-backyard attitude, neighborhood associations often imagined high-density zoning as a pathway to environmental degradation and a detractor from quality of life. The Zilker Park Posse, the first environmental activist group to emerge after Austin Tomorrow, often sought to combat higher-density zoning that would allow for apartments on grounds that would spoil a view or create congestion, but they linked apartments to the deterioration of creeks and to oil pollution from increased traffic and asphalt as well.[24] They thought primarily in terms of quality of life and sought to retain a physical layout that would promote peace and quiet and safety for their families and maintain the natural areas and open spaces in their neighborhoods.[25] The neighborhood level, though, proved overwhelming for many small groups that lacked cohesion and, more importantly, a power base. Neighborhood-based activists often found themselves fragmented and ineffective, fighting an endless series of "small brush fires" as the city itself metaphorically burned.[26] Yet their resistance to density and infill development at the neighborhood level was also a tacit complicity with the very sprawl that most threatened the environment. Caught between resisting higher density and simultaneously resisting growth on the periphery, neighborhood groups found themselves in a position where they had to either resist growth entirely or advocate for growth in other neighborhoods. As such, groups as well as individuals often framed amorphous "growth" as the primary issue facing citizens.[27] Given the options, the logical choice was to turn attention to those who were facilitating growth and profiting from it: real estate developers, powerful politicians, and businesspeople interested in growth.

But after 1974, development was occurring in a way that blurred neighborhood boundaries and threatened to change the fundamental nature of the city. The building of large roads and the deterioration of urban creeks provided the initial impetus for a citywide antidevelopment coalition that transcended neighborhoods. Roads and creeks both connect otherwise distant places because they run between neighborhoods, simultaneously joining them and creating boundaries between them. Urban creeks are important sites of recreation and water but also sites of trash and pollution brought from one neighborhood to the next. In Austin, and especially West Austin, roads and creeks are also connected because concrete deflects pollutants from cars and creates an impervious cover that drains those pollutants into creeks and does so more rapidly than soil, which filters them out. Building materials used to construct roads also pollute waterways, and asphalt used to pave parking lots constitutes one of the most damaging types of impervious cover. Many Austinites also saw highways and attendant billboards as aesthetic blemishes on the landscape, markers of the relationship between rampant growth and the commercialization of nature.[28] But the most fundamental concern was that large highways also dictate growth by allowing people to travel to newer areas more efficiently; this also meant that development was much more likely to be profitable in areas that were connected to downtown and other centers of population via roads.

The threat produced by highways through western Austin and the potential deterioration of urban creeks all over West and South Austin were attended by a more existential problem, the rapidly expanding boundaries of the city (discussed in chapter 8) and the liberal annexation laws established in Texas. The 1970s saw the city absorb over forty square miles of land, more than half the total area at the beginning of the decade; most of the annexed land was outside of Austin Tomorrow's recommended growth corridor.[29] From the city's perspective, annexation was a tool to combat the biggest problem in older northeastern and midwestern cities, the flight of capital and people to independent suburbs. Annexation would ensure that new developments on the periphery would share in Austin residents' economic burden if these developments were to enjoy their resources. While, initially, opponents of annexation framed their debate as economic (why should we raise bonds for utility delivery, for example, to neighborhoods that do not yet pay taxes?), by the late 1970s the issue took on an environmental tone as people began to notice development in formerly undeveloped areas as well as problems with their water.

By the early 1980s the loose connections among creeks, MoPac, and annexation tightened. In the 1970s, Westside neighborhood groups bitterly fought MoPac, which they perceived as a decidedly local concern because the road cut a 300-foot-wide swath through neighborhoods, displacing people and creating a barrier, at some places over fifty feet high. They argued that the highway benefited nonresidents at the expense of residents

MoPac construction at Thirty-Fifth Street. The large expressway cut a broad swath through West Austin in the early 1980s and became a source of outrage for West Austin environmentalists. PICA 02343, Austin History Center, Austin Public Library.

and created a massive eyesore, made worse by annoying billboards, on the otherwise picturesque landscape of West Austin.[30] By the late 1970s, as plans emerged to extend the highway south of the Colorado River, more vociferous environmental groups were already connecting MoPac to environmental pollution in the Edwards Aquifer and Barton Creek. A study of the Barton Creek watershed commissioned in 1975 found that rapid increases in impervious cover near the creek had negative effects for the water and adjacent land.[31] Perhaps more importantly, the potential extension of MoPac connected land on the Edwards Aquifer recharge zone to other developed areas and almost certainly meant explosive development above the sensitive groundwater. Motorola had already located its new facility in Oak Hill, a lightly developed neighborhood in Southwest Austin in 1974. That year, real estate developers also began construction on Barton Creek Country

Club, a 2,200-acre site in far southwestern Austin along Barton Creek, with three golf courses, a large convention center, and a hotel.

By 1983, however, smaller, more environmentally focused groups including the Zilker Park Posse and Save Austin's Neighborhoods and Environment, a coalition that included Latino activists, began to gain support because of increased construction in Southwest Austin and the apparent complicity of the municipal government. Two developments proved pivotal. The first was Barton Creek Square Mall, developed in the late 1970s by national mall builder Melvin Simon, who chose the site based on its beautiful surroundings and view of downtown. Yet for many residents it was the blatant disregard for the natural landscape that generated concern. Contractors dynamited and bulldozed the top off the property's hill, flattening it to make room for the largest mall in Texas, which opened in 1981, and its asphalt parking lots. The second was the largest planned community to be proffered for development in Austin's history. In 1982 real estate developer Gary Bradley and three partners purchased a 2,700-acre tract of land called Circle C Ranch, directly in the path of the MoPac extension and over the aquifer's vital recharge zone, where rainwater was absorbed directly into the aquifer. The following year the city council approved the creation of four municipal utility districts (MUDs) that would provide water and sewerage to Circle C and agreed to annex the development within ten years.[32]

The buzz generated by the MCC decision, the siting of Motorola on the aquifer, and the large highways, malls, and tract developments that dotted the hilly landscape portended ominously for members of Austin's environmental community. But it was also important to characterize the threat in as wide a way as possible to generate grassroots support. To achieve this goal they broadened their stance in an effort to make environmental threat feel like something that touched as many lives as possible. We Care Austin, for example, actively encouraged its participants to think about a broadly defined community well-being as part of environmental awareness in its early years in the 1970s.[33] When discussing coalition building, We Care Austin leader Jean Mather understood that developing broad support for environmental awareness meant making the issue personal. "People will organize themselves around common interests, concerns, and goals," she wrote. "They are always highly motivated when their homes and neighborhoods or the future of their children are in question."[34] The Zilker Park Posse reframed its initial arguments about the deterioration of Barton Creek from an issue that affected one neighborhood to an issue that affected the aquifer and, therefore, all Austinites in some way.[35] And an irresponsibly built physical environment was threatening, especially since by the 1980s new subdivisions were being created in anticipation of population growth. Austin residents already had places to live and did not think new subdivisions were positive in and of themselves. Only the developers profited, and the city council was subsidizing them with tax-generated funds. To cast the widest net, though, environmentalists argued in as fundamental a way as possible that the environment was being threatened.

For many Austinites, nothing proved to be a more fundamental threat to their well-being than polluted water. Developers and development, represented most clearly by Barton Creek Square Mall, with its location, size, complicity with national developers, and brutal physical transformation, became an easy target for environmentalists. The negative effects the mall had on the creek became apparent when Barton Springs was forced to close during the mall's construction because of pollution, thirty-two times in 1981 alone.[36] Sensing that the threat to Barton Springs could be a unifying theme, the Zilker Park Posse quickly tied the two issues together. Their Christmas 1981 newsletter featured an image of the closed pool with a pile of floating debris in the foreground. The insert showed a sign reading "pool closed" and encouraged readers to "boycott Barton Mall." Their first television ads showed footage of bulldozers in Barton Creek, literal machines in the garden symbolizing that real estate interests did not care about Austin's creeks and waterways. Science was important as well. The posse included in their newsletters hydrological charts that informed readers about how water moved from the ground to the aquifer and how impervious cover could harm that process, and they argued that while no one source of pollution could kill the aquifer, many small ones could.[37] Increasingly, the posse traced a battle plan that sought to limit municipal bonds, which funded services to new MUDs on the urban periphery, to curtail development. Not only did bonds fund irresponsible growth; they also increased the financial burden of residents who did not directly benefit from the extension of services. For the developments that inevitably would be approved the posse sought restrictions on the amount of impervious cover that each could contain.[38] And because of rapidly increasing MUDs, annexation, and the pro-growth mentality on the city council, they also began to see the municipal government as a willing partner in the destruction of the aquifer and Barton Springs.

The Zilker Park Posse, as well as some other smaller groups, were able to demonstrate that the built environment could have dramatic and disastrous effects on water and natural areas in Austin. They made the environment something that seemed connected to the well-being of residents and transcended neighborhoods. They identified rampant, unplanned development, and the political and economic coalition that supported it, as the major threat to the city's quality of life by accusing the council of ignoring growth guidelines.[39] Starting in the 1970s, but especially after the MCC announcement, they included technology firms in their arguments and linked semiconductor production with pollution.[40] Through 1983, they were able to amass a broad coalition of voters that often defeated pro-growth candidates and struck down bond initiatives. They pressured the city council to write a development ordinance for the Barton Creek watershed, which became one of the most aggressive water-quality ordinances in the country and strictly limited the amount of impervious cover permissible in developments over the recharge zone.[41] In 1985 they led the defeat of pro-development mayor Ron Mullen and installed Frank Cooksey in his place; they also seated on the city council three people who had participated in neighborhood environmental groups. And they set the parameters of debate that would dominate Austin's environmental politics into the 1990s: bonds for utility service to outlying areas and limitations on density and impervious cover in new subdivisions.[42]

On a rhetorical level, protecting nature against the externalities generated by human greed, on the one hand, and population growth, on the other, proved a solid strategy. Yet even electoral victories were essentially meaningless for environmentalists in Texas, where variances could easily be applied to ordinances to facilitate development. While the Barton Creek Ordinance was strong, ordinances for other creeks were not or did not exist at all until 1985 when the Comprehensive Watershed Ordinance was passed by the more environmentally friendly council. But as Austin boomed and money flowed, close to 80 percent of projects received a zoning variance that allowed developers to circumvent the strict standards codified by the ordinance. The city still had no legal jurisdiction over MUDs, nor did the ordinances apply to land outside the city, so MUDs remained the developments of choice and were essentially impervious to legal control. By 1987, when the Texas legislature passed a bill that outlawed cities from altering water-quality standards once a development was begun, environmentalism, though ideologically and politically robust, seemed in freefall. The next year, the massive Circle C development, directly above the recharge zone, put its first houses up for sale.

Ironically, it was a sharp economic downturn and concomitant real estate bust that rejuvenated mainstream environmentalism and presented a new global threat to Austin's natural environment and sense of place. Severe overbuilding mixed with a national recession ended the building boom in Austin and crippled the city's real estate industry. In March 1987 the city's largest home builder, Nash Phillips/Copus Inc., filed for bankruptcy, as did former governor John Connally and former lieutenant governor Ben Barnes, who were the most prominent real estate developers in the early 1980s and owners of a large tract of undeveloped land over the aquifer. Average home prices dropped by only 6 percent, but over forty-four hundred residential properties were foreclosed on in Travis County, and almost as many commercial properties were taken over by lenders. Gary Bradley, developer of the Circle C property, filed for bankruptcy. The office market was equally as poor. The downtown rental market dropped in value by half between 1986 and 1987, and One Congress Plaza, the largest downtown office building, was 80 percent empty. By December 1987 Austin had the most overbuilt office market in the country, with 34 percent vacancy, leading the *Wall Street Journal* to call the city's real estate market "an absolute disaster."[43]

Global multinational corporation Freeport-McMoRan stepped in to clean up the fragments in 1988. The corporation swiftly purchased over three thousand acres of land in the Barton Creek watershed, loaned Bradley money to keep his property, and sold the Barton Springs Country Club property to ClubCorp, one of the largest developers of private golf clubs in the United States. Shortly thereafter, Freeport-McMoRan CEO Jim Bob Moffett proposed the largest-ever development in Austin, the Barton Creek Planned Unit Development (PUD) containing 2,500 homes, 1,900 apartment units, 3.3 million square feet of office and retail space, and four golf courses on 4,400 acres, overlooking Barton Creek just seven miles upstream from the springs.[44] While the environmental threat was obvious—the increase in traffic alone would pose immediate pollution concerns, not to mention the pesticides, wastewater, and solid refuse—Freeport-McMoRan and Moffett also presented a more ideological threat that promised to finally transform Austin, once and for all. Here was Moffett, a former football player, oil tycoon, and the sixth-highest-paid executive in the United States, leading Freeport-McMoRan, a global multinational corporation known for rampant environmental and human rights abuses in its mineral extraction enterprises across the globe, looking to similarly raze Austin's natural landscape for profit. University of Texas president and proponent of high tech William Cunningham assisted Moffett on the Freeport-McMoRan board. And, because of the real estate downturn and Freeport-McMoRan's economic might, they were able to buy up huge swaths of foreclosed land from desperate banks and developers and count golf courses as open space.[45]

The corporation provided the perfect symbol of what Austin would become if grassroots support did not materialize expeditiously: a careless destroyer of the environment managed by people from other places who did not care about or understand Austin, its people, and its sense of place. Signs of degradation were already abundant. Barton Springs was still inexplicably closed from time to time. Population growth continued. And by 1990 the golden-cheeked warbler was added to the endangered species list on an emergency basis.

One week before the city council hearing that would determine the fate of the PUD, the small alternative weekly *Austin Chronicle* ran an article by environmental and political activist Daryl Slusher showing a skull and crossbones on a beach ball and warning Austinites, "If you don't read this issue we'll poison Barton Springs." The effect was immediate. Hundreds of people with little previous interest in environmental activism descended on the springs the night before the hearing. On the day of the hearing, hundreds more gathered on the street in front of city council chambers, encouraging one another to "take back our city" to the honking approval of passing cars. Over seven hundred people signed up to talk at the city council hearing, which quickly turned from a municipal hearing into an antidevelopment bacchanalian festival featuring poetry reading to the council, home videos taken at Barton Springs, and technical reports detailing pollution and Freeport-McMoRan's plans to sell its land and quickly leave Austin. Moffett, who confidently assured the audience that as a geologist and a great university student he knew more about Barton Creek than anyone else in the room, was greeted with uproarious laughter several times throughout his speech. The seemingly endless line of Barton Springs defenders kept going until close to six o'clock in the morning, when the weary council finally voted unanimously to disallow the PUD. Environmentalist and sociologist William Scott Swearingen argues that this hearing, and the protection of Barton Springs, was the most important moment for environmentalism in Austin: the creation of a lasting grassroots coalition that would defend Barton Springs at all cost and ensure that Austin would remain a city defined by environmentalism.[46]

Austin's environmental leaders used the grassroots energy generated by the PUD hearing as a platform to launch a systematic defense of Barton Springs. In 1991 We Care Austin leader Mary Arnold delivered a statement to the Texas Water Commission claiming that the city was obligated to protect the springs from pollution.[47] Lawyer and activist Bill Bunch wrote a position paper that cast the springs as the center of the city, arguing that its fragile ecology was particularly susceptible to pollution because it absorbed runoff quickly and because there was little soil on the ground above it. The Comprehensive Water Ordinance, Bunch argued, was simply not enough given the ease with which the city granted variances. He attacked Freeport-McMoRan's emphasis on private property rights, arguing that the city had a greater obligation to protect public water than to ensure developers' rights.[48]

In the lead-up to the 1992 municipal elections, SOS drafted a new ordinance that sought to protect Barton Springs and the aquifer from any further development. The alliance collected over thirty thousand signatures in an effort to get the ordinance put on the ballot as a citizen's initiative; the pro-development city council stalled the hearing but was ultimately powerless to stop the ordinance. It passed the referendum easily, by a two-to-one margin. It established much more stringent protections against potential degradation and more firm regulations on developments in the watershed. Zoning variances, the principal means by which developers circumvented previous watershed ordinances, were outlawed. Impervious cover was limited to 15–25 percent of a development's surface area. Restrictions were put on how close to the creek structures could be built. And comprehensive testing to ensure non-degradation (new developments cannot affect water quality) was mandated. Developers fought the ordinance in the courts with some success, but the Texas Supreme Court upheld the city's right to curtail development to maintain water quality because it was in the public's interest.[49]

As a symbol of Austin's collective environmental spirit, Barton Springs was a metaphorical line in the sand for thousands of residents. The springs demarcated the limits of Austin's 1980s growth explosion; through the 1990s, SOS's grassroots appeal engendered significant changes in the city's ideology that, to that point, had acquiesced to developers. Many environmental groups portrayed the pool as an unchanging piece of nature, something that reflected Austin itself and, if destructed, threatened the city's identity. Yet if Barton Springs was cast by environmentalists as something that brought the city's residents together, it was also something that demarcated the limits of the city's collective identity. Although the ordinance passed easily, even winning more conservative districts in outlying western and northern Austin,[50] it was opposed by eight precincts in the solidly minority eastern and northeastern portions of the city. Just one member of

SOS was African American. And despite its own vigorous environmental campaign, PODER was not shy about claiming that it was not invited into the environmental-developer debate that characterized Westside environmental politics.[51]

Environmental Impact: PODER and the Hybrid Landscape

Like earlier environmentalist groups and the SOS Alliance, Eastside environmentalists argued that they were defending the environment against irresponsible development, pollution, and threats to their quality of life. Like early conservationists in Austin, they drew on the emerging national environmental justice movements, particularly the Southwest Network for Environmental and Economic Justice (SNEEJ), and saw themselves as part of a regional and, at times, global struggle against oppression. They gained national notoriety for their work in exposing environmental racism. In many respects the ways they configured these issues were "environmental" in a similar way as well: drinking water, high tech, and overly developed built landscapes were decidedly environmental issues. But to this list they added concerns that reflected the history of structural and institutional inequality, as well as geography, that minority communities had endured for generations in Austin: economic discrimination, health disparities, and land use zoning caused by institutional racism and capitalist urban planning. By the late 1990s, they framed environmental injustice as a civil rights issue. As an outcome, their definition of environment was more apt to include human-based environments and social concerns than that of mainstream environmentalists, who rendered the environment as something that humans used at times but were also destroying.

Their arguments situated environmental justice within a broad social and historical context of continued discrimination that allowed them to challenge existing structures from a civil rights perspective. By constructing the environment as a hybrid landscape, one where natural and built reinforced one another and combined to undermine minorities' health and access to jobs, education, and recreation, PODER was able to forcefully demonstrate how environmental racism was an important community issue. PODER had great success in transforming the Eastside by changing zoning laws, getting industrial and other bothersome and hazardous facilities shut down, and demonstrating the historical patterns of discrimination that minorities endured. Yet it was externalities generated by its victories—higher real estate values that threatened community cohesion—that proved the most difficult for PODER to overcome. While pollution and industry seemed like environmental problems to Anglo environmentalists and politicians, gentrification was a social issue and, in the minds of many, an important component of the cure to Austin's environmental troubles.

PODER organized in reaction to the siting of SEMATECH's semiconductor processing plant in 1991, but it framed its concerns about high-tech pollution using history and geography that made SEMATECH, and later other high-tech facilities, appear as part of a larger problem.[52] One of the group's first public statements called on Congress to review the environmental effects that SEMATECH had on local communities.[53] The initial argument drew on environmental justice studies of Silicon Valley that found high-tech production had damaging effects on water, soil, and air. In Travis County, PODER found, manufacturing facilities emitted over 8 million pounds of toxic chemical waste in 1992. High-tech facilities also generated a high amount of wastewater; in East Austin, electronics manufacturers increased their discharges by 36 percent from 1990 to 1991.[54] It found that seven of the top ten polluters in Austin were high-tech companies; five of those were located in East Austin.[55] After learning of a huge spill of cleaning solvents at a Motorola facility outside Phoenix, PODER began to investigate spills at high-tech facilities in Austin.[56] It found a number of toxic spills at the AMD facility in 1992 and urged the city to undertake quality tests in the affected areas; the city often used tax-generated money to pay for cleanup as well.[57] Water played a large role in PODER's understanding of environmental threat, as it had for Westside environmentalists, and the group linked water with high-tech development. The city's water use increased by 52 percent from 1980 to 1994; high-tech facilities were responsible for over 10 percent of the overall total, the largest percentage of any sector. Water was a scarce resource being used up rapidly by high-tech facilities, which also produced a substantial amount of wastewater; PODER recorded a 36 percent increase from 1990 to 1991 alone. And, PODER pointed out, all three of Austin's wastewater treatment facilities were located in East Austin.[58]

Unlike the tank farm controversy, however, PODER's approach to questioning high tech reached beyond environmental hazards by interrogating the city's stance on industrial development and its economic benefits to citizens. PODER leaders, well of aware of the Eastside's historical lineage as an industrial receptacle, identified and attacked the city's stance on high tech as an economic engine for minority residents. For minorities, planning injustices went as far back as planning itself in Austin. And while after the 1970s planners and politicians discussed Eastside development as a community improvement issue, rather than as the development of empty space, their plans rarely envisioned anything much different for the Eastside. The most recent manifestation was a 1988 ordinance that declared most of East Austin south of Manor Road an "enterprise zone," which made a variety of subsidies available for firms locating there.[59] Enterprise zones, the city argued, would have the double benefit of attracting industry to Austin and providing jobs close to neighborhoods with higher-than-average levels of poverty and lower-than-average levels of education. The zones, though, were broadly defined and did not take note of existing land uses or zoning. The plan contained no buffer areas between neighborhoods and industrial sites. As in previous plans, the city envisioned East Austin as one-dimensional, an undifferentiated plot of land ready to be made more profitable via industrial relocation.[60]

PODER was quick to point out both the historical significance of enterprise zones and the faulty logic undergirding them. Citing George Kozmetsky's vision for Austin in *Creating the Technopolis,* they framed high tech as an economic development strategy meant to foster growth from outside, predominantly among scientists, researchers, and businesspeople who stood to profit from regional growth.[61] Conversely, and despite the rhetoric of distributive benefits embedded in Kozmetsky's writings and in the city's stance, electronics manufacturing was not a boon for Austin's minorities. In fact, they argued, it was a detriment. In Montopolis, site of SEMATECH, Motorola, and AMD and over 90 percent minority, unemployment increased drastically between 1980 and 1990, from just under 11 percent to just under 16 percent. Among Montopolis residents the poverty level grew more than in any other Austin neighborhood during the 1980s, reaching 38 percent in 1990. Less than 1 percent of residents had a college degree, and 43 percent had less than a ninth-grade education; in contrast, high-tech firms in Austin disproportionately employed educated white men in higher-paying positions. PODER pointed to city planning documents that ironically argued that quality of life and education were important components of locational choices for high-tech companies, yet those statistics did not deter firms from choosing Montopolis as a site—as long as skilled employees could have access to those benefits in other places.[62]

Because the benefits of high tech did not accrue evenly, and because local residents were more vulnerable to the environmental hazards high tech presented, PODER argued that the subsidies the city gave to relocating firms were discriminatory and forced taxpayers to fund environmental degradation. Motorola, for example received over $18 million in tax abatements from 1990 to 1994. In total, the city abated roughly $42 million in taxes from 1990 to 1994 for thirteen high-tech companies operating in East Austin.[63] As part of its generous incentive package, the city gave electronics giant Samsung $121 million in tax abatements when the company relocated to Austin in 1996.[64] To make up for lost revenue and offset costs of extending service, the city increased utility rates for customers, effectively making citizens bear the cost of the subsidies. In consort with SNEEJ, PODER representatives wrote a paper arguing that subsidies for high-tech development were a form of environmental racism and corporate welfare endemic to many communities that housed high-tech agglomerations. Costs associated with high-tech firms were not borne by the firms themselves but rather were passed on to low-skilled workers, local residents, the environment, and the community as a whole.[65] For PODER and other activists involved with SNEEJ, environmental inequities were indistinguishable from socioeconomic inequities generated by policy that favored business at the expense of residents and taxpayers.

In linking environmental inequities with economic and social inequities inherent in high tech, PODER laid the groundwork for recasting the environment as a civil and human rights issue. In response to Tokyo Electron's announcement of its relocation to Montopolis, on October 31, 1995, on Día de los Muertos, PODER filed a complaint with the EPA alleging that the amount of environmental hazards generated by high-tech companies in Montopolis constituted a civil rights violation. While there were numerous other high-tech facilities that polluted air and water in Austin, none were as clustered or in immediate proximity to residences as the four largest in Montopolis and none reflected a similar legacy of siting bias. Rooting its arguments in both environmental and social inequity, PODER charged that the city and the Texas Natural Resource Conservation Commission violated residents' civil rights by systematically failing to address discrepancies in the siting of hazardous facilities and not testing to determine the environmental and health impacts of those facilities.[66] While the EPA declined to hear the case, the civil rights argument linked environmental hazards to historical decisions about land use in East Austin and allowed PODER to concentrate on transforming municipal law to improve its communities.

PODER increasingly focused on zoning and industrial siting as the primary culprits and targeted entities that symbolized the lack of attention that Austin politicians historically paid to the Eastside. The tank farm, discussed in the opening of the chapter, formed the basis for PODER's first large attack of the city's zoning practices. Because the farm symbolized significant environmental and health threats to people beyond the Eastside, and it benefited large corporations, the cause drew resources from all over Austin. Illegal dumping was also a long-standing problem in East Austin, where Eastside businesses and others dumped waste on vacant land, oftentimes in close proximity to residences, whose occupants experienced higher-than-average incidences of maladies, such as lead poisoning. The practice dated to at least the 1950s and was another example of the widespread view that anywhere in East Austin could serve as an industrial receptacle.[67]

Finally, the city itself was also responsible for creating environmental hazards on the Eastside because of its siting of a garbage truck cleaning facility in an African American residential area. The East Austin Strategy Team, a majority African American group, began to question the city's placement of the facility in 1993, when a storm brought debris and industrial runoff from the site into Boggy Creek and private backyards. The city and the Texas Water Commission issued citations to the facility, but the Travis County appraiser also devalued dozens of properties near the site and along the creek.[68] In 1993 the East Austin Environmental Initiative was founded in response to an issue brought up during the tank farm controversy. The initiative drew support from both sides of town and included a new water-quality monitoring program and a plan that regulated industry discharge into storm sewers.[69] Yet other industrial facilities that PODER attacked were seen by some Westside progressives as either locally beneficial or not important enough to invest in. In all instances, PODER drew on the city's history of zoning discrimination to make its case. It emphasized the location of facilities that served greater Austin and asked why Eastside residents were consistently subjected to industrial facilities while Westsiders were not. Its arguments reveal the group's complex understanding of what constituted the environment and environmental discrimination and also demonstrate that Austin's Westside progressives were not as willing to support PODER when they perceived that industrial facilities were benefits to the city or the natural environment. By the mid-1990s, frustrated Eastside activists were openly questioning the positive effects of environmental sustainability on their communities.

The most publicized site of PODER's protest was the Holly Power Plant, located on the northern bank of Town Lake close to the center of a majority Latino residential neighborhood. Conceived of in the 1950s and built throughout the 1960s, the Holly plant was the largest Austin power generating facility when it was constructed, able to generate 570 megawatts of power, over 20 percent of the total capability. It was sited close to the Longhorn Dam, the final dam in the Highland Lakes system, to ensure an adequate supply of cooling water. The city claimed eminent domain to evict approximately two hundred residents and demolish fifty privately owned houses to build the plant.[70]

The demolished houses, though, were part of a relatively dense neighborhood that had been in existence since the turn of the century, at first a white neighborhood that slowly changed to Latino, especially on the floodplain where the plant was located. By the 1950s most residents within two blocks of the facility were Latino; many cited the plant as a source of pollution, health problems, and daily irritations. PODER first criticized the plant in response to a large fire on the grounds in 1993. It drew on numerous accounts of fires and chemical spills that marred the neighborhood and subjected vulnerable residents to increased hazards. Before the 1970s the city did not record disturbances at the plant. But the first official report, filed in 1974, documented a spill of between ten thousand and twenty thousand gallons of fuel.[71] In 1976 a City of Austin study found that the plant created more noise than any other facility in the city, including Mueller Airport.[72] In 1985 a fuel line ruptured, causing another devastating spill. Seven spills were reported from 1991 to 1993, one resulting in a fire. For years the plant was a source of constant minor pollution and a bane to residents' quality of life. The noise shook the foundations of homes, made it impossible to open windows or enjoy outside activities, and occasionally covered the neighborhood in soot. It stood within three blocks of three parks and two schools.[73] Yet because of its age the Holly plant was not subject to Texas air-quality laws, which meant that change would have to happen via grassroots support.[74]

In 1993 the city council passed a resolution to establish a public process to minimize the plant's impact on residents. PODER's response was to document the disparities that living near the Holly plant created for residents to demonstrate that environmental racism was a civil rights as well as a health issue. The goal, PODER wrote, was to decommission the plant as expeditiously as possible. Drawing once again on SNEEJ precedents that used social surveys and scientific testing to demonstrate environmental racism, Sylvia Herrera and Susana Almanza canvassed the immediate area with questionnaires for residents living within two thousand feet of the Holly plant. Four cases of lymphoma and three cases of breast cancer were reported among the 105 households surveyed, a much higher percentage than among the general public.[75] PODER called on the Texas Natural Resource Conservation Commission to carry out air pollution tests around the plant; the commission and the Austin Electric Utility Department agreed to undertake the study. The report found "significant air pollution," including higher incidences of nitrogen oxide, particulate matter, and electromagnetic fields.[76] By summer of 1994, PODER had convinced a number of Austin environmentalists to support shutting the plant down. In a letter to Austin mayor Bruce Todd, the Lone Star Sierra Club argued that closing the plant would reduce pollution, ozone, and electromagnetic fields and be cheaper in the long run.[77] The city council agreed. In 1994 the council began the process of finding alternative power sources for the city, commissioned a $1.2 million project to put silencers in the plant, and initiated soundproofing programs in the neighborhoods. In January 1995 the council passed a resolution to phase retirement of the plant. Of the four power generating units in the facility, two would be closed by 1998 and two more by 2005.[78]

Inspired by the apparent victory in the power plant dispute, PODER turned its attention to other industrial facilities located in Eastside residential districts. The siting of industrial facilities, long a bane of Eastside residents, allowed PODER to forcefully address the historical impact of zoning on the Eastside and set a course for higher levels of resident control over land use. It framed land use and zoning as the symptom of a larger social and economic issue: the built environment itself was an instrument of oppression, not just a nuisance or a health hazard but an architecture that excluded minorities from prosperity. While Westside environmentalists were often open to supporting issues of health and pollution, the way they imagined the environment precluded issues of socioeconomic disparity. And, in some cases, it was their understanding of what constituted environmentally friendly practices that contributed to socioeconomic malaise on the Eastside.

In reports they wrote and newspaper articles where they were quoted, PODER representatives as well as leaders from other Eastside community groups never tired of emphasizing history. Where Westside environmentalists constructed a historical narrative of social decline linked to environmental deterioration, effectively constructing an environmental jeremiad for the city, Eastsiders articulated a much different subjectivity, one that emphasized a continuous pattern of exclusion. There was no paradise to decline from in their story. While many minorities had used Barton Springs and Zilker Park since integration in the 1960s, both were enmeshed in a history of discrimination and alienation that was part of minorities' larger sense of exclusion. This was not just a lack of access to natural areas, where black residents living on the Westside in Clarksville or Wheatsville had to walk across town to Rosewood Pool rather than swim in nearby Deep Eddy.[79] Alienation was also generated by a series of promises unfulfilled, from public housing to urban renewal to a series of economic development initiatives that promised Eastsiders growth but rarely delivered.[80] While Westside environmentalists meant well, their own sense of the city's history as well as its sacred places was so radically different that they often contributed to alienation.

The firmest linkage between historical zoning discrepancies and the contemporary environmental landscape came in the form of recycling. Recycling, since the 1970s an important issue to many environmentalists in Austin and nationwide, was also something PODER framed as a symbol of the chasm separating west from east and used as a jumping-off point to fundamentally alter how land use was determined in the city. Among the many industrial facilities located in residential neighborhoods on the Eastside, the Balcones Recycling plant and the Browning-Ferris Industries (BFI) plant were two of the most pronounced. Situated on East Sixth Street, one of the Eastside's main commercial corridors, the BFI plant was built on the former site of a lumberyard and backed up to a densely built residential neighborhood. Much larger and more intense than the lumberyard, the plant recycled over one hundred tons of paper and cardboard and two tons of aluminum per day. And, PODER pointed out, there was not a public hearing when the Balcones Recycling plant opened because when properties zoned for industry changed hands there was no legal need for a zoning hearing or for a public announcement.[81] PODER linked this to historical zoning practices in order to demonstrate why zoning disincentivized development in East Austin and discouraged real estate appreciation. Even in areas that were predominantly residential, and even in places where homes were in good condition, the threat of industry moving in was always present because of the city's long-standing system of cumulative zoning, where an area zoned for more intense activities could still house less intensive activities. Thus many of the statutes enacted in the city's first zoning plan in 1931 still determined land use in East Austin. For PODER and other groups, the threat of industry also went far in explaining why new residential construction was limited almost entirely to federal housing projects or owner-constructed houses throughout most of the area's history.[82]

For PODER and other Eastside activists, the BFI plant, located in a less central but equally residential neighborhood, Govalle, symbolized the environmental hazards of recycling and also the ways that mainstream environmentalism itself could produce externalities that disproportionately affected minorities. It took a large fire at the plant in July 1996 to bring to light what "neighbors had complained about for years,"[83] the noise, truck traffic, and bug infestations and the refuse that ended up in residents' yards. Yet, for the city, the BFI plant was of utmost importance to the recycling program itself because it was the only facility able to process every type of material that wound up in recycling containers. Without it, the city's solid waste director warned, the entire recycling program would be in jeopardy.[84]

Little changed until the following May when Govalle residents protested the postponement of a city council vote to roll back the facility's limited industrial zoning to limited office zoning, a change that would forbid recycling at the location. Residents charged the plant with creating roach and rat infestations in addition to noise and solid waste pollution, and they charged the environmental community with a lack of caring about the plight of Eastsiders living near industry. Six weeks later a resident came to the city council vote with a recycling bin. The point, she said, was that she was refusing to recycle because she cared about her neighborhood and thought recycling had negative effects on it. BFI argued that the plant added value to the neighborhood because it employed many local workers and to the Austin community by recycling and also by facilitating compact, new urbanism-style neighborhoods.[85] The newly elected city council voted to roll back zoning on the BFI property in July 1997.[86]

PODER was unusually successful in winning so many environmental justice battles so swiftly. The group, along with other African American community activist coalitions, played skillfully on the city's history of environmental abuses and use of East Austin as both dumping ground and industrial area to convince politicians that their communities were at risk. PODER representatives created a narrative that linked planning decisions from the past with health problems in the present. They showed how the built environment reflected their position as subalterns and how they bore the burden of Austin's growth: power plants, recycling, tank farms, and high-tech facilities were manifestations of the city's economic growth or environmental culture, aspects of Austin that Eastsiders benefited from less than other Austinites.

Unlike the tank farms, the Holly Power Plant remained a problem for Eastsiders long after resolutions and zoning rollbacks were passed by the city council. Because the plant produced 22 percent of the city's power, it was almost impossible to shut it down expeditiously. A group of five former city council members, including strong supporters of SOS, formed Citizens Against Shutting Holly in 1997, arguing that the economic costs for residents would be too great.[87] In 1999, after the date to begin the shutdown passed, PODER and El Concilio filed a complaint with the EPA. In 1999 and 2001, fires broke out; the 2001 fire occurred on Halloween evening, and the city failed to notify residents and their trick-or-treating children.[88]

Divided History, Divided Ethic of Place

The 1990s proved a victorious decade for environmentalists on both sides of IH-35. The deleterious effects of widespread economic, demographic, and spatial growth prompted activists to demonstrate the ramifications of unchecked growth, leading to grassroots protests and political change. In 1997 Austinites elected the Green Council, headed by dynamic young mayor Kirk Watson and composed of environmentally friendly members, including journalist Daryl Slusher. Council members quickly pledged support to both environmental factions. On the Eastside, one of their first actions was to roll back zoning on the BFI recycling plant. From a broad perspective, they recognized that zoning changes were the most pragmatic way to make the Eastside more livable, healthy, and enjoyable for residents. PODER drew on a recent report that found that new development in East Austin was stymied by mixed-used, interspersed residences and industries and irregular lots, all of which were "freely" mixed for decades.[89] The new council supported an overlay district for the central Eastside, essentially allowing neighborhood plans created by residents and community groups to dictate new zoning laws under a broad, citywide framework.[90] On the Westside, the new council pledged support for SOS initiatives, Barton Springs, and other undeveloped natural areas in the city's western hinterland. The council's new Smart Growth Initiative (discussed in the Epilogue) promised to do what environmentalists had proposed for years: funnel the city's development into the urban core using a mix of incentives and subsidies and curtail development in Austin's western hills by implementing strict regulations or buying open space to preserve it.

In the 1990s, as in previous decades, the two sides of town were brought together periodically. The tank farm, the Holly Power Plant, and high-tech growth drew ire from many groups and people, and PODER often supported SOS. Yet the geographic and socioeconomic gap that defined race relations in Austin proved more divisive. While mainstream environmentalists rallied around the relationship between humans and nature, subaltern groups defended their communities from the built environment. The way they used the past differed. Mainstream environmentalists invoked a nostalgic past where humans and nature lived in harmony, a pristine landscape spoiled by greed and growth. Subalterns drew on a past where nature was already undermined by the logic of discriminatory policy, a messy landscape where urbanity had already curtailed their relationship with nature and undermined their quality of life. By the late 1990s, the divergence became obvious. The zoning changes that PODER fought for became a centerpiece of the city's new environmental philosophy, forged in part by SOS's grassroots popularity, which sought to appease developers and environmentalists by focusing growth on the urban core, increasing density, and redeveloping older neighborhoods. For the Eastside, a vibrant development initiative had finally come.

Epilogue

From Garden to City on a Hill: The Emergence of Green Urbanity

In 1998 the Greater Austin Chamber of Commerce announced a new strategy for urban and regional growth called Next Century Economy. Coming on the heels of the 1997 victory of Austin's Green Council[1] and new mayor Kirk Watson's implementation of the EPA-endorsed Smart Growth Initiative, Next Century Economy endorsed a path for Austin that reflected a growth ideology much different from previous decades. While the emphasis was still on regional cooperation and the high-tech industry, the document also portrayed environmental sustainability as an economic benefit that could facilitate and improve economic growth.[2] The positive tone echoed the Green Council's excitement at the prospect of a compromise among the city's developers, politicians, and environmentalists after four decades of animosity. Watson proved to be a lynchpin; his political savvy got the chamber of commerce, the Real Estate Council of Austin, and the SOS Alliance to agree that they could all get what they wanted by simply adopting a new ideology and embracing Austin's future as a more urban place. In a sense, all those groups had valued natural space—some for its economic potential, others for its peaceful qualities, and others because it represented their ideas about the place they lived. All of them, it seemed, finally had to accept that Austin actually was a city and begin to treat it as such.

Upon election, Watson and the Green Council began a campaign to institute Smart Growth, a national urban planning movement that encouraged blending quality-of-life issues with economic development initiatives to create sustainable communities. Smart Growth was an attempt to create policies that promoted and rewarded the implementation of New Urban designs: pedestrian-friendly, mixed-use, transit-oriented, filled with open spaces, and properly dense according to local guidelines in order to create human-scale, sustainable communities. At its core, the movement sought to mitigate the damaging social and environmental effects of suburban sprawl.[3] In Austin, smart growth was seen as a means to protect the environment as a place of pristine beauty and recreation for citizens while mitigating, but not destroying, economic and demographic growth by funneling it into already-existing areas of the city using subsidies and infrastructural improvements. For environmentalists, the chance to protect Barton Springs and the open space beyond city limits proved extremely attractive. The chamber and the real estate council responded positively because the plan promised to incentivize centralized development without messy legal issues rather than penalize peripheral development. And politicians hoped to increase tax revenue and slow down peripheral growth, where service extension was costly.[4] In 1999 the chamber of commerce, the city council, the SOS Alliance, and the Real Estate Council of Austin reached an agreement that realized the importance of environmental management to the city's future, and all parties agreed to take care of it.[5] In some sense, Roberta Crenshaw's advice that attractive, usable places confer economic benefits had finally been taken by Austin's business and political elites. The initiatives coincided with a national turn toward more urban-friendly policies that encouraged infill development and higher levels of density after decades of suburban-dominated policy. Since the 1990s, many commentators have noted the revaluation of urbanity and a return of capital to urban cores.

The Green Council developed two bond packages to institutionalize a Smart Growth Initiative proposal, its vision for the city, in 1998. The initiative sought to protect Balcones Canyonlands Preserve to the northwest of the city, constrain development along urban creeks, and institute protective measures over much of the aquifer. To promote responsible development, it advocated neighborhood-based development groups that would actively participate in developmental issues, while also forcing builders to conform to determined sustainable building guidelines outlined by the initiative. Most importantly, it sought to channel building into specific areas of the city. Similar to the Austin Tomorrow recommendations twenty years prior, the Smart Growth Initiative proposal created three geographic zones that determined where development was most and least appropriate. The first, the drinking water protection zone, which conformed to western watersheds that fed the aquifer or supplied water directly, was the least desirable zone for development. It included almost everything west of Lamar Boulevard stretching south into Hays County and north to Austin's city limits. The second was the desired development zone, which made up the rest of the city proper and some outside environs exclusive of the central area. Finally, the third, the urban desired development zone, roughly bounded by Lamar on the west, State Road 45 and U.S. 183 to the north, U.S. 183 to the east, and a combination of the Colorado River, IH-35, and U.S. 290 to the south to include central South Austin, was based on funneling development inside urban watersheds.[6] Over the next year, citizens voted to fund close to $800 million in bonds to purchase land and make infrastructural improvements in the urban desired development zone.

Downtown, the $103 million that the city offered generated a frenzy of deals within two years of the bond passage. Planners imagined that downtown redevelopment would attract members of the "creative class," young high-tech workers and neoartisanal producers who would be attracted to the amenities and cultural opportunities a redeveloped downtown could offer.[7] Three large tech companies, Intel, Vignette, and Computer Sciences Corporation (CSC), announced plans to build new downtown facilities; CSC's facility would be incorporated into the new Austin City Hall along Cesar Chavez Street. A mixed-use facility, subsidized with $23 million in public money, was announced on the IH-35 frontage road between Eleventh and Fifteenth Streets. AMLI, one of the largest real estate corporations in Austin, agreed to build a 250-unit apartment complex adjacent to the CSC–City Hall complex. By 2000 the various projects promised to bring two thousand jobs downtown and dozens of unused lots were under construction.[8]

In the new century, Austin has become a city notable for its environmental and economic success, a rare blend of spectacular growth, quality of life, and environmental protections. Popular sources routinely characterize the city as one of the most economically robust, culturally significant, hip, environmentally friendly, and livable cities in the country.[9] In particular, Austin's reputation as a green, environmentally sustainable city is paramount. In the wake of the bond packages, the City of Austin, Travis County, the LCRA, and other private groups jointly purchased or set aside over thirty thousand acres of undeveloped land in perpetuity extending from northwestern Austin into unincorporated areas in Travis County. The land helps to keep development away from the Barton Creek watershed and the Bull Creek watershed in Northwest Austin. The seventy-one-square-mile park is now used primarily as a habitat for twenty-seven endangered species.[10] Since 2000, the city has purchased another twenty thousand acres of land on the Edwards Aquifer recharge zone using bonds. By employing conservation easements, where the city essentially pays private owners to leave land undeveloped, thousands more acres will remain undeveloped.[11] Although the SOS Alliance has faded from prominence and the health of Barton Springs remains perpetually tenuous, Austin has become a leader in numerous aspects of sustainability. The city plans to be carbon neutral by 2020, and city-run Austin Energy is the nation's leading seller of renewable energy.[12] Austin also offers generous rebates for rainwater collection, green insulation, low-flow toilets, solar panels, and many other environmentally friendly upgrades. The American Planning Association gave Imagine Austin, the city's newest comprehensive plan that emphasizes New Urbanism concepts and green development, its Sustainable Plan Award for 2014.[13] National publications laud Austin as a "clean tech hub," and academics have held the city up as an example of community-driven sustainability in recent years.[14] From an environmental perspective, Austin has become a progressive leader among U.S. cities, and increasingly a model of sustainability, in a state and region more notable for environmental degradation than for environmental achievement. From an economic perspective, in the competitive neoliberal environment, environmental amenities and awards give Austin an advantage over other cities.

The widespread zoning changes and quality-of-life improvements won by PODER and other environmental justice groups in the 1990s made the Eastside cleaner and more livable. The Green Council immediately passed a series of zoning rollbacks to finally attempt to bring stability and order to the Eastside neighborhoods that contained mixed and noncompatible uses. Watson facilitated a zoning truce between BFI and its Govalle neighbors that induced a standing ovation at a city council meeting.[15] Yet environmental improvements generated externalities that, similar to twentieth-century Austin, have had deleterious effects on Austin's most vulnerable residents. Those victories also led to dramatic increases in real estate values and property taxes that threatened the very communities that PODER, the Organization of Central East Austin Neighborhoods (OCEAN), and other Eastsiders worked so hard to save.

In accordance with the Smart Growth Initiative and with historic city planning ideology, starting in 1997 the Watson council encouraged neighborhoods to create their own development plans that the planning commission would then implement through zoning and other ordinances. One of the three trial neighborhood plans that began the process and would serve as models for future plans was the East Cesar Chavez Neighborhood Plan, which covered the area bounded by IH-35, Seventh Street, Chicon Street, and Town Lake, directly adjacent to downtown Austin. The council and planning commission correctly assumed that the neighborhood would be in demand for redevelopment because of its proximity to downtown and inexpensive real estate, and the plan could serve as a test case for other plans on the central Eastside. After three years of debate the plan was finally adopted in 2000. It established firm and ubiquitous zoning regulations for the neighborhood that directly facilitated New Urbanist development. The plan passed a mix of rollback industrial zoning, severely limiting all types of warehouse, recycling, and light industrial space, and infill zoning, which encouraged higher density through garage apartments, aggregate living space, and commercial-residential-office multiuse developments. The entire neighborhood was zoned for mixed-use, which outlawed certain already-existing industrial and commercial spaces in East Cesar Chavez. As in downtown, the city developed a set of benefits for development that it encouraged and restrictions against development it discouraged. To encourage mixed-income housing, the city offered a variety of service-delivery subsidies and fee waivers for builders who included a certain number of low-income units in their developments.[16]

Gentrification began slowly in the 1990s, but the zoning changes and investments hastened the pace dramatically after Smart Growth Initiative policies were adopted.[17] While the average home price in Austin increased by 71 percent from 1992 to 2002, Eastside residents were often least equipped to sustain the increase in taxes, rent, or mortgage payments. In 2002 the Travis County appraiser reported that East Austin property values were increasing faster than anywhere else in the country. In Austin, 72 percent of foreclosures occurred on the Eastside in 2000 and 2001, the year just after neighborhood zoning changes were adopted.[18] In East Cesar Chavez, costs spiraled out of control rapidly. From 1998 to 2004, land values increased by 400 percent on average and property taxes increased by 123 percent. Based on a University of Texas housing study, in 2004 the Austin Human Rights Commission recommended a ninety-day construction moratorium in the East Cesar Chavez neighborhood to further study the effects of redevelopment there.[19]

By 2005 the neighborhood had been largely remade as developers took advantage of low real estate prices and municipal incentives. As real estate prices increased and it became clear that there was demand for condominium and multiuse space in East Cesar Chavez and Holly, the market became more stable and profit more dependable. Very few aggregate living developments included low-income housing units. Even though the city offered a variety of benefits and subsidies for developers that did include a certain percentage of low-income units, almost none participated. In 2000 the average household income in East Cesar Chavez was $27,177, less than 40 percent of the city's median and one of the lowest figures in the city.[20] Even the most modestly priced one-bedroom at Pedernales Lofts, which was considered inexpensive and was one of the developments located the

Changing African American Landscape—Eastern Core *African American Population Concentrations, 2000 and 2010*

Census 2000 Data
Census Blocks

Percentage of the Total Population that is African American

Less than 20%
20% to 40%
40% to 60%
60% to 80%
80% Plus

Census 2010 Data
Census Blocks

Percentage of the Total Population that is African American

Less than 20%
20% to 40%
40% to 60%
60% to 80%
80% Plus

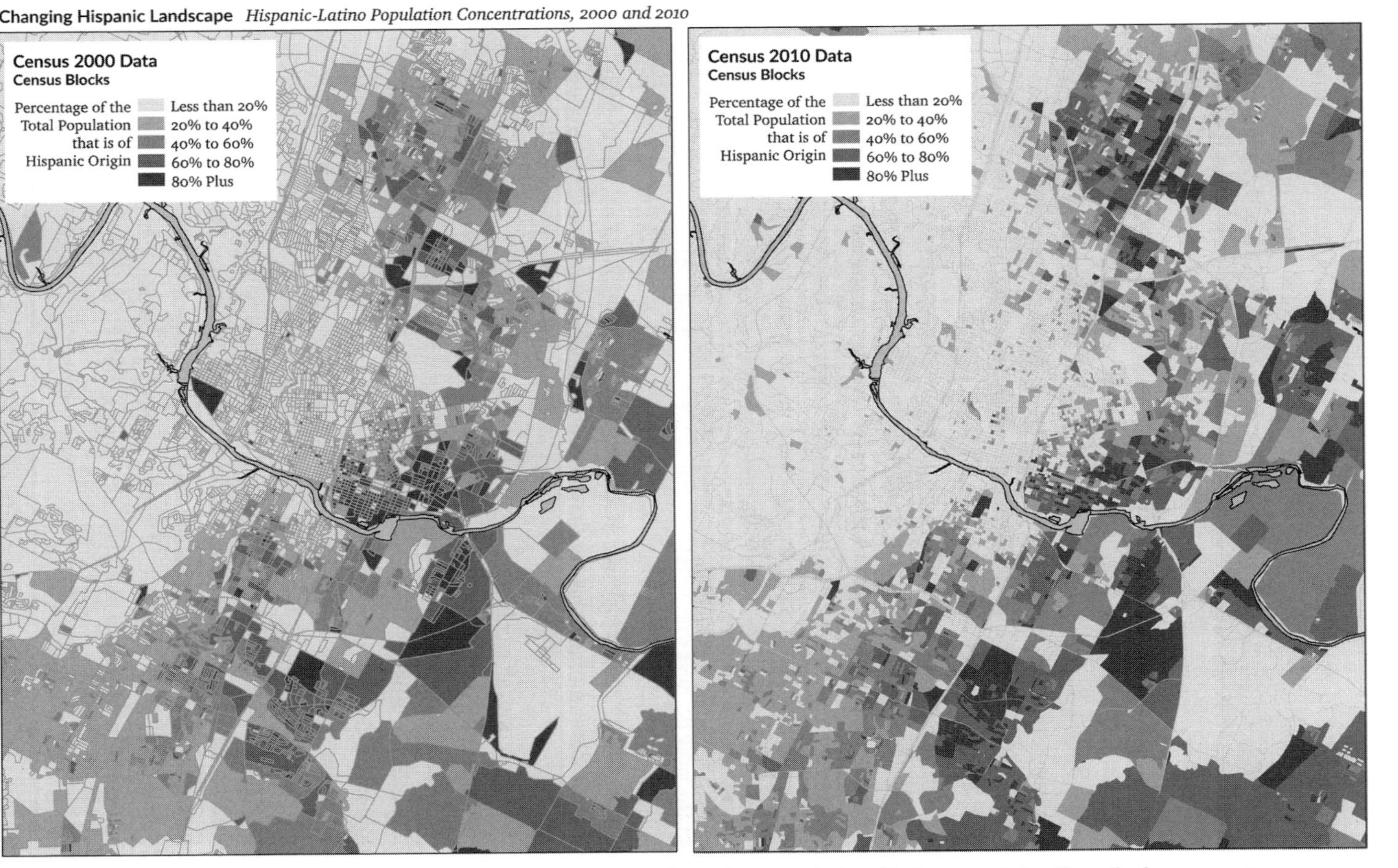

Out-migration of African Americans, and to a lesser degree Latinos, from the central Eastside increased significantly from 2000 to 2010. Courtesy Ryan Robinson and the City of Austin.

farthest east of where real estate was less expensive, were on the market for $154,000 in 2005, a price point that was prohibitive for most East Austin families.[21] Closer to downtown, Waterstreet Lofts, another live/work new urban aggregate living community, was offering one-bedroom apartments for $180,000 in 2005, prompting protests from PODER and other residents at its opening.[22]

PODER leaders were once again at the vanguard of opposition to redevelopment, which they characterized as a social, architectural, and environmental issue. Susana Almanza wrote that "PODER and generations of Mexican-American residents did not ask for the development of new condos or lofts" on the Eastside. Residents in Eastside neighborhoods signed petitions against new zoning measures that allowed for commercial mixed-use zoning characteristic of New Urbanist development.[23] Other Eastsiders thought that the new businesses represented transformation and displacement. One wrote that "most of these projects cater to affluent, white, young professionals" and that "the housing units (aka condos and lofts) are overpriced, and the small coffee shops and businesses (art galleries) don't appeal to us because they were not made to serve us."[24] "Smart Growth equals gentrification," wrote another. "If it didn't it wouldn't work!"[25] PODER was not the only activist group to speak out. OCEAN president Rudolph Williams argued that the "development of high-rent condos and townhomes encircling our neighborhoods" was evidence of the forces that were pushing the "minority community" farther east.[26] For Eastsiders, *sustainability* was yet another slick buzzword that sounded positive but actually had negative effects on community cohesion.

PODER's attempts to curtail gentrification were similarly structured to emphasize how practices considered sustainable were actually facilitating instability for Austin's most vulnerable residents.[27] At its core, PODER's argument rested on a conception of the environment as something in which people and their communities were central. Its attempts to block historic preservation zoning were perhaps the clearest example of this idea. Historic preservation was a centerpiece of the city's efforts to redevelop the Eastside, in large part because the area concentrated so many historical structures relative to the rest of the city due to long-term disinterest from developers. By 2004 Austin offered the most generous subsidies and tax abatements for historic preservation among all U.S. cities.[28] PODER noted that historic zoning had the potential to increase property values by as much as 1,400 percent, a far more severe increase than any other type of zoning change. Pointing out the vast difference in the concept of environment,

Almanza argued that historic zoning honors buildings rather than the people who live in them. Whites, with more resources to research properties and apply for zoning variances, were buying historic structures, getting them H-zoned as historically significant structures, and profiting by taking advantage of subsidies or flipping the properties.[29] Longtime African American activist Dorothy Turner noted the racially tinged irony of H-zoning's popularity as a tool for redevelopment, arguing, "I have a serious problem with white folks coming over here and telling us what is historical. What some white folks think is history is just bad memories for us."[30] Anita Quintanilla pointed to the irony of her former neighborhood becoming nicer and the building of a Mexican American culture center occurring "at the same time that the Mexican community is being torn apart and pushed out due to gentrification."[31] Almanza and Quintanilla persuaded the city to create a task force to determine the costs and benefits of H-zoning. It found that H-zoning facilitated gentrification by accelerating the appreciation of property values and taxes, but it simultaneously forestalled gentrification by preventing the demolition of buildings and preserving neighborhood character. The irony, of course, is that their findings indicated that people were being displaced while buildings were not, which supported Almanza's claim that the city cared about buildings rather than people.[32]

Unlike efforts to address environmental racism in East Austin, PODER's fight against gentrification has done little to stem the tide of displacement, largely because the group has not convinced environmentalists and political leaders that gentrification is an environmental issue.[33] Despite recommendations that run the gamut—from using property tax revenue to subsidize affordable housing for severely vulnerable residents, to instituting rent control or establishing a community land trust—the pace of gentrification has increased. In Census District 1, which includes central East Austin as well as a large portion of historically African American East Austin, the number of African American residents declined by 3,711 (14.5 percent) from 2000 to 2010.[34] Core areas closer to downtown experienced more intense African American population loss.[35] The area is marketed as historically relevant in accord with Austin's new urban tastes, yet the new neighborhood reflects starkly different demographics and consumer preferences. The symbolic reclamation of the East End as a viable part of the city's fabric signaled demographic changes that affected a much larger portion of the central Eastside. The goal of the ARA's (Austin Revitalization Authority) public-private development was to increase investment. The Eleventh Street corridor opened up the area to grassroots gentrification, new single-family homes, condominiums, apartment complexes, and other commercial developments on the central Eastside. Between 2000 and 2010 the former African American neighborhoods underwent intense demographic and economic changes. All four census tracts north of Seventh Street and south of Manor Road adjacent to IH-35 experienced between 18 and 31 percent increases in white population.[36] The same area experienced heavy African American out-migration, averaging a 15.6 percent loss across the four tracts.[37] Overall population on the central Eastside has also grown significantly younger, and the size of households has decreased dramatically as younger people without children have replaced families.[38]

Austin's central core, and especially its historically minority neighborhoods on the Eastside, becomes more sustainable as social and economic capital becomes more fixed on its landscape. What is changing is not the socioeconomic disparity but rather the geography and architecture of exclusion. Shaping investment patterns toward a more urban landscape to save natural areas on the periphery from development has worked well for developers, politicians, and environmentalists because urban life has proved attractive to consumers and hence has provided a new place where capital flows into. Real estate values and property taxes have increased by over 40 percent in Austin since 2000, and in some areas the increase has been much more dramatic. Increased costs have not interrupted growth (rather the opposite is true), and fragile ecosystems on the periphery have been saved.

Yet minorities continue to bear the burden of environmental improvements and the commodification of sustainable urban design. It is useful here to reiterate some of the statistics that I discussed in the Introduction. African Americans have lost population share in Austin every decade since 1920 and experienced a real decline from 2000 to 2010, even as the city as a whole grew by 20 percent. Austin is the only fast-growing major city in the United States that is losing African American population. It is one of the few metropolitan regions in the United States with a higher percentage of African Americans in suburbs than in the central city—and poverty in Austin's suburbs rose by 143 percent between 2000 and 2011, the second-fastest increase in the United States.[39] Certainly, many minority residents were able to take advantage of rising real estate costs to profit from selling their property or the property owned by relatives. Yet if demographic trends continue, Austin's geography will simply become inverted from the decades-long era of suburbanization in Austin and elsewhere.

Minorities that remain continue to experience a socioeconomic lag similar to what they experienced during most of the twentieth century, averaging half the household income of whites. The city's population as a whole has a poverty rate of around 23 percent, a majority of whom are people of color. In some Eastside neighborhoods, poverty was 2,000 percent greater among African Americans than among whites in 2013.[40] In 2015 Richard Florida's Martin Prosperity Institute of the University of Toronto named Austin–Round Rock as the most economically segregated metropolitan area in the country.[41] These statistics indicate that, despite economic growth and an overall strong quality of life, Austin is not a particularly sustainable place for its historically disadvantaged residents. Socioeconomic bifurcation results from historical discrimination and a political geography that aligns minorities with undesirable urban functions and spaces. As urbanity becomes popular in Austin, the people long associated with negative urbanism are forced out.

Sustainability's Prospect

Austin has in many ways become a city on a hill, in large part because it has been successful in the physical blending of city and garden. Its creative and environmental legacy along with its success as a tech center has become a model for progressive nonindustrial urban growth. In the postindustrial era defined by global systems of production and rising competition for resources, many American cities are attempting to grow in this way. The award-winning 2012 comprehensive plan for Imagine Austin locates sustainability as the central policy goal for the city over the next quarter century. The region's robust business and demographic growth, even during the prolonged economic malaise of 2008 to 2012, speaks to its sustainability as an emerging metropolis. For environmentalists, sustainability indicates an increasing quality of life via protected natural spaces and water, as well as the related effort to increase density and green building in the urban core. They are also invested in Austin's unique lifestyle, especially the benefits of having a larger number of natural spaces integrated into the urban fabric or a short distance away. Yet nature and culture are still seen as integrated in this way. The Austin Recreation Commission slogan signals this ideal: "Cultural Places, Natural Spaces." The emphasis has obviously paid off.[42]

Yet being sustainable must also mean creating a new conceptualization of the environment that recognizes the historical unevenness and discrimination that Austin's development has necessitated and understanding that history has a profound impact on today's social and environmental landscapes. The failures to produce truly cross-cultural dialogue, in planning, in environmental consciousness, in the battles for the city's future, are all rooted in a history that is acknowledged but that does not actually inform the present. Whites were able to live where they wanted, enjoy natural areas, and count on being a part of the city's consistent economic growth. Minorities had far fewer opportunities and were often moved, contained, or dispersed based on decisions they had little or no part in. Gentrification is thus best viewed as the most recent example of possessive whiteness in a line running back to the nineteenth century. History and geography defined peoples' relationships to the environment and set the parameters for their ability to create meaning via a relationship to place.

Viewed from this perspective, the relationship between humans and their natural world—whether it be Barton Springs, parks, the Colorado River, or undeveloped cedar forests—is not sufficient to create a holistic, supportive, and integrative sense of belonging in a modern city. Protecting Barton Springs and other natural treasures and indispensable resources in and around Austin is vitally important. The springs are a source of community, a place that inspires both artistic and scientific exploration, and an oasis from the continually expanding urbanity surrounding them. The parks and trails along the riverfront, fought for by Roberta Crenshaw, are similarly islands in the land, used by bikers, strollers, and all kinds of other people including minorities. The long swath of reservoirs and dams that connects the city to its western hinterlands still provides water, electricity, flood control, and numerous social and recreational benefits for people. The picturesque hills and creeks captivate people and provide solitude and calm in the midst of the city. William Scott Swearingen and the mainstream environmentalists are correct to note that places matter and that people do develop human-ecology attachments that help to define where they live.

The connection to place is important, and Austin certainly generates uncommon love of place, *topophilia*. The availability of natural amenities is central to what makes Austin special for so many people. Gardens contain nature, but that nature is augmented by humans. Even the most pristine natural areas have histories fraught with human intervention and social meaning that frame how they are valued. The movement of capital from the western hills to the Eastside similarly demonstrates a shift in values, one where previously unvalued areas are suddenly at the center of investment plans, municipal subsidies, and commodification strategies. Vulnerable residents might see this as an opportunity to profit, as a threat to their community stability, or as another example of the failure of the city to look out for their interests. But it is again the place that is valued—the urbanity, the historical importance, the fact that building there is not going to destroy the environment, the profitability—more so than the people. Environment must mean a place where diverse people have a say, where communities can remain intact, and where a balance is struck, not just between people and their natural places but among groups that seek to share in the benefits and responsibilities of urban life equally.

In Austin and elsewhere, popular notions of sustainability still seem to apply predominantly to the natural environment and, despite convincing arguments to the contrary (including the Austin Recreation Commission), nature and culture, or environment and society, are still largely viewed as separate entities with separate policy prescriptions. Austin's history deeply challenges that notion and forces us to rethink how environment, race, and ideologies of growth reinforced and shaped one another. The new emphasis on urbanity in twenty-first-century Austin exposes the limits of sustainability and also shows why incorporating historical understanding of how places emerge and change is vital to creating more just cities. As such, it is of utmost importance to cast off the idea that the pristine natural world is somehow unsullied or that it can be imagined as something that exists outside of human relationships and struggles. The natural world is something that has been valued predominantly because it is easy to enjoy and it reflects collective social experience, creating a deep sense of meaning. One of the most attractive aspects of environmentalism is that people get to speak for birds and tree and pools, which cannot speak for themselves. Yet its meaning is something that is continually reconstructed and reconstituted. To imagine it as integrated into a larger human ecosystem, as something man-made and imbued with the same politics and messiness as a building or a neighborhood or a school or race relations, may be difficult. Yet those who lacked access to those places, or whose lives and communities were subordinated to their interests, were forced to view them from that very perspective. Those who were oppressed by the environment viewed it holistically, and I suggest that this is a good starting point for imagining what a new, environmental consciousness less invested in possessive whiteness might look like in Austin and elsewhere.

A new concept of sustainability means thinking about the economy in new ways where more sectors of the labor market are invited to participate. In Austin, mainstream environmentalists have often portrayed their battle as one of environment versus growth. Swearingen argues that Austin environmentalists attacked the growth machine's claim that economic expansion brought jobs on the grounds that most environmentalists were middle-class, white-collar workers often employed by the state, university, or city.[43] In doing so they staked a precarious position, and one reminiscent of Austin's 1950s growth advocates: a middle class in between greedy developers, on the one hand, and those blue-collar mid- and low-skilled workers who subsisted because of growth, on the other hand. Looked at this way, their ability to not include social and economic issues as part of an environmental discourse appears axiomatic. Yet a hybrid sustainability makes more sense here. Neoliberalism and globalization mean that the federal government takes less responsibility for stimulating job growth and simultaneously that stable mid- and lower-skill work has migrated to developing countries. The American middle class continues to struggle; average income grows more bifurcated every year. In this landscape, more responsible growth will necessitate a revaluation of local production systems that value blue-collar work as well as high-tech and neoartisanal forms of production. While "industry without smokestacks" served Austin well, it also led to policies that devalued participation from minorities and encouraged growth from the outside. Policy makers need to work to generate jobs for all kinds of people, not just for those that manage sites of consumption that cater to the tastes of creative workers and others with disposable income. A strong emphasis on local chains of production—not just in foodways or art, but in manufacturing and services—is a great first step toward more equitable urban life. Those that have been excluded from robust growth must be able to participate. Austin is in a wonderful position here. Its residents spend more disposable income per capita than residents of any other American city. This demand for consumer goods can be better captured by more local chains of production and consumption.

The new sustainability must deeply consider the importance of housing and neighborhood stability as fundamental components of a city's livability. Austin, like many other cities, has long had problems with producing enough affordable housing despite federal programs and subsidies for private developers. Today neoliberal policy means that cities are increasingly forced to create solutions to housing shortages but are also dedicating more resources to attracting investment than ever before. As the city becomes more desirable, housing costs will continue to increase. The city is already extremely segregated by class with a high level of poverty, and unchecked gentrification promises to push even more marginalized residents out of the urban core. It also saw an across-the-board increase of 40 percent in property taxes from 2008 to 2014. Numerous choices to stabilize communities exist. Rent controls, land trusts, direct subsidies, tax freezes, levies on big-box stores, and higher taxes on short-term rentals have helped in other places. What is necessary is a commitment from the city and political and social leaders to imagining housing, jobs, and community cohesion as indispensable components of a holistically livable environment.

The new sustainability must also acknowledge that capital improvements need to be distributed more equitably to make up for deep historical discrepancies in siting decisions. While industry went into East Austin, infrastructure went into West Austin. The dam system on the Colorado River was by far the most important infrastructural improvement in the city's history. Owners of land in West Austin and in adjacent areas near the watershed universally benefited, as did business owners, developers, and others. Conversely, East Austin's waterways, and especially the Colorado, remained largely unimproved well into the 1980s. Boggy Creek, the Eastside's largest waterway, was channelized rather than beautified when it was improved in the 1980s. East Austin is also lower than West Austin, meaning that it is the hydrological drain for the city and contains places extremely prone to flooding.[44] The earliest Mexican American communities on the Eastside were on the floodplain between Holly Street and the Colorado River, an area that remained completely unimproved until 1960. Given this history, it is equitable to guide resources toward communities that have been historically underfunded and devalued.

Holistic sustainability means making our sense of community more inclusive. Cities are essentially imagined communities, disparate people pulled together or pushed apart by their ideas of what a city should be and what values it should reflect. While anyone may be invited to join a movement or to enjoy a place, people with different histories will not experience the same depth of feeling and, in some instances, may feel that movements and places have been part of their marginalization. Defining a city using a human-natural relationship that emphasizes a few important places encourages a sense of belonging that necessarily excludes people who do not feel the same way about those places, whose histories have forced them to value different spaces. Places are deeply contextualized. Understanding, acknowledging, and using their histories to create a hybrid sense of place, one that reflects beauty as well as messiness, can help to make more people feel that they belong.

Notes

Introduction

1. One story reported from CBS is Anthony Mason, "Austin, Texas Leads Nation in Job Growth," *CBS Evening News*, January 8, 2011, accessed August 29, 2011, http://www.cbsnews.com/stories/2011/01/07/eveningnews/main7224063.shtml. Kiplinger's in 2011 rated Austin the number one city for personal business over the next decade, http://www.kiplinger.com/article/business/T006-C000-S002-10-best-cities-for-the-next-decade.html, accessed December 30, 2016. *Forbes* also considers Austin the top U.S. growth pole among standard metropolitan areas with over one million residents. See Joel Kotkin, "The Next Big Boom Towns in the U.S.," *Forbes*, July 11, 2011, accessed July 12, 2011, http://finance.yahoo.com/career-work/article/113083/next-big-boom-towns-forbes; Francesca Levy, "Cities Where the Recession Is Easing," *Forbes*, March 3, 2010, accessed March 9, 2010, http://realestate.yahoo.com/promo/cities-where-the-recession-is-easing.html. *Time* ran an article on Austin's success during the recession as well. See Barbara Kiviat, "The Workforce: Where Will the New Jobs Come From?," *Time*, March 19, 2010, accessed August 3, 2010, http://www.time.com/time/magazine/article/0,9171,1973292,00.html.

2. U.S. Bureau of the Census, "Ten U.S. Cities Now Have One Million People or More," May 21, 2015, accessed September 15, 2015, http://www.census.gov/newsroom/press-releases/2015/cb15-89.html; Erin Carlyle, "America's Fastest-Growing Cities 2016," *Forbes*, accessed May 26, 2016, http://www3.forbes.com/business/americas-fastest-growing-cities-2016/21.

3. Florida, *Cities and the Creative Class.*

4. Mother Nature Network, "Top Ten Green U.S. Cities," October 16, 2009, accessed July 27, 2015, http://www.mnn.com/health/allergies/photos/top-10-green-us-cities/what-makes-a-city-green.

5. A list of many of Austin's environmental awards and green initiatives is available at JW Properties, "Austin, a Green City," accessed September 15, 2015, http://jwproperties.net/default.asp.pg-AustinaGreenCity. For trails, see "The Best Fitness-Walking Cities," *Prevention*, April 2007, 118.

6. City of Austin, "Imagine Austin Wins National Sustainability Award," accessed July 27, 2015, https://www.austintexas.gov/news/imagine-austin-wins-national-sustainability-award.

7. Bryan Walsh, "Red State, Green State: How Austin Has Become America's Clean-Tech Hub," *Time*, January 16, 2012, accessed July 27, 2015, http://content

.time.com/time/magazine/article/0,9171,2103780,00.html; Moore, *Sustainable City*. For a fairly comprehensive list of what the city has done and is planning, see Howard Witt, "Austin's Green Ambitions," *Seattle Times*, October 5, 2007.

8. City-Data.com, "Austin, Texas (TX) Poverty Rate Data," accessed July 15, 2013, http://www.city-data.com/poverty/poverty-Austin-Texas.html. For suburban poverty, see Kneebone and Berube, *Confronting Suburban Poverty*.

9. Florida and Mellander, *Segregated City*.

10. Throughout the book the term *minority* refers to Austin's historically disadvantaged residents of Latino and African American heritage. While in recent decades many Asian and Indian migrants have settled in Austin, historically their numbers have been small.

11. This argument about cities and capital follows Harvey, *Social Justice*.

12. The term *new urban poverty* was developed by William Julius Wilson in a landmark sociological study of the changing nature of segregation. See Wilson, *When Work Disappears*.

13. Lipsitz, "Possessive Investment," 371; Sugrue, *Urban Crisis*; Hirsch, *Second Ghetto*; Gordon, *Mapping Decline*; Sides, *L.A. City Limits*; Self, *American Babylon*. Some cities, such as Atlanta and Miami, were more like northern cities that experienced in-migration and heightened racial tension over space. See Bayor, *Twentieth-Century Atlanta*; Mohl, "The Second Ghetto"; and Connolly, "Sunbelt Civil Rights." Others saw business coalitions replace old guard Dixiecrats and implement business-friendly policies that intensified segregation. See Scribner, *Renewing Birmingham*; Silver, *Twentieth-Century Richmond*.

14. Studies of urban environmental hazards emphasize race, yet most concentrate on natural disasters. Colton, *Unnatural Metropolis*; Kelman, *River*; Orsi, *Hazardous Metropolis*; Steinberg, *Acts of God*.

15. There is a robust history of urban environmental improvements, yet it often neglects the racial component and focuses on earlier eras. Melosi, *Garbage in the Cities*; Melosi, *Sanitary City*; Schultz and McShane, "To Engineer the Metropolis"; Duffy, *Sanitarians*; Schultz, *Constructing Urban Culture*; Tarr, *The Search*.

16. While some works address racial aspects of suburban and urban mainstream environmental movements, racial discrimination is rarely central. Rome, *Bulldozer*; Sellers, *Crabgrass Crucible*; Klingle, *Emerald City*; Walker, *Country in the City*.

17. Lipsitz, "Possessive Investment," 371.

18. The idea of an envirotechnical system is best articulated by Pritchard, "Environmental History," and Pritchard, *Confluence*. Also see White, *Organic Machine*. Also important is Hughes, *Networks of Power*.

19. Swearingen, *Environmental City*, 176.

20. HoSang, *Racial Propositions*; Lassiter, *Silent Majority*.

21. Two works that root Austin's distinctiveness in citizen opposition to urban change and the protection of Austin's environmental and cultural heritage are Long, *Weird City*, and Swearingen, *Environmental City*. Even before the 1960s, Austin had

a much stronger liberal presence in government than most southern cities did. See especially Orum, *Power, Money, and the People*, 201–225. Austin also had a particularly robust antiwar and countercultural movement in the 1960s and 1970s. See Orum, *Power, Money, and the People*, 267–305; and Rossinow, *Politics of Authenticity*.

22. Tretter, *Shadows*, 112–113.

23. Sugrue, "Crabgrass-Roots Politics"; Sugrue, *Urban Crisis*; Hirsch, *Second Ghetto*; Gordon, *Mapping Decline*; Sides, *L.A. City Limits*; Self, *American Babylon*.

24. Bayor, *Twentieth-Century Atlanta*; Scribner, *Renewing Birmingham*; Silver, *Twentieth-Century Richmond*; Kruse, *White Flight*; Lassiter, *Silent Majority*.

25. For postwar suburbanization, see Jackson, *Crabgrass Frontier*, 231–245; Hayden, *Building Suburbia*; Beauregard, *When America Became Suburban*; Kruse and Sugrue, *New Suburban History*.

26. Self, *American Babylon*.

27. HoSang, *Racial Propositions*, 2.

28. A recent work that parses postwar racial liberalism in California, and exposes its shortcomings, is Brilliant, *Color of America*.

29. For mainstream environmentalism as a component of the postwar liberal regime, see Rome, " 'Give Earth a Chance.' "

30. White, " 'Are You an Environmentalist?,' " 173.

31. Swearingen, *Environmental City*, 64.

32. Ibid., 87.

33. A good synopsis of this history can be found in Joseph E. Taylor III and Matthew Klingle, "Environmentalism's Elite Tinge Has Roots in Movement's History," *Grist*, March 9, 2006, accessed May 25, 2016, http://grist.org/article/klingle.

34. For an excellent analysis of how the knowledge economy has driven growth in Austin's history, see Tretter, *Shadows*.

35. A similar process occurred in other cities with large universities that wanted to grow via research and development. See O'Mara, *Cities of Knowledge*.

36. Florida, *Cities and the Creative Class*; Scott, "Capitalism and Urbanization?"; Scott, "Creative Cities." For Austin, see Smilor, Kozmetsky, and Gibson, *Creating the Technopolis*; Gibson and Rogers, *R&D Collaboration*; Tretter, *Shadows*, 57–76.

37. Portney, *Taking Sustainable Cities Seriously*; Wheeler, *Planning for Sustainability*; Joss, *Sustainable Cities*.

38. Portney, "Sustainability in American Cities."

39. City of Austin, *Imagine Austin Comprehensive Plan*.

40. Most accounts of environmental racism deal with either industrial pollution or agricultural issues. The two classic texts are Pulido, *Environmental and Economic Justice*, and Bullard, *Dumping in Dixie*. Others accounts include Forman, *Promise and Peril*; Szasz, *EcoPopulism*; Bullard, *Unequal Protection*; Lerner, *Diamond*; Cole and Foster, *From the Ground Up*. Industrial cities have figured prominently in this discussion as well. See Pellow, *Garbage Wars*; Sze, *Noxious New York*; Hurley, *Environmental Inequalities*.

41. Olmsted, *Journey through Texas*, 129.

42. Environmental Protection Agency, "Ecoregions," accessed August 6, 2015, http://www.epa.gov/wed/pages/ecoregions/na_eco.htm#Level%20II.

43. Most studies of urban water issues in the American West focus on coastal cities: Los Angeles and, less so, San Francisco. See Righter, *Battle over Hetch Hetchy*; Carle, *California Dream*; Kahrl, *Water and Power*. One inland exception is Kupel, *Fuel for Growth*. For western urbanization more generally, Findlay, *Magic Lands*, and Abbott, *Metropolitan Frontier*, are two standard bearers. The former argues that western cityscapes fit into patterns of development best seen in built spaces such as Disneyland, Stanford Research Park, Sun City, and the Seattle World's Fair.

44. I discuss these geological and meteorological particularities at length in chapter 1.

45. A nice example of how rivers can "create" regional identity is Vogel, "Defining One Pacific Northwest."

46. Especially prescient are Nash, *Federal Landscape*; Wright, *Old South, New South*; Abbott, *New Urban America*; Fones-Wolf, *Selling Free Enterprise*; Schulman, *Cotton Belt to Sunbelt*; Goldfield, *Cotton Fields and Skyscrapers*. None of these works offers more than a mention of Austin.

47. A good example of the Sunbelt's attractive landscape and leisure cultures can be found in Culver, *Frontier of Leisure*.

48. The goal of many cities and states in the South especially was to bring in manufacturing, and many cities that chose that path encountered the same problems associated with heavy industry in the North. See Cobb, *Selling of the South*.

49. Tang and Ren, *Outlier*.

Chapter One

1. "Engineering Triumphs in Texas," *Harper's Weekly*, July 8, 1893; *Scientific American* (cover), August 1896.

2. See Adams, *Damming the Colorado*, 5–10; McDonald, *The Great Dam*; Taylor, *Austin Dam*; " 'Austin Dam' Report—Notes on Water and Light History, Minute Book Entries, 1884–1922," Box 7, Office of the City Clerk Records (hereafter OCC), Austin History Center (hereafter AHC), Austin Public Library, Austin; Burnett, *Flash Floods*, 1–9.

3. For utopianism in planning, see Hardy, "Planning of London"; Schultz, *Constructing Urban Culture*; Hall, *Cities of Tomorrow*; Boyer, *Dreaming the Rational City*.

4. Hays, *Gospel of Efficiency*.

5. Pinchot, *Fight for Conservation*.

6. Kerr, *Seat of Empire*, 8–10; *Austin, Texas, Illustrated*.

7. Olmsted, *Journey through Texas*, 110; Barker, "Description of Texas."

8. Running through discussions of Austin's place as the capital is this theme that the public center of Texas should symbolize the entire state and be owned by

all Texans. See, e.g., "Mrs. Lyndon B. Johnson Remarks," December 10, 1971, Folder "Town Lake Advisory Group—31 July 1998 Guiding Principles," Box 2, OCC, AHC. See also Winkler, "Seat of Government." Stephen F. Austin, for whom the city was named, had a similar impression of the site. See Barker, "Description of Texas," and Adams, *Damming the Colorado*, 6–9.

9. Nash, *Urban Crucible*; Tarr, "The City as an Artifact."

10. Colton, *Unnatural Metropolis*; Kelman, *River.*

11. McKinstry, *Colorado Navigator.*

12. Adams, *Damming the Colorado*, 5.

13. Winkler, "Seat of Government."

14. Lower Colorado River Authority (LCRA), "Flood Safety," accessed July 9, 2014, http://floodsafety.com/texas/regional_info/regional_info/lcra_zone.htm#a; Jordan, *Texas*; Dumble, *Geological Survey*; Student Geology Society, *Guidebook*; Burnett, *Flash Floods.*

15. E. H. Johnson, "Edwards Plateau," *Handbook of Texas Online*, accessed July 9, 2014, http://www.tshaonline.org/handbook/online/articles/rxe01; Bureau of Economic Geology, "Geologic Wonders of Texas," accessed July 9, 2014, http://www.beg.utexas.edu/UTopia/centtex/centtex_what.html; Austin Public Library, "West Travis Country," accessed July 10, 2014, http://www.austinlibrary.com/ahc/outside/west.htm.

16. United States Census Bureau, "Texas—Race and Hispanic Origin for Selected Large Cities and Other Places: Earliest Census to 1990," accessed July 10, 2014, http://www.census.gov/population/www/documentation/twps0076/TXtab.pdf.

17. Quotations from "Engineering Triumphs." For university-related economic growth, see John A. Lomax, "The University of Texas and Its Relation to Austin," *Austin Progress* 1, no. 4 (October 1913).

18. *Austin and Travis County*; *Austin, the Capital of Texas.*

19. American National Bank, "Citizen of No Mean City."

20. The late nineteenth century saw a growing belief in the restorative power of climate and open space. See Olmsted, "Public Parks"; Sellers, *Crabgrass Crucible*, 13; Nash, *Inescapable Ecologies.*

21. This organization would later become the Austin Chamber of Commerce.

22. "Engineering Triumphs"; Austin Commercial Club, *Austin, Texas*, 13, 21. See also Austin, Texas, Board of Trade, "Health, Beauty, Prosperity: Something of Interest concerning the Great Capital of Texas" (Austin, 1890).

23. The city ran dry numerous times, as late as 1885. " 'Austin Dam' Report—Notes on Water and Light History"; "A Brief History of the Austin Dam," Folder "Austin Dam Correspondence, 1938," Box 5, OCC, AHC.

24. Burnett, *Flash Floods*, 299–300.

25. Adams, *Damming the Colorado*, 9–10; Long, *Flood to Faucet.*

26. Alexander P. Wooldridge, *Austin Statesman*, January 1, 1888; Orum, *Power, Money, and the People*, 33–37. For the way technology reflected democratic culture and American greatness, see Nye, *American Technological Sublime.*

27. Alexander P. Wooldridge, "Our Water Power," *Austin Statesman*, June 25, 1889.

28. For regional support of envirotechnical systems, see White, *Organic Machine*. Scholars have portrayed dam building at the center of a debate over the proper place of nature and wilderness. One side tends to argue that untainted nature or wilderness is a positive spiritual force with inherent value. The other side argues that nature should be augmented to better serve human needs. See Nash, *Wilderness*. Robert W. Righter argues that this battle is often actually about what regions benefit from particular improvements, not about wilderness and civilization. Righter, *Battle over Hetch Hetchy*.

29. Burnett, *Flash Floods*; "Brief History of the Austin Dam"; Austin Commercial Club, *Austin, Texas*. For thinking about nature and cities symbiotically, see Spirn, *Granite Garden*.

30. "General Lightfoot," Minutes of Austin City Chamber of Commerce, July 15, 1915, AHC.

31. Austin National Bank, "Citizen of No Mean City."

32. Austin Chamber of Commerce, *Progressive Austin* (1916).

33. Nye, *American Technological Sublime*; Marx, *Machine in the Garden*. For the size of the dam and its significance, see "Brief History of the Austin Dam"; Burnett, *Flash Floods*, 1–2; *Scientific American*, August 1896.

34. "Engineering Triumphs"; Austin Commercial Club, *Austin, Texas*.

35. The theme of "the machine in the garden" was commonplace in nineteenth-century literature. See Marx, *Machine in the Garden*.

36. Burnett, *Flash Floods*, 1–4, quotation on 3; "Brief History of the Austin Dam"; Austin Chamber of Commerce, *Progressive Austin* (1916); "The Great Dam across the Colorado River at Austin, Texas," *Scientific American*, September 24, 1892.

37. "Brief History of the Austin Dam"; " 'Austin Dam' Report—Notes on Water and Light History"; Allen Hazen to the City Council, May 16, 1896, Folder " 'Austin Dam' Report—Notes on Water and Light History, Minute Book Entries, 1884–1922," OCC, AHC; Adams, *Damming the Colorado*, 9–12.

38. See Adams, *Damming the Colorado*, 5–10; McDonald, *The Great Dam* ; Taylor, *Austin Dam*; "Death and Ruin at Austin," *San Antonio Express*, April 8, 1900; *Austin Semi-Weekly Statesman*, April 11, 1900.

39. Taylor, *Austin Dam*; Mead, "Report on the Dam."

40. "Complete the Austin Dam," May 1917; Councilman and Superintendent of Parks and Public Property to Harrison B. Freeman, August 20, 1915; and Guy A. Collett to the Mayor and City Council, May 15, 1916, all found in Folder " 'Austin Dam' Correspondence, 1913–1919 (Bulk 1916–1919)," Box 4, OCC, AHC. Citizens wrote in demonstrating frustrations as well. Fleming A. Waters to the City Council, October 21, 1933; Chas. A. Timm to Guiton Morgan, City Manager, October 19, 1933; and J. S. Speir to Guiton Morgan, March 4, 1934, all found in Folder " 'Austin Dam' Correspondence, 1933–1936," Box 5, OCC, AHC.

41. H. C. Patterson to Water and Light Commission, October 31, 1901, Folder " 'Austin Dam' Correspondence, 1913–1919 (Bulk 1916–1919)," Box 4, OCC, AHC.

42. Adams, *Damming the Colorado*, 9–12; "Brief History of the Austin Dam"; A. C. Scott to Harrison B. Freeman, June 3, 1915; S. S. Posey to E. C. Bartholomew, Superintendent, January 27, 1913; "Austin, Texas, March 16, 1912"; and Consulting Engineer to E. C. Bartholomew, March 15, 1915, all found in Folder "'Austin Dam' Correspondence, 1911–1913," Box 4, OCC, AHC.

43. Collett to the Mayor and City Council, May 15, 1916.

44. For a good overview of the literature see Hundley, "Water and the West"; Carle, *California Dream*, Kahrl, *Water and Power*; Righter, *Battle over Hetch Hetchy.*

45. Adams, *Damming the Colorado*, 12–15. For navigation, see Hays, *Gospel of Efficiency*, 5–26.

46. "Brief History of the Austin Dam"; Guy A. Collett to the Mayor and City Council, March 22, 1928, Folder "'Austin Dam' Correspondence, 1922–1932," Box 5, OCC, AHC.

47. Adams, *Damming the Colorado*, 16–23; Banks and Babcock, *Corralling the Colorado*, 28–37.

48. Ickes, *Back to Work*; Smith, *New Deal Liberalism*; Wright, *Old South, New South.*

49. Adams, *Damming the Colorado*, 33; Haw and Schmitt, "Report on Federal Reclamation."

50. Adams, *Damming the Colorado*, 34–37.

51. The best work on the LCRA is John A. Adams's *Damming the Colorado*, which painstakingly details the local politics and actual building of the dams. For a history of the LCRA and the actual building of the Highland Lakes dams, please refer to Adams's book. Also see Banks and Babcock, *Corralling the Colorado*. For text, see "'Austin Dam' Act of Legislature (as Amended) Creating LCRA, 1934," in folder of same name, Box 4, OCC, AHC.

52. Orum, *Power, Money, and the People*, 59.

53. Long, *Flood to Faucet*; Long, *Something Made Austin Grow.*

54. *Austin American-Statesman*, January 1, 1935; "Dams and Lakes of Central Texas Topics of Volume," *Austin American*, August 28, 1937; "New Cen-Tex Lakes Furnish Material for C. of C. Book Here," *Austin American*, August 28, 1937; "Burnett Attorney Asked to Keep Lake Roads Open," *Austin American*, January 22, 1938.

55. Adams, *Damming the Colorado*, 31–42.

56. The dams were originally named Hamilton Dam and Marshall Ford Dam. Each was renamed after a politician who helped engender the LCRA and secure funding for the projects. Buchanan headed the Appropriations Committee, and Mansfield chaired the Rivers and Harbors Committee. The first version of Mansfield Dam was finished in 1939. Another higher dam wall was completed in 1941. Inks Dam and Tom Miller Dam, the second and fourth structures, were completed in 1938 and 1940, respectively. For a narrative of the political machinations necessary to secure funds for Mansfield Dam, see Adams, *Damming the Colorado*, 66–92.

57. Burnett, *Flash Floods*, 79; Adams, *Damming the Colorado*, 43–47.

58. "Special Meeting of the City Council and Citizens Advisory Committee," June 2, 1937; [no title], October 8, 1937; and "Meeting of the City Council and Citizens

Advisory Committee, Thursday, May 13, 1937," all found in Folder " 'Austin Dam' Citizen's Advisory Committee for LCRA, 1937," Box 4, OCC, AHC. Alvin Wirtz to Tom Miller, January 20, 1938, and "Austin Dam Contract, 1938," both found in Folder " 'Austin Dam' Correspondence, 1938," Box 5, OCC, AHC.

59. Lyndon Johnson, "Introduction of the Honorable Harold L. Ickes," Folder "Introduction of Harold Ickes, Dedication of Buchanan Dam, Burnet Co, TX," Box 1, Statements of LBJ, 1927–1937, LBJ Presidential Library (hereafter LBJL), Austin.

60. Ibid.

Chapter Two

1. Wright, *Old South, New South.*

2. Lomax, "University of Texas," 3, 5; Dan Zehr, "History of Austin's Racial Divide in Maps," *Austin American-Statesman Online*, accessed July 29, 2015, http://projects.statesman.com/news/racial-geography; Tretter and Sounny-Slitine, *Austin Restricted.*

3. In industrial cities, and especially after the Civil War, waste disposal and sewer systems were increasingly seen as important components of urban reform and public health. See Melosi, *Garbage in the Cities*; Melosi, *Sanitary City*; Schultz and McShane, "To Engineer the Metropolis"; Duffy, *Sanitarians*; Schultz, *Constructing Urban Culture*; Tarr, *The Search.*

4. Burnett, *Flash Floods*, 22–27; Hamilton, *Social Survey of Austin.*

5. Austin Chamber of Commerce, *Austin, the Friendly City.*

6. McDonald, *Racial Dynamics*, 18–31, 182–200; Foley, *White Scourge*; Allen, *Organized Labor*, 145–151.

7. Tretter and Sounny-Slitine, *Austin Restricted*; Austin Chamber of Commerce, *Austin, the Friendly City.*

8. Loose materials, Folder "Barton Springs—History," Box 1, Walter E. Long Papers, AHC; Walter E. Long, "Austin History," Folder "Austin History," Box 1, Long Papers.

9. Hamilton, *Social Survey of Austin*, 32–37.

10. Austin National Bank, "Citizen of No Mean City."

11. McDonald, *Racial Dynamics*, 105.

12. Hamilton, *Social Survey of Austin*, 56–57.

13. Jackson, "East Austin," 56.

14. Accurately gauging the number of Mexicans in Austin before 1930 is difficult because Mexicans were considered white both in the census and by Anglo Austinites, and whites often assumed that Mexicans were migratory workers who would return to Mexico and were not permanent residents. See McDonald, *Racial Dynamics*, 18–31; Austin Human Relations Commission, *Housing Patterns*, 16–18.

15. Foley, *White Scourge*; McDonald, *Racial Dynamics*; Kantrowitz, *White Supremacy*; Roediger, *Wages of Whiteness.*

16. U.S. Bureau of the Census, *Fifteenth Census of the United States, Unemployment* (Washington, DC: Government Printing Office, 1931), 1:952.

17. McDonald, *Racial Dynamics*, 182–193; Austin Commercial Club, *Austin, Texas.*

18. Freeman, *East Austin*; Jackson, "East Austin," 63–72.

19. Tretter and Sounny-Slitine, *Austin Restricted.*

20. Freeman, *East Austin*; McDonald, *Racial Dynamics*, 106–107, 150. U.S. Bureau of the Census, *Twelfth Census of the United States* (Washington, DC: Government Printing Office, 1900).

21. Hamilton, *Social Survey of Austin*; Jackson, "East Austin," 63; McDonald, *Racial Dynamics*, 201.

22. Freeman, *East Austin*; McDonald, *Racial Dynamics*; Orum, *Power, Money, and the People*, 177–182.

23. The predominant newspaper for the African American community was the *Austin Herald.* For churches, see McDonald, *Racial Dynamics*, 118–122; Orum, *Power, Money, and the People*, 185–187.

24. U.S. Bureau of the Census, *Twelfth Census.*

25. McDonald, *Racial Dynamics*, 18–26; Foley, introduction to *White Scourge*; Edwin Waller, "Plan for the City of Austin" (map, 1839), Perry-Castaneda Library Map Collection, accessed January 3, 2017 https://www.tsl.texas.gov/arc/maps/images/map0926d.jpg; Hamilton, *Social Survey of Austin*; East Austin Chicano Economic Development Corporation, *Barrio Unido Neighborhood Plan.*

26. Melosi, *Garbage in the Cities*; Melosi, *Sanitary City*; Monkkonen, *America Becomes Urban*; Tarr, *The Search.*

27. Melosi, *Garbage in the Cities*, 8–14; Nelson, *Scientific Management*; Park, Burgess, and McKenzie, *The City*; Riis, *The Other Half*; Teaford, *Unheralded Triumph*; Tarr, *The Search.*

28. Boyer, *Urban Masses and Moral Order*; Teaford, *Unheralded Triumph*; McGerr, *Fierce Discontent*; Flanagan, *America Reformed.*

29. Melosi, *Garbage in the Cities*, 42–65; Schultz and McShane, "To Engineer the Metropolis"; Waring, *Modern Methods*; Waring, *Street Cleaning.*

30. Riis, *The Other Half*; Sinclair, *The Jungle*; Addams, *Democracy and Social Ethics*. For studies of urban poverty and health, see, e.g., *Charities and the Commons.*

31. Olmsted, "Public Parks," quote on page 304.

32. Smith, *The Plan of Chicago*; Howard, *Garden Cities of To-morrow.*

33. Wright, *Old South, New South*; Cobb, *Industrialization and Southern Society*; Carson, *The Coming of Industry.*

34. Gillette, *Civitas by Design*; Boyer, *Dreaming the Rational City.*

35. Martin V. Melosi's *Garbage in the Cities* remains the standard on the meaning of garbage in the Progressive Era. Also see Melosi, *Sanitary City*, 103–116.

36. Most works on environmental justice focus on events and issues that have occurred since the 1960s, and most likewise focus on industrial waste and pollution, although some deal with concerns of the nineteenth and early twentieth centuries.

One such work is David Naguib Pellow's *Garbage Wars*. See also Bullard, *Dumping in Dixie*; Szasz, *EcoPopulism*; Lerner, *Diamond*; Pulido, *Environmental and Economic Justice*; and Sze, *Noxious New York*.

37. Hamilton, *Social Survey of Austin*, 2–5, 14; Hamilton, *Social Survey of Austin #2*.

38. Hamilton, *Social Survey of Austin*, 4–7.

39. "Acute Suffering Result of Flood," *Austin American*, April 24, 1915; Burnett, *Flash Floods*, 22–27.

40. Hamilton, *Social Survey of Austin*, 9.

41. Even Hamilton was quick to judge the habits of poor blacks. He faulted them for the unsanitary conditions in the city's many slaughterhouses, their need to use discarded food and other garbage, and their personal hygiene habits. Hamilton, *Social Survey of Austin*, 32–41.

42. Hamilton, *Social Survey of Austin*, 9–13, quotations on 12.

43. Ibid., 32–45, 58–60.

44. Ibid., 54–55.

45. Ibid., 53–54.

46. Ibid.

47. Ibid., 57–60, quotation on 60.

48. Hamilton, *Social Survey of Austin #2*, 27–29.

49. Ibid., 36–37.

50. Tretter and Sounny-Slitine, *Austin Restricted*.

51. *Austin Statesman*, August 14, 1915; McDonald, *Racial Dynamics*, 102–112; Jackson, "East Austin," 61–64.

52. Austin Commercial Club, *Austin, Texas*.

Chapter Three

1. The relationship between urban planning and civic virtue was established during the Progressive Era. See Bachin, *Building the South Side*. Other studies of Progressive Era urban planning and ideology include Kirschner, *Paradox of Professionalism*, and Boyer, *Dreaming the Rational City*.

2. For Southern progressivism, see Kirby, *Darkness at the Dawning*; Link, *The Paradox*.

3. Tretter and Sounny-Slitine, *Austin Restricted*; also see Delaney, *Race, Place, and the Law*; Gotham, *Uneven Development*; Vose, *Caucasians Only*. For the relationship among racial and other types of property restrictions, see Fogelson, *Bourgeois Nightmares*.

4. "City Plan Supplement," *Sunday Morning News*, February 12, 1928.

5. McDonald, *Racial Dynamics*, 18–31; Kennedy, *Over Here*; Grossman, *Land of Hope*; Trotter, *The Great Migration*; Foley, *White Scourge*.

6. McDonald, *Racial Dynamics*, 18–31; Bureau of Research in the Social Sciences, *Population Mobility*. McDonald estimates that lax counts caused the Latino popula-

tion to be undercounted. He argues that Austin was likely 10 percent Latino by 1930.

7. See Spear, *Black Chicago*; Grossman, *Land of Hope*; Osofsky, *Harlem*; Tuttle, *Race Riot*; McWhirter, *Red Summer*.

8. McDonald, *Racial Dynamics*, 301–331, 251–259; Orum, *Power, Money, and the People*, 362; NAACP, *Mobbing of John R. Shillady*.

9. Brownie Bradford, "We Feed Them, Then Teach Them English," *Austin Statesman*, October 19, 1928, and J. W. Markham, "Mexican Youths Given Special Attention," *American Statesman*, November 13, 1936, both found in Folder 4, Box 5, Camacho Family Papers, AHC.

10. McDonald, *Racial Dynamics*, 31–54.

11. *Austin, Texas, Illustrated*.

12. James Pinkerton, "Struggle of Blacks Traced in Austin History," *Austin American-Statesman*, October 8, 1984, "AF—Subdivision—East Austin S6090," General Folder 2, Austin Files, AHC.

13. Jackson, "East Austin," 62–67; McDonald, *Racial Dynamics*, 106–109; Tretter and Sounny-Slitine, *Austin Restricted*.

14. Melosi, *Sanitary City*, 108–110; Tretter and Sounny-Slitine, *Austin Restricted*.

15. *Austin Daily Statesman*, December 29, 1908; Bridges, *Morning Glories*, 60–61.

16. Bridges, *Morning Glories*, 63; Long to Will P. Jones, June 13, 1927, Folder "City Manager Form of Government, 1922–1927," Box 18, Long Papers.

17. Austin Chamber of Commerce to George Roark, May 20, 1922, and "City Problems" (n.p., n.d.), both found in Folder "City Manager, 1920–1927," Box 18, Long Papers; Long, *Something Made Austin Grow*.

18. Long, "Austin History."

19. "'Unpolitical' and Uneconomic," *Austin Statesman*, May 24, 1922, and "City Manager Geo. Roark to Tell Austin Wednesday How Beaumont Plan Works" (n.p., n.d.), both found in Folder "City Manager, 1920–1927," Box 18, Long Papers.

20. *Austin American*, August 4, 1924, quoted in Bridges, *Morning Glories*, 113.

21. "City Manager Fight May Be Settled within Next Month," *Austin Statesman*, September 27, 1925, Folder "City Manager Form of Government, 1922–1927," Box 18, Long Papers; "Unless City Manager Is Given Solid Charter Backing Plan Is Meaningless, Says Geo. Roark" (n.p., n.d.), Folder "City Manager, 1920–1927," Box 18, Long Papers.

22. Long to Will P. Jones, July 13, 1927, Folder "City Manager Form of Government, 1922–1927," Box 18, Long Papers.

23. Long, "Austin History"; Long to P. W. McFadden, September 17, 1926, Folder "City Planning, 1927," Box 19, Long Papers.

24. Schmidt, *Back to Nature*.

25. In the 1910s Austin mayor and longtime civic advocate Alexander P. Wooldridge parked a strip of land eight blocks long between the northbound and southbound lanes of East Avenue, the street that both carried north-south through traffic in Austin and increasingly made up the boundary between Anglo and minority

Austin. He also dedicated one city block as a park at Guadalupe and Tenth Streets. See "Mr. Louis P. Head," April 13, 1928, Folder "City Planning, 1926, 1927, 1928," Box 19, Long Papers. For the traveler's comment, see "A Tourist to Chamber of Commerce," April 2, 1928, Folder "City Planning, 1926, 1927, 1928," Box 19, Long Papers.

26. "Austin, under City Manager for 4-1/2 Years, Tells Results," *Dallas Morning News*, February 22, 1931, Folder "City Plan, 1931," Box 20, Long Papers; Walter E. Long, "Barton Springs History" (n.d.), Folder "Barton Springs History," Box 1, Long Papers; "Suggestions Made to the Austin Chamber of Commerce for a City Plan," Folder "City Plan, Sept. 1926–Oct. 1927," Box 19, Long Papers.

27. "Everybody's Capital" (n.p., n.d.), Folder "City Planning, 1927," Box 19, Long Papers.

28. Austin Chamber of Commerce, *Progressive Austin* (1916).

29. *Austin, Texas, Illustrated*; "Everybody's Capital."

30. Lomax, "University of Texas."

31. American National Bank, "Citizen of No Mean City"; "Everybody's Capital"; Austin Chamber of Commerce to Tudie Thornton, November 12, 1926, Folder "City Planning, 1927," Box 19, Long Papers.

32. "Suggestions Made to the Austin Chamber of Commerce."

33. Ibid.

34. Sonia Hurt, in her history of zoning in the United States, argues that the practice reflects "spatial individualism" despite curtailing what individuals can do with their land. Zoning also allowed for segregation to become more rigid, despite the illegality of racial zoning from the outset. Hurt, *Zoned in the USA*; Hoyt, *Land Values in Chicago*; Weiss, *Community Builders*.

35. Houston, for example, never implemented land use zoning.

36. Long to McFadden, September 17, 1926; John E. Suratt to Long, November 5, 1926, Folder "City Plan, Sept. 1926–Oct. 1927," Box 19, Long Papers; City Planning Publishing Company to Long, April 29, 1927, Folder "City Planning, 1926, 1927, 1928," Box 19, Long Papers; "Simplifying Zoning," Folder "City Planning, 1915," Box 18, Long Papers.

37. Austin Commercial Club, *Austin, Texas*, 5.

38. Koch and Fowler, *City Plan*, 3.

39. Ibid., 20–33.

40. Ibid.

41. Ibid., 20–36, quotations on 28 and 25, respectively. For neighborhoods, Pemberton Heights, immediately west of Shoal Creek between Twenty-Fourth and Twenty-Ninth Streets, was begun in 1927 and grew quickly through the 1930s. Bryker Woods, directly to the north, was developed in the 1930s. According to Tretter and Sounny-Slitine, from the 1890s through the 1940s nearly every subdivision in West Austin north of Fifteenth Street and south of Thirty-Fifth Street had a racially restrictive covenant adopted. Tretter and Sounny-Slitine, *Austin Restricted*, 59.

42. Long to Mead, December 1, 1926, and E. A. Wood to Long, November 1, 1926, both found in Folder "City Planning, 1926–1928," Box 19, Long Papers.

43. Kessler Planning Associates, "Need City Planning Legislation" (memo, 1927), Folder "City Plan, Sept. 1926–Oct. 1927," Box 19, Long Papers.

44. Koch and Fowler, *City Plan*, 57.

45. Ibid.

46. There was already only one African American–designated high school in Austin, for example, built in 1913 on the Eastside.

47. Koch and Fowler, *City Plan*, 8–20, 53–57.

48. Ibid, 44.

49. Mexicans had been slowly leaving Little Mexico since the mid-1910s, likely induced by lower rents in East Austin and increasingly by animosity to their presence adjacent to downtown in a neighborhood that could easily rise in exchange value. Koch and Fowler make this point as well. See ibid., 44–45.

50. Ibid., 47.

51. Ibid., 4–6.

52. Ibid., 34–37. What little incentive might have existed to build houses on the Eastside was negated when the Home Owners' Loan Corporation refused to support mortgage loans anywhere on the Eastside in 1935.

53. Bridges, *Morning Glories*, 75; Martin, "Municipal Electorate." Roscoe C. Martin essentially states in his article that the city government was controlled by landowning taxpayers.

54. "Austin, under City Manager"; "Bond Issue 1928," Folder "Bond Issue 1928," Box 13, Long Papers.

55. Jackson, "East Austin," 66–68, 92–93.

56. The city average, of course, was the sum of all Austin, so the Westside was likely even lower. See McDonald, *Racial Dynamics*, 113–114; Jackson, "East Austin," 98–99.

57. Tretter and Sounny-Slitine, *Austin Restricted*, 17–19.

58. Jackson, "East Austin," 75, 84, 98, 105; Austin Housing Authority, *Report*; Johnson, *Tarnish on the Violet Crown*.

59. Koch and Fowler, *City Plan*, 47–50; Austin City Council, "City of Austin, Texas Use District Map," 1939, courtesy AHC.

60. Dale Carrington, "Mrs. Zamarripa Says East Austin 'Victim of Poor Zoning,'" *La Fuerza*, April 11, 1974.

61. *Austin, Texas, Illustrated*.

Chapter Four

1. See, e.g., Raymond Brooks, "National, State Leaders to Dedicate Wirtz Dam," *Austin American-Statesman*, June 15, 1952; Brooks, "Dam Ceremony Seen by Six Thousand," *Austin American*, June 16, 1952; Brooks, "Colorado Put to Work for the Public," *Austin American*, June 17, 1952. For the Shivers quotation, see Banks and Babcock, *Corralling the Colorado*, 180.

2. Brooks, "Colorado Put to Work."

3. Wirtz to Lyndon Johnson, August 9, 1944, LBJ Archives, Selected Names File, Alvin Wirtz, Box 37, LBJL.

4. Hays, *Beauty*, 13–39.

5. Cohen, *Consumer's Republic*; Potter, *People of Plenty*; Donaldson, *Abundance and Anxiety*; Lindsey, *Age of Abundance*.

6. Culver, *Frontier of Leisure*. Most works that tie urban development to regional hinterlands can be traced back to William Cronon's *Nature's Metropolis*. These include Walker, *Country in the City*; Catton, *National Park, City Playground*; Klingle, *Emerald City*; Vogel, "Defining One Pacific Northwest"; Rawson, *Eden on the Charles*. For suburbs, see Rome, *Bulldozer*; Kruse and Sugrue, introduction to *New Suburban History*; Sellers, *Crabgrass Crucible*.

7. Most studies of urban water issues in the American West focus on coastal cities: Los Angeles and, less so, San Francisco. See Righter, *Battle over Hetch Hetchy*; Carle, *California Dream*; Kahrl, *Water and Power*. One inland exception is Kupel, *Fuel for Growth*. For western urbanization more generally, two standard bearers are Findlay, *Magic Lands*, and Abbott, *Metropolitan Frontier*. *Magic Lands* argues that western cityscapes fit into patterns of development best seen in built spaces such as Disneyland, Stanford Research Park, Sun City, and the Seattle World's Fair.

8. Pritchard and Zeller, "Industrialization"; Pritchard, *Confluence*.

9. The idea that "landscape" constitutes both the physical characteristics of a place and the human imprint on it goes back to Carl O. Sauer, one of the first geographers to investigate landscape in the 1920s. See Sauer, "Morphology of the Landscape."

10. For other examples of dam systems allowing for more regional consciousness (in the Northwest), see White, *Organic Machine*, and Vogel, "Defining One Pacific Northwest."

11. Johnson, "Introduction of the Honorable Harold L. Ickes."

12. Graff, *Dallas Myth*; Feagin, *Free Enterprise City*.

13. White, *Organic Machine*, 59–64.

14. Of the two hundred largest counties in the United States, Travis County had the lowest percentage of its workforce employed in manufacturing well into the 1960s.

15. John Urry calls this the "tourist gaze," a manner of looking into the important places as well as the ordinary lives of people within those places. Urry, *The Tourist Gaze*.

16. Busch, "Building 'A City.'"

17. National Parks Service, LCRA, and Texas State Parks Board, *Highland Lakes of Texas*, available in Box 183, Papers of Lyndon Baines Johnson (hereafter LBJ Papers), House of Representatives, 1937–1949, LBJL. The survey was undertaken over approximately three years, from 1936 to 1939, as part of the Park, Parkway, and Recreational Area Study Act of June 1936. Maps, histories, and other materials were used as reference, and public officials and private citizens were consulted. See ibid., 15.

18. Mansfield Dam was built in two separate stages. The first stage was completed in 1939 and the second in 1941.

19. National Park Service, LCRA, and Texas State Parks Board, *Highland Lakes of Texas*, 2–4.

20. Ibid., 30. The study suggests that Texas's population will grow from 6 million in 1940 to 8.5 million residents by 1960.

21. LCRA, "Highland Lakes of Texas," 9–10.

22. LCRA, "Highland Lakes of Texas," Ibid., 31, 39, 51; quotation on 51.

23. John E. Babcock, "'Playground of the Nation' Is Dream for Central Texas' Lake Chain Area" (n.p., n.d.), Folder "AF—Highland Lakes HO800, " AHC.

24. Miles and Miles, *Consuming Cities*.

25. "Master Plan for Dollar Flow," *Austin in Action*, May 1965, 19.

26. Rugh, *Are We There Yet?*

27. Bill Brown, "Trends to Affect Austin," *Austin in Action* 1, no. 11 (April 1960). For income statistics, see Baughn, *Texas Commercial Banking*, 8.

28. Wrobel, *Promised Lands*.

29. Austin Chamber of Commerce, *Austin Area Lakes*; National Park Service, LCRA, and Texas State Parks Board, *Highland Lakes of Texas*.

30. Austin Chamber of Commerce, *Austin in a Nutshell* (1942).

31. Gottdiener and Lagopoulos, *The City and the Sign*.

32. Nye, *American Technological Sublime*. For Las Vegas and the Hoover Dam, see Rothman, "Selling the Meaning of Place," 550.

33. *LCRA News*, September 1947.

34. Austin Chamber of Commerce, *Lakes in the Austin Area*.

35. Shaffer, *See America First*.

36. Banks and Babcock, *Corralling the Colorado*, 143–144.

37. "Operation Waterlift," *Texas Parade*, February 1950; Banks and Babcock, *Corralling the Colorado*, 170–171. For decentralization, see Federal Civil Defense Administration, "Industrial Survival," November 8, 1956, 5, and Office of Civil and Defense Mobilization, "Ten Steps to Industrial Dispersal," April 1960, both found in Folder "Civil Defense—Industry," Box CDL3 2005-111/2, J. Neils Thompson Papers, BCAH; see also Wood, "Industry *Must* Prepare."

38. John P. McKenzie, "Highland Lakes Country a 'Vacation Wonderland,'" *Austin Statesman*, December 16, 1964, Folder "AF—Highland Lakes HO800 Misc.," Box "AF-HO800—General," AHC.

39. Ibid.; "'Vacation in Your Own Back Yard' Really Applies to Central Texas," *Austin American-Statesman*, August 27, 1961; "Lakes Rate among Top Attractions," *Austin American*, May 15, 1956; "Highland Lakes: Centex Mecca," *Austin American*, August 12, 1964.

40. Glaston to Miller, January 8, 1958, Folder "Jan.–Feb., 1958," Box "FPF.10B; Austin Mayors; Miller, Robert Thomas; Correspondence, Jan. 1958–Dec. 1960," AHC.

41. *Top Spot for Fun* series, LCRA Corporate Archives, Austin.

42. "L.C.R.A. a Boon in -," *Austin in Action* 3, no. 2 (July 1961).

43. Brooks, "Colorado Put to Work." For lake resorts, see Highland Lakes Development Association, "Texas' Top Vacation Spot," *Top Spot for Fun* (n.d.), Folder "Lakes and Recreation," LCRA Corporate Archives.

44. Similar suburban developments are described in Culver, *Frontier of Leisure*, 213–217.

45. "The Lure of the Lakes," *Austin in Action* 3, no. 2 (July 1961): 7.

46. "Barnes and Jones' Highland Lakes Homesites," *Austin in Action* 6, no. 1 (April 1964).

47. "Living in Lago Vista," *Top Spot for Fun—and Industry*, Spring 1959; Barnes and Jones, "Highland Splendor at Your Doorstep" (brochure, n.d.), courtesy LCRA Corporate Archives.

48. Dwight L. Williams, "Planning of Research and Development Work" (pamphlet, n.d.), Folder "Planning, 1944," Box 21, Long Papers.

49. "A Golden Hue to Highland Lake Blue," *Austin in Action* 1, no. 2 (July 1959): 8–11; *Austin in Action* 2, no. 4 (September 1960). For the increase in boat sales, see Glaston to Miller, January 8, 1958; "Commodore, the Amazing Paddleboat That Seems to Have Churned out of Another Era," *Austin in Action* 3, no. 2 (July 1961).

50. "Lure of the Lakes," 5.

51. Rothman, "Stumbling towards the Millennium," 140; Rothman, "Selling the Meaning of Place," 525.

52. "Aqua Festival Had Civic Start," *Austin American*, August 16, 1971.

53. "Delegates Enjoy Aqua Festival Bonus," *Texas Public Employee*, August 1973, Subject File "Aqua Festival," AHC.

54. Vance E. Murray, "1965 Austin Aqua Festival," *Municipal Perspective* (1965), and "Summary" (1964), both found in Subject File "Aqua Festival," AHC.

55. "First Annual Austin Aqua Festival" (brochure, 1962); "The Seventh Annual Austin Aqua Festival," *Holiday Inn Magazine*," July 1968, Subject File "Aqua Festival," AHC.

56. "First Annual Austin Aqua Festival"; "Municipal Perspective" (pamphlet, July 1966); Bob Inderman, "Rio Noche Night Water Parade Thrills 150,000," *Austin American-Statesman*, August 10, 1968; Ginger Banks, "Thousands Witness Aqua Fest Parade," *Austin American-Statesman*, August 2, 1969.

57. "Municipal Perspective"; "Aqua Band Contest on Sunday," *Austin American-Statesman*, July 27, 1972; and "Aqua Band Contest Altered," *Austin American-Statesman*, July 20, 1972, all found in Subject File "Aqua Festival," AHC.

58. "1969 Austin Aqua Festival" (brochure), Subject File "Aqua Festival," AHC; Crispin James, "Bergstrom's Aerofest Draws Eighty Thousand Viewers," *Austin American-Statesman*, August 8, 1971; Rick Timmons, "Color Austin Aqua" (n.p., July 26, 1972), Subject File "Aqua Festival," AHC. For resistance to integration, see Southern Union Gas Company, "A Study of the Housing Market" (report, 1971), Vertical File "Austin, Texas—Housing and Real Estate (Travis County)," BCAH; and Orum, *Power, Money, and the People*, particularly 249–266.

59. "Plans for Admirals Ball Made at Party," *Austin Statesman*, June 29, 1966, Subject File "Aqua Festival," AHC. The ball's program in 1972 had an Indian theme revolving around images of the god Krishna. See "The Admirals Club" (program, 1972), Subject File "Aqua Festival," AHC.

60. "For Payrolls, the Big Push," *Austin in Action*, April 1966; "Centex Soars in State's Biggest Jump," *Austin Statesman*, July 25, 1968.

61. "For New Business, New Residents, What Do We Have to Offer?," *Austin in Action*, November 1964.

Chapter Five

1. Austin Area Economic Development Foundation (AAEDF), "Austin and Industry," *Austin and Industry* 4, no. 1 (April 1950); Wood, "Outline." For decentralization, see Wood, "Industry *Must* Prepare"; Office of Civil and Defense Mobilization, "Ten Steps to Industrial Dispersal." Fear of economic collapse was widespread after World War II. The Austin Chamber of Commerce received a document on the topic from the Chamber of Commerce USA, Hahn, *Deficit Spending and Free Enterprise*.

2. The dominant narrative depicts the postwar decades as an era of significant urban decline. Suburbanization, capital and white flight, and deindustrialization left cities with a depleted tax base, strained budgets, and poor services by the 1970s and 1980s. See, e.g., Sugrue, *Urban Crisis*; McKee, *The Problem of Jobs*; Sides, *L.A. City Limits*; Bluestone and Harrison, *Deindustrialization of America*; Cowie and Heathcott, *Beyond the Ruins*.

3. O'Mara, *Cities of Knowledge*; Geiger, *Research and Relevant Knowledge*; Leslie, *Cold War*; Hall and Markusen, *Silicon Landscapes*; Wang, *American Science*.

4. While urban historians have tended to focus on the decline of large industrial cities in the postwar era, scholars have also documented the regional shift to the South and West during that period. It is generally known as the Sunbelt shift and is attributed to a number of factors including federal investments, cheaper production costs, state and local subsidies, better weather, and federal policies that encouraged decentralization. Nash, *Federal Landscape*; Mohl, *Searching for the Sunbelt*; Nickerson and Dochuk, *Sunbelt Rising*; Abbott, *New Urban America*; Newman, *The American South*; Schulman, *Cotton Belt to Sunbelt*; Vance, *All These People*.

5. Wood, "Industry *Must* Prepare"; Federal Civil Defense Administration, "Industrial Survival," 5; Office of Civil and Defense Mobilization, "Ten Steps to Industrial Dispersal"; Howard, *Hidden Welfare State*; O'Mara, *Cities of Knowledge*.

6. Scribner, *Renewing Birmingham*; Silver, *Twentieth-Century Richmond*; Bayor, *Twentieth-Century Atlanta*.

7. The chamber of commerce and other boosters were acutely aware that marketing the city's unique amenities, or what one publication called the "merits of Austin as a place," was a viable strategy for growth and one preferable to subsidies for industry. See AAEDF, "The Foundation's Principal Objective Is the Planned and

Diversified Economic Expansion of Austin, Consistent with the High Character of the City," *Austin and Industry* 3, no. 1 (May 1950).

8. For postwar suburbanization, see Jackson, *Crabgrass Frontier*, 231–245; Hayden, *Building Suburbia*; Cohen, "Town Center."

9. Walter E. Long to Frank M. Soule, August 13, 1945, Folder "Industrial Inquiries, 1944–1945," Box 36, Long Papers; "Inventory of Miscellaneous Items," Folder "1944 Post War City Planning 'Inventory,'" Box 21, Long Papers; Long to Herman Hettinger, November 16, 1944, Folder "Planning, 1944," Box 21, Long Papers.

10. "Recent Industrial Advance in Texas," *Texas Business Review* 18, no. 12 (January 1945).

11. Austin Chamber of Commerce to Lyndon Johnson, November 11, 1941; Ray E. Lee to Johnson, September 30, 1941; Porter A. Whaley to Johnson, October 9, 1941; Johnson to Taylor Glass, October 13, 1941; "Sept. 27, 1941"; Walter A. Dickerson to Johnson, October 11, 1941, all found in Folder "1941 Defense Austin Magnesium Plant," Box 213, LBJ Papers, House of Representatives, 1937–1949, LBJL.

12. "Austin—Tomorrow," speech at Austin Chamber of Commerce annual meeting, 1947, Folder "Austin—Tomorrow," Box 12, Long Papers.

13. Moore, "Planned or Unplanned Growth."

14. Dwight L. Williams, "Planning of Research and Development Work" (pamphlet, n.d.), Folder "Planning, 1944," Box 21, Long Papers; AAEDF, "Austin and Industry." For histories of the rise of scientists during the Cold War, see Wang, *American Science*; Kevles, *The Physicists*.

15. Austin Planning Commission, *The Austin Master Plan*, 11; Wood, "Outline."

16. AAEDF, "The Foundation's Principal Objectives"; Minutes of the Industrial Bureau Meeting, August 22, 1944, Folder "Industrial Bureau, 1944 (Jan.–June)," Box 38, Long Papers.

17. Wood, "Outline," 2–8.

18. For its first meeting in April 1948 the foundation drew up a short addendum to Wood's "Outline" that listed the foundation's chief aims and responsibilities. AAEDF, *Annual Report, 1949*; AAEDF, *Austin and Industry* 1, no. 3 (January 20, 1949).

19. *Austin and Industry* 1, no. 1 (October 28, 1948); *Austin and Industry* 1, no. 2 (December 5, 1948); *Austin and Industry* 2, no. 6 (February 19, 1950).

20. *Austin and Industry* 3, no. 1 (May 1950): 4–8; "Payrolls without Smokestacks," *Texas Parade*, November 1950, was circulated by the AAEDF.

21. Federal Reserve Bank of Dallas, "Austin," *Monthly Business Review*, December 1, 1951, Vertical File "Austin, Texas—Business," BCAH.

22. The Bureau of Business Research and the Bureau of Economic Geology were foremost among University of Texas research entities, and most of their work analyzed business prospects in Texas industries, especially oil. A. B. Cox to J. W. Calhoun, April 5, 1938, Folder "Industrial and Commercial Research Council, 1937–38," Box VF8-B.b; Cox to H. Y. Benedict, January 10, 1933, Folder "Bureau of Business Research, 1932–33," Box VF7-A.b; Waldo B. Little to F. A. Buechel, May 25, 1939, Folder "Industrial and Commercial Research Council,

1937–38," Box VF8-B.b, all found in University of Texas President's Office Records, 1907–1968, BCAH.

23. Gaines to John H. Bickett Jr., May 18, 1943; W. A. Cunningham to J. A. Burdine, November 10, 1943; W. R. Woolrich to Homer Rainey, September 27, 1941; Texas Research Corporation Minutes, February 28, 1942; C. E. MacQuigg and W. R. Woolrich, "Memorandum on Engineering Experiment Stations Available for Government Research," June 5, 1942, all found in Folder "Texas Research Corporation, 1943–44," Box VF8-B.b, University of Texas President's Office Records, 1907–1968, BCAH.

24. Called the War Research Laboratory from its inception in 1942 until 1945, when its name was changed to Military Physics Research Lab.

25. Austin Chamber of Commerce, *Austin Invites You*; Carl L. Covington to Homer Rainey, November 12, 1942, Folder "War Research Laboratory, 1942–44," Box VF 8-B.b, University of Texas President's Office Records, 1907–1968, BCAH.

26. Tretter, *Shadows*, chap. 2.

27. T. S. Painter to C. L. Andrews, January 28, 1946, Folder "War Projects—Austin Magnesium Plant #2, Corr., January, 1946," Box 220, LBJ Papers, House of Representatives, 1937–1949, LBJL.

28. Granberry to Johnson, November 8, 1945, Folder "FHA—University of Texas," Box 274, LBJ Papers, House of Representatives, 1937–1949, LBJL; Guiton Morgan to Marshall W. Amos, March 29, 1946, Folder "War Projects—Austin Magnesium Plant #2, Corr., January, 1946," Box 220, LBJ Papers, House of Representatives, 1937–1949, LBJL.

29. A. O. Greist to Sam H. Husbands, October 22, 1945, and Johnson to Lucius C. Andrews, August 22, 1945, both found in Folder "War Projects—Austin Magnesium Plant," Box 220, LBJ Papers, House of Representatives, 1937–1949, LBJL.

30. For clarity I will refer to the facility as the Balcones Research Center (BRC) even though it was called the Off-Campus Research Center and the Texas Memorial Research Laboratories before being named the BRC in 1954 and is today named after longtime U.S. congressman J. J. Pickle.

31. Raymond Brooks, "WAA Gives Magnesium Plant to Austin," *Austin American*, April 23, 1949. Thompson to Painter, March 28, 1949, and "University Assured Magnesium Plant" (n.p., April 1949), both found in Folder "Off-Campus Research Center, 1948–49," Box VF28-C.a, University of Texas President's Office Records, 1907–1968, BCAH.

32. Leslie, *Cold War*.

33. J. Neils Thompson, "Proposed Development of the Memorial Research Laboratories at the Austin Magnesium Plant" (unpublished paper, 1949), Vertical File "Balcones Research Center, University of Texas," BCAH.

34. J. Neils Thompson, "Integrating Sponsored Research into the University Research Program" (unpublished paper, 1949); "UT Prof Fears Trend to Mediocrity," *Austin American*, October 22, 1950.

35. Thompson, "Integrating Sponsored Research," 5–13.

36. Ibid., 10–14.

37. This last point regarding a return to craft production and a revaluation of labor is taken from Piore and Sabel, *Second Industrial Divide*.

38. "Director of Research at UT Says Center Ready for War Switch," *Austin American-Statesman*, August 13, 1950; "Texas U Device Aiding Defense," *Dallas News*, July 23, 1950; "UT Research Aids Defense," *Daily Texan*, February 18, 1951; "Off Campus UT Research Pursues War Trend," *Austin American-Statesman*, July 29, 1951; Thompson to J. C. Dolley, June 27, 1951, Folder "Off-Campus Research Center, 1951–52," Box VF28-C.b, University of Texas President's Office Records, 1907–1968, BCAH.

39. W. George Parks to Thompson, May 29, 1951, Folder "Off-Campus Research Center, 1950–51"; Thompson to Logan Wilson, January 18, 1954, Folder "Balcones Research Center, 1953–54"; J. Neils Thompson, "Administrations and Functions of the Balcones Research Center" (memorandum, January 1954), Folder "Balcones Research Center, 1953–54," all found in Box VF28-C.b, University of Texas President's Office Records, 1907–1968, BCAH.

40. Thompson to Logan Wilson, March 31, 1953, Folder "Balcones Research Center, 1952–53"; "Institute for Advanced Engineering Created to Help Old Grads Catch Up," *Engineering-Science News* 2, no. 2 (March–April 1954), Folder "Balcones Research Center, 1953–54"; Thompson to Judge James P. Hart, January 28, 1953, Folder "Balcones Research Center, 1952–53," all found in Box VF28-C.b, University of Texas President's Office Records, 1907–1968, BCAH.

41. Norman Hackerman to Wilson Stone et al., August 22, 1966, Folder "I—Special Programs OGSR [Office of Graduate Studies and Research] Projects," Box VF33-B.a, University of Texas President's Office Records, 1907–1968, BCAH.

42. J. Neils Thompson, "U.S. Department of Commerce State Science and Technology Conference" (unpublished report, 1964), Folder "Balcones Research Center, 1960–68," Box VF38-D.a, University of Texas President's Office Records, 1907–1968, BCAH.

43. Ibid.

44. Ibid. For developing social relationships based around technological work, see Turner, *From Counterculture to Cyberculture*.

45. "Growing Research Activity 'Catches on' Here," *Austin in Action* 3, no. 5 (October 1961): 8.

46. Originally called Associated Consultants and Engineers before changing to Texas Research Associates in 1960 and Tracor in 1962.

47. Austin Chamber of Commerce, *Austin Invites You*.

48. TRACOR, "Texas Research Associates," pamphlet (Austin, n.d.), Vertical File "TRACOR," BCAH. The pamphlet is most likely from around 1959 or 1960. "The New Breed: Richard Lane," *Texas Business and Industry*, August 1969, Vertical File "TRACOR," BCAH.

49. "The New Breed"; "TRACOR: Annual Report, 1965," Vertical File "TRACOR," BCAH.

50. "TRACOR . . . Diversified Brain Trust," *Austin in Action*, November 1964; "Research Activity 'Catches On.'"

51. Richard N. Lane, "How Are Chances? For a Booming Scientific Complex Here?" *Austin in Action*, April 1965.

52. "TRACOR: Annual Report, 1964"; "TRACOR: Annual Report, 1965"; "TRACOR: Annual Report, 1967"; Scott Armstrong, "McBee of TRACOR Riding High in High-Tech Saddle," *Christian Science Monitor*, March 31, 1982, all found in Vertical File "TRACOR," BCAH.

53. Dale B. Elmore to Lester Palmer, November 29, 1966, Folder "Park Board," Box 1, Roberta Crenshaw Papers, AHC. Orum, *Power, Money, and the People*, 245–248.

54. For the comparison to Silicon Valley regarding Department of Defense expenditures, see Lécuyer, *Making Silicon Valley*, 6–8. For TRACOR history, see "The New Breed."

55. See, e.g., Vic Mathias, "Should Our Growth Continue?," *Austin in Action* 2, no. 8 (January 1961): 14–17; Austin Department of Planning, "Basic Data about Austin and Travis County" (report, 1955), Vertical File "Austin, Texas—City Planning (I)," BCAH.

56. John Ferguson, "One Big Damn Subdivision," *Texas Observer*, November 16, 1973; Larry BeSaw, "Apartments, Autos Expand Role in Austin's Lifestyle," *Austin American*, September 6, 1973; Larry BeSaw, "Time Alters Shopping Patterns," *Austin American*, September 7, 1973; "Building Permits Breaking Records," *Austin American*, September 6, 1967, and "Austin Building Sixteenth in Nation," *Austin Home Builder*, March 7, 1968, both found in Folder "General Texas Austin—1968," Box 95-112-203, Papers of J. J. Pickle, BCAH.

57. "Centex Economy Soars in State's Biggest Jump," *Austin Statesman*, July 25, 1968, Folder "General Texas Austin—1968," Box 95-112-203, Pickle Papers.

58. "Goal Area: Economics," "Goal Area: Population," "Goal Area: Transportation," "Goal Area: Land Use." All figures come from data gathered by the Austin Planning Department in the early 1970s as part of the initial stages of redeveloping the Austin City Plan (the latest version of the plan was adopted in 1961). Austin Planning Department, n.d., Folder "AF City Planning C4170 (3) Austin Tomorrow (before 1974)," Vertical File "Austin Tomorrow," AHC.

59. Vic Mathias, "Local Population More for Saying than Believing," in "Austin Progress Report," special edition, *Austin Statesman*, February 28, 1969, Vertical File "Austin, TX—Business (1—General)," BCAH.

60. "Employment in Travis County Sets New Record," in "Austin Progress Report," special edition, *Austin Statesman*, February 28, 1969, Vertical File "Austin, TX—Business (1—General)," BCAH.

61. "Austin's Industry Puts Capital into High Gear," in "Austin Progress Report," special edition, *Austin Statesman*, February 28, 1969, Vertical File "Austin, TX—Business (1—General)," BCAH; Chris Whitcraft, "Austin Economy Hits New Heights," *Austin American-Statesman*, October 19, 1968, Folder "General Texas Austin—1968," Box 95-112-203, Pickle Papers.

62. “Budget, like City, Continues to Grow,” in “Austin Progress Report,” special edition, *Austin Statesman*, February 28, 1969, Vertical File “Austin, TX—Business (1—General),” BCAH.

63. “Looking Forward to Another Record Breaking Year in 1969!,” in “Austin Progress Report,” special edition, *Austin Statesman*, February 28, 1969, Vertical File “Austin, TX—Business (1—General),” BCAH; Economic Development Fund of the Austin Chamber of Commerce, “Investors Dividend Report” (n.d.), Vertical File “Austin, TX—Industries,” BCAH.

64. “Austin's Industry Puts Capital into High Gear,” in “Austin Progress Report,” special edition, *Austin Statesman*, February 28, 1969, Vertical File “Austin, TX—Business (1—General),” BCAH; “Building Permits Breaking Records,” *Austin American*, September 6, 1967, Folder “General Texas Austin—1968,” Box 95-112-203, Pickle Papers; “Austin Building Sixteenth in Nation.”

65. “Bigger Government Yields Big Austin,” special edition, in “Austin Progress Report,” special report, *Austin Statesman*, February 28, 1969, Vertical File “Austin, TX—Business (1—General),” BCAH.

66. Southern Union Gas Company, “A Study of the Housing Market” (report, 1971), Vertical File “Austin, Texas—Housing and Real Estate (Travis County),” BCAH.

67. Dorothy Blodgett, “Downhill for Downtown?,” *Austin in Action* 2, no. 2 (July 1960): 8. See also Orum, *Power, Money, and the People*, 241–244. Developer Joe Crow initially developed all three malls because he thought that there simply were not enough stores in Austin.

68. Arbingast, Horton, and Ryan, “Austin, Texas R&D Nucleus”; J. Neils Thompson, “Report of Industrial Parks Tasks Committee to the Economic Development Council” (n.d.), Folder “II(b) Austin Chamber of Commerce, 1960–1962,” Box VF34-E.b, University of Texas President's Office Records, 1907–1968, BCAH; Bridges, *Morning Glories*, 152–159.

69. Quoted in “Research Activity ‘Catches On,’ ” 8.

70. Ibid.

71. Arbingast, Horton, and Ryan, “Austin, Texas R&D Nucleus.”

72. Ibid.

Chapter Six

1. “The Complete Text of Austin's Fair Housing Ordinance and the Federal Fair Housing Act,” *Austin Statesman*, May 24,1968, Vertical File “TX Cities—Austin—Housing,” BCAH. Also see Orum, *Power, Money, and the People*, 261–266.

2. Puett to Pickle, June 4, 1968, Folder “General Texas Austin—1968,” Box 95-112-203, Pickle Papers.

3. Orum, *Power, Money, and the People*, 264–266; L. C. Todd to J. J. Pickle, April 30, 1968, Folder “General Texas Austin—1968,” Box 95-112-203, Pickle Papers.

4. Bayor, *Twentieth-Century Atlanta*; Mohl, "The Second Ghetto"; Connolly, "Sunbelt Civil Rights"; Scribner, *Renewing Birmingham*; Silver, *Twentieth-Century Richmond*; Gordon, *Mapping Decline*.

5. There was one incident, in 1919, where a fight occurred in which an NAACP representative, John R. Shillady, died as a result of his injuries, but overt racially motivated violence was relatively rare in Austin. Orum, *Power, Money, and the People*, 203.

6. See, e.g., former Austin City Council member Charles Urdy in an interview with KLRU, accessed August 31, 2011, http://www.klru.org/austinnow/archives/gentrification/index.php. Neighborhood restaurateur Ben Wash, interview with author, March 30, 2007, transcribed as part of the Southern Foodways Alliance, "Texas Barbecue Trail," accessed December 2, 2010, http://www.southernbbqtrail.com/bens_longbranch.shtml. Orum, *Power, Money, and the People*, 184–186.

7. Quoted in James Pinkerton, "Urban Renewal," *Austin American-Statesman*, 1984, "AF—Subdivisions—East Austin S6090," General Folder 2, Austin Files, AHC.

8. Orum, *Power, Money, and the People*, 192–194.

9. Ibid., 132–135.

10. R. H. Davidson to Tom Miller, October 4, 1955, Folder "FPF.10; Miller, R. T.; Correspondence Oct.–Dec. 1955," Box "Miller, May 1955–March 1956," Robert Thomas Miller Papers, AHC; Otis L. Bush to Mayor Tom Miller, August 3, 1957, Folder "FPF.10B; Miller, R. T.; Correspondence June–Sept. 1957," Box "Miller, Apr. 1956–Dec. 1957," Miller Papers.

11. Arthur DeWitty to Tom Miller, Mayor, December 18, 1956, Folder "FPF.10B; Miller, R. T.; Correspondence Oct.–Dec. 1956," Box "Miller, Apr. 1956–Dec. 1957," Miller Papers.

12. Connie A. Miller to Congressman J. J. Pickle, January 17, 1966, and February 26, 1966, Folder "General Texas—Dept. Public Welfare," Box 95-112-104, Pickle Papers; "New Patterns of Traffic Flow," *Austin in Action* 3, no. 7 (December 1961).

13. Ethel Limuel to J. J. Pickle, March 19, 1966, Folder "General—Texas—Austin," Box 95-112-104, Pickle Papers.

14. Morgan V. Smith Jr. to John G. Tower, July 29, 1968, Folder "General Texas Austin—1968," Box 95-112-203, Pickle Papers.

15. T. W. Fourqurean to Ben White, October 30, 1956, Folder "FPF.10B; Miller, R. T.; Correspondence Oct.–Dec. 1956," Box "Miller, Apr. 1956–Dec. 1957," Miller Papers.

16. Orum, *Power, Money, and the People*, 190.

17. While African Americans were institutionally segregated, Latinos (almost all Mexican Americans at the time) were legally classified as "white." But de facto segregation remained strong between Latinos and whites.

18. Austin Housing Authority, "From Slums through Public Housing."

19. Anthony R. Orum, in *Power, Money, and the People* (254), relates a story in which a white adult sexually assaulted a twelve-year-old African American girl, and the police detained her, sent her away for evaluation, and then deemed her unfit to testify against her attacker.

20. City Department of Urban Renewal, "Slum Districts" (pamphlet, n.d.), Vertical File "Austin, Texas—Industry (Cities)," BCAH. This pamphlet appears to have been published in 1958. It draws on census data from 1950 but uses neighborhood boundaries from 1955. Austin Planning Commission, *The Austin Master Plan* (report, 1958), 26.

21. Carol Guistine, "The 'Other Side' Is Beginning to Cross the Tracks While Staying Right Where It Is," monthly supplement, *Daily Texan*, April 1963, Folder "(1) General, 1959–1964," Subject File "AF—Urban Renewal Project and Program U5000," AHC; Feagin and Jackson, "Delivery of Services." For the decline in home ownership, see Orum, *Power, Money, and the People*, 187–188. Orum claims that East Austin had a 55 percent home ownership rate in 1949, far higher than the Guistine piece reported in 1963.

22. Tretter, *Shadows of a Sunbelt City.*

23. Orum, *Power, Money, and the People*, 170–171.

24. Elsworth Mayer, "Five Areas Up for Renewal," *Austin in Action* 7, no. 8 (March 1966); "Rx for Cities: Urban Renewal" (pamphlet, n.d.), Vertical File "Austin, Texas—Industry (Cities)," BCAH. This pamphlet, almost certainly from 1958 or 1959, contains pro-urban renewal rhetoric—"A vote *for* urban renewal is a vote *against* slums and blight"—as well as statistics about urban renewal. The vote to create the urban renewal agency was extremely close.

25. Austin City Planning Commission, "The Austin Plan," 3–4. Although the planning commission put together the report, much of the information and analysis was produced by Pacific Planning and Research.

26. Ibid., 6; Fred Day, "New Industry Needn't Impair City's Beauty and Liveability," *Austin in Action* 2, no. 2 (July 1960).

27. Austin City Planning Commission, "The Austin Plan," 11; "Summary of Texas Urban Renewal Law—House Bill 70," Vertical File "Austin, Texas—City Planning (I)," BCAH; Orum, *Power, Money, and the People*, 171.

28. Amy Smith, "One More Detour on Holly Street," *Austin Chronicle*, January 21, 2011, http://www.austinchronicle.com/news/2011-01-21/one-more-detour-on-holly-street.

29. See, e.g., photo of 1188 Graham Street, Folder 21-2, Box 21, Appraisal Associates of Austin Records, AHC.

30. Southern Union Gas Company, "A Study of the Housing Market" (report, 1971), Vertical File "Austin, Texas—Housing and Real Estate (Travis County)," BCAH.

31. Alice Wightman, "Council Watch—Getting Screwed, All around Town: Eastside and Westside," *Austin Chronicle*, July 31, 1992, Folder "Tank Farm Activities, 1992–1993," Box 74, People in Defense of Earth and Her Resources (PODER) Records, AHC; City of Austin Electric Utility Department, "Holly Power Plant Closure Options: RFPs for Replacement Resources" (report, June 21, 1994), Folder "HPP Closure Options," Box 46, PODER Records. The particulars of environmental racism in the city are discussed in detail in chapter 9.

32. Feagin and Jackson, "Delivery of Services."

33. W. W. Collins to J. J. Pickle," June 7, 1966, Folder "Urban Renewal Administration—Department of Housing and Urban Development," Box 95-112-66, Pickle Papers.

34. "Summary of Texas Urban Renewal Law House Bill 70," Vertical File "Austin, Texas—City Planning (I)," BCAH.

35. Austin Urban Renewal Agency, "The Workable Program for Community Development," (pamphlet, 1964), Folder "(1) General, 1959–1964," Subject File "AF—Urban Renewal Project and Program U5000," AHC.

36. Mayer, "Up for Renewal"; Diana Dworin, "Austin Housing Complex Is Neglected, Decaying," *Austin American-Statesman*, November 17, 1996, Vertical File "Austin, Texas—Housing and Real Estate (Travis County)," BCAH.

37. Pinkerton, "Urban Removal."

38. Pickle to Ella Louise Davis, September 9, 1968, Folder "General Texas Austin—City of Austin Urban Renewal Agency," Box 95-112-203, Pickle Papers.

39. Todd to Pickle, April 30, 1968.

40. Leon M. Lurie to Scott, March 21, 1968, and Scott to J. J. Pickle, March 28, 1968, both found in Folder "General Texas Austin—City of Austin Urban Renewal Agency," Box 95-112-203, Pickle Papers.

41. *HOC Newsletter*, November 14, 1969, Folder "(2) General, 1965–," Subject File "AF—Urban Renewal Project and Program U5000," AHC.

42. In 1968 Austin was the sixty-seventh-largest metropolitan area in the United States but ranked sixteenth in the value of construction permits. "Austin's Industry Puts Capital into High Gear"; "Building Permits Breaking Records"; "Austin Building Sixteenth in Nation."

43. Austin Human Relations Commission, *Housing Patterns*, 127.

44. Southern Union Gas Company, "The Housing Market"; Austin Urban Renewal Agency, "Annual Report, 1969–70" (1970), Folder "(1) General 1965–," Subject File "AF—Urban Renewal Project and Program U5000," AHC.

45. Austin Human Relations Commission, *Housing Patterns*, 157, 170.

46. East Austin Chicano Economic Development Corporation, *Barrio Unido Neighborhood Plan*.

47. Southern Union Gas Company, "The Housing Market."

48. For household income spent on housing and for single-family homes built, see Austin Human Relations Commission, *Housing Patterns*, 65, 124–127; quotation on 127.

49. Southern Union Gas Company, "The Housing Market."

50. David Harvey, "The New Imperialism."

51. City of Austin, *Block Grant Application*, 32.

52. Federal Reserve Bank of San Francisco, "Austin, Texas: The East Austin Neighborhood" (n.d.), accessed October 4, 2010, http://www.frbsf.org/cpreport/docs/austin_tx.pdf. Statistics appear to be taken from census data. In 1988, for example, *Inc.* magazine named Austin the best city for business in the United States. Austin Human Relations Commission, *Housing Patterns*, 90.

53. City of Austin, *Block Grant Application*, 29. The Brackenridge and University East tracts were parts of areas with 9.6 percent and 8.4 percent substandard housing, respectively. They had 14.1 percent and 42.6 percent minority populations, respectively, as well, indicating a small migration of more affluent minorities out of the traditional minority neighborhoods by 1977.

54. Austin Human Relations Commission, *Housing Patterns*, 102–113.

55. Ibid., 180.

56. Even in relatively integrated neighborhoods, like the area around East First and Chicon Streets, custom dictated that schools remain rigidly segregated between whites and Latinos. In that neighborhood, whites attended Metz Elementary and Hispanics attended Zavala well into the 1960s.

57. *United States of America v. Texas Education Agency*, 467 F.2d 848 (1972), accessed September 4, 2011, http://law.justia.com/cases/federal/district-courts/FSupp/347/1138/1404362/; *Blackshear Residents Organization v. Housing Authority of the City of Austin*, 347 F. Supp. 1138 (1972), accessed September 4, 2011, http://174.123.24.242/leagle/xmlResult.aspx?xmldoc=19721485347FSupp1138_11300.xml&docbase=CSLWAR1-1950-1985.

58. *United States of America v. Texas Education Agency.*

59. For busing in Boston, see Formisano, *Boston against Busing.*

60. All forthcoming quotations are from an interview of Lynch conducted in 1973 by Charles Edwin Davis and given in his dissertation. See Davis, "United States v. Texas Education Agency." For Austin busing, also see Davidson, *Race and Class*, 236; Cunningham, *Cowboy Conservatism*, 213–215.

61. Davis, "United States v. Texas Education Agency," 187.

62. Ibid., 189.

63. Ibid. Will Davis was the head of AISD.

64. Ibid.

65. Elvia R. Arriola, "Austin Schools Project" (1998), accessed October 20, 2011, http://www.womenontheborder.org/AUSTINschools.htm.

66. Ibid.

67. East Austin Chicano Economic Development Corporation, *Barrio Unido Neighborhood Plan.*

68. Rossinow, *Politics of Authenticity.*

69. United States Census Bureau, "Travis County Census Tracts" (May 1980).

70. City of Austin, "Block Grant Application," 35.

71. Federal Reserve Bank of San Francisco, "Austin, Texas." The figure for the rest of the city, 20.5 percent, most likely includes university students.

72. Southern Union Gas Company, "Market Analysis: A Study of the Housing Market" (report, 1971), Vertical File "Austin, Texas—Housing and Real Estate (Travis County)," BCAH, 3–4; Northcutt, "Austin." The statistics are taken from Myers, *Quality of Life.*

73. "For New Business, New Residents, What Do We Have to Offer?," *Austin in Action*, November 1964.

Chapter Seven

1. For a description of the incident, see Rice, *Distant Publics*, 1–3. For the quotation, see "Copy to Jim Berry, City Editor" (from Mrs. Fagan Dickson), October 23, 1969, Folder "Waller Creek," Box 1, Crenshaw Papers. Ed Smith, "'Ax Erwin' Rally Attended by Three Hundred" (n.p.), November 5, 1969, Folder "Waller Creek," Box 1, Crenshaw Papers.

2. Two works that demonstrate how environmentalism flourished in the United States after World War II largely because of suburban grown and metropolitan decentralization are Rome, *Bulldozer*, and Sellers, *Crabgrass Crucible*. See also Hays, *Beauty*, 1–99. Early scholars looked to conservation-minded organizations, such as the Sierra Club, as the core of the movement. See Nash, *Wilderness*.

3. See, e.g., Packard, *Waste Makers*; Galbraith, *Affluent Society*; Carson, *Silent Spring*; Rome, *Bulldozer*; Hays, *Beauty*.

4. Swearingen, *Environmental City*.

5. Lessoff, *Where Texas Meets the Sea*; Feagin, *Free Enterprise City*; Abbott, *Metropolitan Frontier*; Abbott, *New Urban America*; Shermer, *Sunbelt Capitalism*.

6. Swearingen, *Environmental City*; Moore, *Sustainable City*.

7. Logan and Molotch, *Urban Fortunes*; Busch, "Perils of Popular Planning."

8. Rome, *Bulldozer*, 119–152; Jackson, *Crabgrass Frontier*, 231–245; Sellers, *Crabgrass Crucible*. For a contemporary take, see Whyte, *Last Landscape*.

9. Peek to Herring, November 14, 1957; Webster to Herring, November 15, 1957; Gardner to Herring, November 15, 1957; Isley to Herring, November 18, 1957, all found in Folder "October–November 1957," Box "Robert Thomas Miller, April 1956–December 1957," Miller Papers.

10. Ruby Ripperton and Mrs. G. M. McNeilly to Herring, November 17, 1957; Webster to Herring, November 15, 1957; Mrs. R. Q. Underwood to the Austin City Council, November 18, 1957; Richard G. Underwood to Herring, November 16, 1957, all found in Box "Robert Thomas Miller, April 1956–December 1957," Miller Papers.

11. See Swearingen, *Environmental City*, particularly chap. 2.

12. Carl M. Rosenquist to the City Council, May 30, 1960, Folder "FPF.10B; Miller, Robert Tom; Corr. January–June, 1960," Box "1958–1960," Miller Papers. See also Walter E. Long to Tom Miller, June 20, 1960, Folder "FPF.10B; Miller, Robert Tom; Corr. January–June, 1960," Box "1958–1960," Miller Papers; Mrs. Herman Jones to Austin American Statesman, June 13, 1960, Folder "FPF.10B; Miller, Robert Tom; Corr. January–June, 1960," Box "1958–1960," Miller Papers; Willis R. Bodine to Tom Miller, July 9, 1960, Folder "FPF.10B; Miller, Robert Tom; Corr. July–August, 1960," Box "1958–1960," Miller Papers; Roberta P. Dickson to Tom Miller, November 20, 1967, Folder "FPF.10B; Miller, Robert Tom; Corr. Sept.–Oct., 1957," Miller Papers.

13. "McMurtry—New," October 12, 1964, Folder "Town Lake Concept—PARD," Box 1, Crenshaw Papers.

14. Wayne Henneberger to Tom Miller, September 25, 1959, Folder "Corr. June–Dec. 1959," Box "FPF.10, Austin Mayors, Correspondence, 1958–1960," Miller Papers;

Lou Unfried to the City Council, December 1, 1955, Folder, "Miller, RT; Corr. Oct.–Dec. 1955," Box "FPF.10; Austin Mayors; Miller, Robert Thomas; Corr., 1947; May 1955–March 1956," Miller Papers; "Preliminary Notes: City of Austin Proposed Budget for Fiscal 1956," Folder "Miller, RT; Corr. Oct.–Dec. 1955," Box "FPF.10; Austin Mayors; Miller, Robert Thomas; Corr., 1947; May 1955–March 1956," Miller Papers; Bill Glaston to Tom Miller, January 8, 1958, Folder "Jan.–Feb. 1958," Box "FPF.10; Austin Mayors; Miller, Robert Thomas; Correspondence, Jan. 1958–Dec. 1960," Miller Papers.

15. Charles V. Hill to the Mayor and Members of the City Council, October 17, 1967, Folder "Town Lake Concept—PARD," Box 1, Crenshaw Papers.

16. Quoted in Roberta P. Dickson to Walter Wendlandt, March 1, 1968, Folder, "Town Lake Policy Zoning," Box 1, Crenshaw Papers. See also Roberta P. Dickson to Stuart King and Alan Y. Taniguchi, December 7, 1962, Folder "Park Board—Town Lake," Box 1, Crenshaw Papers. For downtown deterioration, see Blodgett, "Downhill for Downtown?"

17. Dickson to King and Taniguchi, December 7, 1962.

18. Untitled document, Folder "P&R Dept. II," Box 1, Crenshaw Papers.

19. P. D. Greer, "Parks Need," *Austin Statesman*, August 2, 1965, Folder, "PARD Board—P&R Dept.," Box 1, Crenshaw Papers.

20. Parks and Recreation Board to the City Council of Austin, June 5, 1963, Folder "Little Texas Proposal," Box 1, Crenshaw Papers.

21. "To Polish a Jewel," *Austin in Action*, January 1964, Folder "Town Lake Lighting—Private Offices to Assist," Box 1, Crenshaw Papers.

22. See, e.g., Rome, *Bulldozer*, 153–188; McHarg, *Design with Nature*.

23. "Triple Award Goes to Mrs. Dickson," *Austin American*, January 21, 1972, loose in Box 1, Crenshaw Papers. For examples of national environmental awareness, see Rome, *Bulldozer*; Sellers, *Crabgrass Crucible*.

24. Sellers, *Crabgrass Crucible*, 251–254.

25. "A Report on Some Developments following the White House Conference on Natural Beauty," September 1965, Folder "Legislation—General," Box 1, Crenshaw Papers.

26. "Speights," July 7, 1968, Folder "Town Lake Concept—PARD," Box 1, Crenshaw Papers.

27. Swearingen, *Environmental City*, 48–50.

28. Roberta P. Dickson to Palmer, June 25, 1965, Folder "PARD Board," Box 1, Crenshaw Papers. For other citizens supporting antidevelopment measures, see Irving R. Ravel to Mrs. Fagan Dickson, November 30, 1964, and Henrietta Jacobsen to Palmer, October 23, 1966, both found in Folder "PARD Board," Box 1, Crenshaw Papers.

29. See, e.g., Mayor Akin's speech, "A Greater Austin through Teamwork," delivered February 21, 1967, Folder "Little Texas Proposal," Box 1, Crenshaw Papers.

30. Swearingen, *Environmental City*, 53.

31. "Little Texas Facts" (memorandum, 1966), Folder "Little Texas Proposal," Box 1, Crenshaw Papers.

32. "Park Land Donations Vetoed," *Austin American*, December 9, 1966, and "Parks Question Answered by City," *Austin American-Statesman*, December 10, 1966, both found in Folder "Town Lake Concept—PARD," Box 1, Crenshaw Papers.

33. "Austin's 'Lake of Lights,'" *Austin American-Statesman*, August 13, 1967, Folder "Town Lake Concept—PARD," Box 1, Crenshaw Papers.

34. Mary Jane Garza, "Tracking the MACC: A Brief History of Austin's Latino Culture Center," *Austin Chronicle*, June 12, 1998, accessed May 11, 2011, http://www.austinchronicle.com/arts/1998-06-12/523624.

35. "Request for Ordinance/Resolution," December 15, 1967; "Summary of Meeting and Action on Amendment," January 10, 1967; untitled document, June 4, 1964; John A. Bowman to the Mayor and Members of the City Council of Austin, January 4, 1967, all found in Folder "Town Lake Concept—PARD," Box 1, Crenshaw Papers.

36. "Report of Meeting on the Proposed City Lake," November 3, 1958, Folder "FPF.10B; Miller, Robert Tom; Corr.," Box "FPF.10; Austin Mayors; Miller, Robert Thomas; Correspondence, Jan. 1958–Dec. 1960," Miller Papers; The debate continued throughout the 1960s. See Rosenquist to the City Council, May 30, 1960; Loren Winship to Buch Head, June 1, 1960, and Mrs. Gifford White to Mrs. Fagan Dickson, February 9, 1966, Folder "Town Lake Motor Boats," Box 1, Crenshaw Papers.

37. Hernandez, "Defending the Barrio."

38. Ibid.

39. Ibid., 128.

40. Ibid., 129.

41. East Austin Survival Task Force, "East Austin Survival Task Force," Folder "AF—Neighborhood Groups N1900 (27) East Austin Survival Task Force," Austin Files, AHC.

42. Hernandez wrote that three Mexican American youths were killed by Austin police between 1972 and 1974; one youth was shot in the back of the head after a burglary and found in possession of bread and milk. Hernandez, "Defending the Barrio," 129.

43. Garza, "Tracking the MACC."

44. Adam Rome deftly illustrates how the idea and language of land use had become commonplace among environmentalists by 1970. See Rome, *Bulldozer*, 221–253.

45. For suburban culture and postwar environmentalism, see Rome, *Bulldozer*; Carson, *Silent Spring*. For the Austin movement, see Rice, *Distant Publics*; Swearingen, *Environmental City*; Dickson, "Copy to Jim Berry, City Editor"; Leslie Donovan and Bill Cryer, "Trees Fall at UT Despite Protesters," *Austin Statesman*, October 23, 1969; Glen Castlebury, "Erwin and Texas Mood" (n.p., n.d.), Folder "Waller Creek," Crenshaw Papers.

46. Molotch, "Growth Machine." Molotch defines the "growth coalition" as an influential group of businesspeople who have power to influence political decision makers in a way that benefits them economically. Gottdiener, *Urban Space*; Boyer,

Dreaming the Rational City. Boyer's work focuses on early city planners, from the 1900s to the 1950s, whom she finds incapable of planning a truly rational city in the interests of urban dwellers. Burke, *Participatory Approach*.

47. Lefebvre, "The Right to the City"; Barker et al., "Non-plan"; Hatch, *Social Architecture*; Bolan, "Emerging Views of Planning"; Burke, "Citizen Participation Strategies"; City of Tacoma Planning Department, *Citizen Involvement in Community Improvement* (Tacoma: City of Tacoma Planning Department, 1971); Godschalk, *Participation, Planning, and Exchange*; Godschalk and Mills, "Planning through Urban Activities"; Mann, "Planning-Participation."

48. Ferguson, "One Big Damn Subdivision"; BeSaw, "Apartments, Autos"; BeSaw, "Time Alters Shopping Patterns"; "Building Permits Breaking Records"; "Austin Building Sixteenth in Nation."

49. "Building Permits Break Records"; "Centex Economy Soars"; "Travis County Sets New Record"; Dickson to Palmer, June 25, 1965.

50. City of Austin Department of Planning, "Austin Tomorrow—Today" (n.d.), 8. Folder "AF City Planning C4170 (3) Austin Tomorrow (1974–)," Subject File "Austin Tomorrow," AHC. The planning department published numerous fliers, handouts, and brochures to publicize Austin Tomorrow starting around 1972.

51. City of Austin Department of Planning, "Ten Thousand Citizens Are Deciding the Future of Austin," newspaper insert, February, 1974. The quotation is from "Lillie Asks Goals from West Austin," *Austin American-Statesman*, October 16, 1973.

52. Michael Norman, "The Austin Experiment," *Sunday Times Adviser* (Trenton, NJ), August 24, 1975, Folder "AF City Planning C4170 (3) Austin Tomorrow (1974–)," Subject File "Austin Tomorrow," AHC.

53. Wood quoted in Walmsley, "City Planning," 60. The interview took place on December 4, 1974.

54. "Three Groups Hesitate to Back City Plan," *Austin American-Statesman*, June 20, 1973; "Letter Confuses City Plan," *Daily Texan*, June 19, 1973.

55. "NLM President Outlines Austin Tomorrow's Flaw," *Austin American-Statesman*, December 20, 1973.

56. "City Plan Involves Students," *Daily Texan*, November 12, 1974; " 'Tomorrow' Panel Gains Twelve Persons in Council Action," *Austin American-Statesman*, September 27, 1974.

57. "New 'Tomorrow' Attracts Neighbors, Lobby," *Austin American-Statesman*, February 19, 1974.

58. The *Austin American-Statesman* covered almost every phase III neighborhood meeting between February 19 and May 1, 1974. See, e.g., "Refunds, Growth Hottest Topics," *Austin American-Statesman*, April 6, 1974. The *Daily Texan* covered many as well. See especially "Guest Viewpoint—Lake Austin: Wilding No More," *Daily Texan*, April 8, 1974.

59. "Austin Growth Viewed as Curse or Blessing," *Austin American-Statesman*, March 20, 1974.

60. Wray Weddell, "Wray Weddell," *Austin Citizen*, January 25, 1974, and February 5, 1974.

61. "Goals Program Snags on Meeting Standards," *Austin American-Statesman*, December 4, 1973; "East Side Austin Tomorrow Attracts Few," *Austin American-Statesman*, March 5, 1974; "Austin Tomorrow Notes Problems," *Daily Texan*, December 5, 1973. For *Tribune* support, see "Local Citizens Can Help Plan East Austin's Future," cited in Walmsley, "City Planning," 99.

62. "East Austinites Complain at 'Tomorrow' Meeting," *Austin American-Statesman*, April 29, 1974.

63. Ibid.; "East Austin Absentee Ownership Allegedly Zoning Problem Cause," *Austin American-Statesman*, March 19, 1974; "Tomorrow Suggestions Voiced in Spanish, English," *Austin American-Statesman*, April 10, 1974. For quality of life, see City of Austin, *Block Grant Application*.

64. Austin Department of Planning, "Austin Tomorrow Ongoing Committee: Your Austin 1978," Folder "AF City Planning C4170 (3) Austin Tomorrow (1974–)," Subject File "Austin Tomorrow," AHC.

65. Quoted in Walmsley, "City Planning," 106.

66. Quoted in ibid., 116.

67. Ibid.

68. Busch, "Building 'A City.' "

69. City of Austin, Department of Planning, *Austin Tomorrow Comprehensive Plan*, 5.

70. Ibid., 7. For environmental groups, see Swearingen, *Environmental City*, chap. 2.

71. With the help of Lady Bird Johnson, Crenshaw and the city secured federal funding to improve the area under the banner Town Lake Beautification Project throughout the 1970s.

72. Swearingen, *Environmental City*, 57–59.

73. Ibid., 205; Marty Toohey, "City Plan in 1970s Reflected What Austinites Wanted, But Not All of It Was Implemented," *Austin American-Statesman*, September 25, 2010; Melody Davison to Letters Editor, U.S. News and World Report, June 18, 1981, Austin File "C4170 (30) 1974–," AHC.

74. Hays, *Beauty*, 23.

Chapter Eight

1. "Executive Summary: The Texas Incentive for Austin," *On Campus*, June 1983, Vertical File "MCC," BCAH; David Butts, "Flawn Envisions Fruitful UT, MCC Link," *Daily Texan*, May 30,1983, Vertical File "MCC," BCAH; "Statement for MCC Project," n.d., Folder "Chicago MCC Trips," Box 1, MCC Recruitment Records, AHC.

2. Quoted in Gibson and Rogers, *R&D Collaboration*, 169.

3. See, e.g., interviews in McCann, "Inequality and Politics"; City of Austin, *Imagine Austin Comprehensive Plan*; Florida, *Rise of the Creative Class*. Peter Hall

makes the point that in the 1980s prior industrial development and architecture could have impeded high-tech growth and been unattractive for high-tech workers. Hall, "The Fifth Kondratieff."

4. See, e.g., Kirk Ladendorf, "Rebooting: Consortium Ponders High-Tech Mission, Future," *Austin American-Statesman*, September 16, 1990, Subject File "MCC," BCAH.

5. Preer, *Emergence of Technopolis*, 55.

6. Tretter, *Shadows*, chap. 3.

7. Hall, "Fifth Kondratieff."

8. Scott, "Capitalism and Urbanization?"; Saxenian, *Regional Advantage*; Oakey, "High Technology Industries."

9. Regarding Austin and the university, see, e.g., Kozmetsky, Williams, and Williams, *New Wealth*, especially 121–130 (a laudatory essay about Austin's growth); Smilor, Kozmetsky, and Gibson, "Sustaining the Technopolis."

10. Marylin Bender, "Deep in the Heart of Texas, 'Real-World' MBA's," *New York Times*, April 25, 1971.

11. Ibid.

12. Gerald McLeod, "In the Private Interest: The Institute for Constructive Capitalism," *Change* 10, no. 11 (December 1978): 14.

13. Bluestone and Harrison, *Deindustrialization of America*; Soja, Morales, and Wolff, "Urban Restructuring."

14. Russell Mitchell, "Business Elite of Texas Will Play a Major Role on Technology Council," *Austin American-Statesman*, July 16, 1984, Folder "Impact Analysis," Box 1, MCC Recruitment Records.

15. Diane E. Downing, "Thinking for the Future: The Promise of MCC," *Austin*, August 1983, 103–110, and Jeffrey M. Guinn, "Consortium Gambling All Its Chips," *Fort Worth Star-Telegram*, June 26, 1983, both found in Vertical File "MCC," BCAH.

16. Kahn and Farley, *Impact of MCC*, 2–4.

17. Quoted in Dwight B. Davis, "R&D Consortia: Pooling Industries' Resources," *High Technology*, October 1985, Folder "MCC," Box 1, MCC Recruitment Records.

18. Downing, "Thinking for the Future"; W. R. Deener III, "Firms Study Austin as Site for Venture," *Dallas Morning News*, April 13, 1983, Vertical File "MCC," BCAH; "How to Expand R&D Cooperation," *Business Week*, April 11, 1983, Folder "MCC/SRC Background," Box 1, MCC Recruitment Records.

19. University administrators clearly saw working conditions and living conditions as equal components of quality of life. Improving university facilities for knowledge workers was part of what they considered an attractive lifestyle. See "A Development Plan for the Balcones Research Center" (unpublished report, University of Texas at Austin, 1989), 17–20, Folder "Balcones (Pickle) Research Center," Box 99-119/2, Robert Ovetz Papers, BCAH.

20. Gibson and Rogers, *R&D Collaboration*, 122–126. Quotation in "What MCC Means to Texas" (n.d.), Folder "Impact Analysis," Box 1, MCC Recruitment Records.

21. See, e.g., Jeffrey M. Guinn, "Computers 'n' Cactus: Fighting an Image," *Fort Worth Star-Telegram*, June 26, 1983, Vertical File "MCC," BCAH; Downing, "Thinking for the Future."

22. "Great Academic Stature at UT Blueprinted for 1970," *Austin in Action* 3, no. 4 (September 1961). For Austin's marketing campaign, see "Big Campaign Beckons Business," *Austin in Action*, April 1965; Tom Miller to J. J. Suran, December 3, 1958, and Joe E. Armstrong to Tom Miller, December 3, 1958, both found in Folder "Miller, Robert Thomas; Corr. July–December 1958," Box "FPF.10B; Austin, Mayors; Miller, Robert Thomas, Correspondence 1958–1960" AHC.

23. "Development Plan for the Balcones Research Center," 6–10; Brenda Bell, "Balcones Industrial Park Concept Is Unique," *Austin Statesman*, December 12, 1975, Folder "Balcones (Pickle) Research Center," Box 99-119/2, Ovetz Papers; Rosanne Mogavero, "Industrial Research Park Projected for Balcones," *Daily Texan*, February 20, 1976, Vertical File "Balcones Research Center, UT, 1946–1986," BCAH.

24. Streetman, a recent hire to the Department of Electrical Engineering in 1982, was one of the university's main representatives on the MCC Task Force.

25. "Development Plan for the Balcones Research Center," 6–10; "No End in Sight to UT Expansion," *Third Coast*, August 1982, and Kerry Gunnels, "Facility to Boost UT Energy Research," *Austin American-Statesman*, December 14, 1981, both found in Folder "Balcones (Pickle) Research Center," Box 99-119/2, Ovetz Papers.

26. "Executive Summary: The Texas Incentive for Austin."

27. Ibid.; Lotte Chow, "Pact May Help MCC Spouses," *Daily Texan*, January 10, 1984, Vertical File "MCC," BCAH; "MCC Relocation Assistance Proposal" (n.d.), Folder "MCC Relocation Off.," Box 1, MCC Recruitment Records. For a truncated version of Austin's attempts to woo MCC, see Orum, *Power, Money, and the People*, 312–316; Gibson and Rogers, *R&D Collaboration*, particularly 139–173.

28. For technopolis-related theories of labor, see Preer, *Emergence of Technopolis*, 54–64.

29. Herbert I. Fusfeld, "The Bridge between University and Industry," *Science* 209, no. 4453 (July 11, 1980): 221.

30. Konecci et al., *Commercializing Technology Resources*, 253–254.

31. Ovetz, "Entrepreneurialization," 46–48.

32. Arnie Weissmann, "Contract Warriors," *Third Coast*, June 1986, Folder "UT Military Research," Box 99-119/9, Ovetz Papers.

33. Ovetz, "Entrepreneurialization."

34. For secondary circuits of capital and their relation to changes in the mode of production, see Harvey, *Urbanization of Capital*, chap. 1.

35. Historically, Texas law has encouraged municipal annexation, generally ruling that "home rule" cities, those with more than five thousand residents, can annex smaller cities without the latter's consent. In 1963 the legislature made this into law. Texas is also one of the most lenient states in the United States regarding extraterritorial jurisdiction, where the city can provide services, apply zoning laws,

and generally govern in adjacent land it does not legally control. See "Annexation Law," Folder "3-16," Box 3, Jean Mather Papers, AHC.

36. City of Austin, *Imagine Austin Comprehensive Plan*, 32–33; "Annexation Law"; League of Women Voters, "The Economics of Annexation" (unpublished report, April 1984), Folder "3-16," Box 3, Mather Papers.

37. The process by which MUDs were annexed in Austin is explained nicely in Swearingen, *Environmental City*, 92–94, and in League of Women Voters, "Economics of Annexation."

38. City of Austin, *Imagine Austin Comprehensive Plan*, 32.

39. Northcutt, "Austin."

40. Paul Schnitt, "Austin Home Process Top Texas Cities, Survey Says," *Austin American-Statesman*, October 28, 1984, and Kelly Hodge, "Can the Boom Last?," *Austin*, August 1984, both found in Vertical File "Austin, Texas—Housing and Real Estate (Travis County)," BCAH.

41. Michael McCullar, "3M Building the Future Right Now," *Austin American-Statesman*, August 21, 1988, Vertical File "Austin, Texas—Industries, 1939–1996," BCAH.

42. Michael McCullar, "From the Outside to the Inside," *Austin American-Statesman*, August 21, 1988, Vertical File "Austin, Texas—Industries, 1939–1996," BCAH.

43. Ibid.

44. "Corporate Offices on the Hills," *Austin Business Executive*, March 1984, Subject File "Austin, Texas—Housing and Real Estate (Travis County)," BCAH.

45. Hodge, "Can the Boom Last?"; Joe Bienvenu, "Austin Emerging as Major High Tech Center," *Southwest Real Estate News*, January 1986, and Crispin Ruiz, "Office Buildings and Industrial Parks: A Report," *Austin*, 1982, 7a–21a, both found in Vertical File "Austin, Texas—Housing and Real Estate (Travis County)," BCAH.

46. Lon Brooks and Ruth Spiegel, "The City's Changing Land Use Patterns," *Austin Business Executive*, March 1984, 55–58, Subject File "Austin, Texas—Housing and Real Estate (Travis County)," BCAH.

47. Ruiz, "Office Buildings," 10a.

48. Ibid., 10a–12a.

49. Ibid., 12a.

50. Swearingen, *Environmental City*, 206.

51. Brooks and Spiegel, "Changing Land Use."

52. Ibid.

53. Roger Duncan, "Slow Growth Proposal" (unpublished report, 1984), Folder "3-13," Box 3, Mather Papers.

54. Quoted in James Pinkerton, "Owners, Tenants Await Change," *Austin American-Statesman*, 1984, "AF—Subdivisions—East Austin S6090," General Folder 2, Austin Files, AHC.

55. Ibid.

56. Ruiz, "Office Buildings," 21a.

57. Damond Benningfield, "Chipping In," *Austin*, January 1989, Folder "Sematech, 1988–1993," Box 99-119/2, Ovetz Papers; Thomas C. Hayes, "Is Austin the Next Silicon Valley?" *New York Times*, January 13, 1988, Vertical File "Austin, TX—Industries," BCAH; Elizabeth Whitney, "How Austin Snared Sematech: The View from the Other Side," *Austin American-Statesman*, July 24, 1988, and Damond Benningfield, "New Chips on the Block," *Alcalde*, January–February 1989, 26–28, both found in Folder "Sematech, 1988–1993," Box 99-119/2, Ovetz Papers.

58. Benningfield, "New Chips on the Block"; Ashley Price, "Expectations Change as Area Does," *Austin American-Statesman*, July 20, 2010, Subject File "Austin, TX—Neighborhoods and Neighborhood Groups—Misc. (2)," BCAH; PODER, "Montopolis: The New Technopolis" (unpublished report), Folder "Sematech 1994," Box 43, PODER Records, AHC.

59. Neil Orman, "Austin Tops in Patent Growth," *Austin Business Journal*, December 1, 1996, accessed August 17, 2011, http://www.bizjournals.com/austin/stories/1996/12/02/story2.html.

60. City of Austin, "Austin's Economic Future: The Mayor's Task Force on the Economy," (report, 2003), 25; "Blame It on the Typewriter," *Economist.com*, September 23, 2006, accessed January 3, 2017, http://www.economist.com/node/7945787.

Chapter Nine

1. Haddon, "Tank Farm Controversy"; see also "Illness Affects Half of Households near Fuel Terminal," *Austin American-Statesman*, May 10, 1992; PODER, "Community Update on Toxic Fuel Storage Tank Farm Austin Texas"; and PODER, "Tank Farms Survey, March 1992," all three found in Folder "Tank Farms Survey, March 1992," Box 67, PODER Records.

2. Mike Ward and Scott W. Wright, "Air of Concern," *Austin American-Statesman*, March 1, 1992; Mike Ward, "More East Austin Wells Tainted," *Austin American-Statesman*, April 9, 1992; and Mike Ward, "Tank Farm Problems Left off List in Official 'Slip Up,'" *Austin American-Statesman*, March 20, 1992, all found in Folder "Newspaper Clips—Tank Farm 1992," Box 74, PODER Records.

3. PODER, "Human Link Picket to . . . Close the Tank Farm" (flier, n.d.), Folder "Tank Farm Monitoring Committee," Box 68, PODER Records.

4. Mike Ward, "'Greening' Closes Gap in Austin," *Austin American-Statesman*, March 8, 1992, Folder "Newspaper Clips—Tank Farm 1992," Box 74, PODER Records.

5. Ken Kramer to Bill Zeis, June 10, 1992, Folder "Tank Farm Maps," Box 67, PODER Records.

6. Maxey to Antonio Diaz, April 8, 1992, Folder, "Tank Farm 1992 Air Control Board," Box 68, PODER Records; Ward, "Problems Left off List"; Ward, "'Greening' Closes Gap."

7. Ward, "'Greening' Closes Gap"; Scott B. Wright, "Chevron Plans to Close Terminal in East Austin," *Austin American-Statesman*, August 27, 1992, Folder "Tank Farm Articles—Papers Written by PODER," Box 70, PODER Records.

8. Quoted in Kate Van Scoy, "Eastsiders Decry BFI: Residents Say Recycling Plant Constitutes Enviro-Racism," *Austin Chronicle*, May 30, 1997.

9. William Scott Swearingen, a scholar and activist who was part of the mainstream environmental movement in Austin, articulates in his book *Environmental City* (63, 123, 231) the idea that mainstream environmentalists wanted to keep Austin's small-town feel. The fear of "Houstonization" drove much antidevelopment discourse, especially in the 1980s. Jenny Rice in *Distant Publics* (104–116) discusses with Austin writers and other interviewees the worry that the city is becoming too much like Houston and Dallas. Also see Long, *Weird City*. The discourse is evident among neighborhood environmental group records. See, e.g., Save Austin's Neighborhoods and Environment, "Austin—Managing Our Way to a Protected Future" (n.d.), Folder "3-12," Box 3, Mather Papers.

10. Pulido, *Environmental and Economic Justice*. Similar ideas are articulated in Bullard, introduction to *Growing Smarter*. Tretter, in "Contesting Sustainability," makes a similar argument regarding PODER.

11. PODER's response to gentrification is discussed in the Epilogue.

12. Research has shown that poorer minorities are more vulnerable to economic change and that they tend to understand community spatially and tie it to particular places, especially in areas that were heavily segregated. Logan and Molotch, "Homes"; Harvey, "Flexible Accumulation through Urbanization"; Horan, "Community Development."

13. Karvonen, *Politics of Urban Runoff*, 35–41.

14. Walter E. Long, "Barton Springs History," Folder "Barton Springs—History," Box 1, Long Papers.

15. Zilker never actually received money from any of these sales. He essentially gave the land to the school board, which then sold the land to the city. The money the school board received created the Zilker Permanent Fund, used for industrial education for Austin students.

16. Nancy Bishop, "Austin's Zilker Park," *Texas Highways*, August 1997, Folder 46, Box 1, Beverly Sheffield Papers, AHC; Limbacher and Godfrey Architects, *Barton Springs Pool Master Plan*, 25–57.

17. Quoted in Bishop, "Zilker Park."

18. Limbacher and Godfrey Architects, *Barton Springs Pool Master Plan*.

19. Pipkin and Frech, *Barton Springs Eternal*.

20. Swearingen, *Environmental City*, 104.

21. Busch, "Building 'A City.'"

22. Don E. Cox to South River City Citizens Member, March 1975, Folder "1-2," Box 1, Mather Papers; United South Austin, "Meeting Notice and Agenda," November 30, 1984, Folder "1-7," Box 1, Mary Arnold Papers, AHC; entirety of Folder "1-16," Box 1, Arnold Papers; West Austin Neighborhood Association, "Austin To-

morrow Conundrum," 1983, Folders "3-11" and "3-13," Box 3, Mather Papers. The material on the ethnic balance in Southeast Austin referred to Latinos, not African Americans. South Austin Neighborhood East to Austin Housing Authority, February 26, 1977, Folder "3-7," Box 3, Mather Papers.

23. "Comments—Tony Gregg"; Brenda M. Berger to Mary Arnold, February 28, 1985; Marc Warner and Deanna Warner to Planning Commission Member (PCM), February 20, 1985; and Ginger Simmons to PCM, n.d., all found in Folder "3-3," Box 3, Arnold Papers.

24. Mike Thomasson to Member of the Planning Commission, August 29, 1978; Bill Collier, "Zilker Park Posse to Demand Apartment Moratorium," *Austin American-Statesman*, June 19, 1979; Bouldin Creek Neighborhood Association to Mayor Carole McClellan, July 21, 1982, all found in Folder "Zoning Changes in Barton Creek Area 1978–1983," Box 1, Zilker Park Posse Records, AHC.

25. Phillip S. Blackerby and Alexandra K. Blackerby, "Testimony," Folder "105 Zoning Changes in Barton Creek Area, 1978–1983," Box 1, Zilker Park Posse Records.

26. "Austin—Managing Our Way to a Preferred Future" (report, 1984), Folder, "3-7," Box 3, Mather Papers.

27. See, e.g., "Growth: #1 Issue," *Austin Neighborhood Newswatch*, August 1980, Folder "3-5," Box 3, Mather Papers.

28. Maureen McReynolds to Dick Lillie, June 22, 1978, Folder, "105 Zoning Changes in Barton Creek Area, 1978–1983," Box 1, Zilker Park Posse Records.

29. City of Austin, *Imagine Austin Comprehensive Plan*, 32.

30. Swearingen, *Environmental City*, 123.

31. The most visible of these groups was the Zilker Park Posse. See especially Swearingen, *Environmental City*, 121–122.

32. "The Battle for the Springs: A Chronology," *Austin Chronicle*, August 9, 2002; Swearingen, *Environmental City*, 142–144.

33. Jean Mather, "We Care Board" (speech, n.d.), Folder "1-1," Box 1-1, We Care Austin Records, AHC.

34. Ibid.

35. Swearingen, *Environmental City*, 114–116.

36. Paul Burka, "The Battle for Barton Springs," *Texas Monthly*, August 1990, 74.

37. Swearingen, *Environmental City*, 114–119.

38. Zilker Park Posse, "We Won't Be Able to Use Barton Springs Anymore" (pamphlet, n.d.), Folder "1-1," Box 1, Zilker Park Posse Records.

39. Carol Siegenthaler to Mayor Ron Mullen and Members of the City Council, September 26, 1984, Folder "1-5," Box 1, Arnold Papers. The counterculture press in Austin was making similar arguments. See Bud Shrake, "The Screwing Up of Austin," *Texas Monthly*, December 27, 1974.

40. See, e.g., Protect Edwards Aquifer Coalition, "Austin Has Six Electronics Plants and More Are Coming on Line" (unpublished pamphlet, n.d.), Vertical File "Austin, Texas—City Planning—Misc. (2), AHC; "Builders Report Update," June 27, 1984, Folder "1-2," Box 1, Arnold Papers.

41. Swearingen, *Environmental City*, 133–141; "Ordinance #80," April 1, 1980, Folder "1-3," Box 1, Zilker Park Posse Records.

42. Ken Manning and Seth Searcy, "Press Release," (unpublished newspaper article, n.d.), Folder "1-3," Box 1, Zilker Park Posse Records.

43. Kim Tyson, "Real Estate Takes a Fall for the Year," *Austin American-Statesman*, December 27, 1987, Vertical File "Austin, Texas—Housing and Real Estate (Travis County)," BCAH.

44. Burka, "Battle for Barton Springs," 72.

45. Ibid.

46. Swearingen, introduction to *Environmental City*.

47. We Care Austin, "Statement to the Texas Water Commission re Barton Springs Water Quality," June 24, 1991, Folder, "1/1," Box 1, We Care Austin Records.

48. Bunch to Austin environmental leaders interested in adopting a position paper on Barton Springs protection, n.d., Folder "1/1," Box 1, We Care Austin Records. We Care Austin circulated a similar memo on the day of the PUD hearing. Mary Arnold to Members of We Care Austin Board, July 5, 1991, Folder "1/1," Box 1, We Care Austin Records.

49. Karvonen, *Politics of Urban Runoff*, 53–57.

50. Swearingen, *Environmental City*, 168.

51. Moscoco, "PODER's Woes"; Eunice Moscoco, "Eastside Issues Are Priority for 'Green' City Council," *Austin American-Statesman*, July 23, 1997, Folder "PODER/Zoning Articles 7/97," Box 28, PODER Records.

52. PODER, "Toxics in Minority Communities" (pamphlet, July 18, 1991), Folder "Sematech, 1988–1993," Box 99-119/2, Ovetz Papers.

53. "Congressional Review of SEMATECH Urged," *La Prensa*, November 8, 1991, Folder "Motorola—TCE Clean Up, 1991," Box 43, PODER Records.

54. PODER, "Untitled Report" (n.d.), Folder "Sematech 1994," Box 43, PODER Records.

55. Matthew Reeves, "Five Austin Companies Release Most of City's Toxins," *Daily Texan*, July 25, 1991.

56. Anthony Sommer, "TCE Clean Up May Last Fifteen Years, Report Says," *Phoenix Gazette*, March 30, 1984, Folder "Motorola—TCE Clean Up, 1991," Box 43, PODER Records.

57. PODER, press statement, September 30, 1992, Folder "Press Statements, '92," Box 42, PODER Records; Oscar de la Torre, Belinda Garcia, and Rocio Torres, "The City of Austin's Economic Development Strategy and Its Impact on Low-Income Communities" (report, December 13, 1996), Folder, "[Cleanup] AMD UT Study 1-97 Industrial Incentives Package," Box 46, PODER Records.

58. Maria Loya and Samuel Echevarria, "Austin Water: High Tech Industry and Water Viability Issues" (PODER report, November 19, 1996), Folder "Ar.2012.015," Box 46, PODER Records; PODER, "Untitled Report"; Bob Burns, "Texas Tops in Creating Toxic Waste, Environmentalists Say," *Austin American-Statesman*, July 24, 1991, Folder "Motorola—TCE Clean Up, 1991," Box 43, PODER Records.

59. Susana Almanza and Roland Ortiz to Daniel J. Rondeau, October 12, 1995, Folder "CWM—Tokyo Motion to Withdraw 1996," Box 44, PODER Records.

60. Ibid.

61. PODER, "Untitled Report."

62. Ibid.; De la Torre, Garcia, and Torres, "Austin's Economic Development Strategy."

63. Sylvia Ledesma, "HiTech Giants: From 'Silicon Valley' to 'Silicon Hills" (report draft, March 1, 1995), Folder "High Tech Giants Report 1995," Box 42, PODER Records.

64. Loya and Echevarria, "Austin Water."

65. SNEEG, "Water: The Life Blood of the Earth, Executive Summary," February 1997, Folder " 'Water: The Life Blood of the Earth,' CRT-SNEEG 2-97," Box 45, PODER Records.

66. PODER to Daniel J. Rondeau, October 11, 1995, Folder "Civil Rights Correspondence 1996," Box 44, PODER Records. The specific allegations are outlined in "EPA File No. 1R-95-R6," Folder "TNRCC Response to EPA-Tokyo EPA Title VI Complaint 4/3/99," Box 44, PODER Records.

67. City of Austin, "Eastside Environmental News," 6, no. 1 (December 2001), Folder "Mobil Oil Press Clips, 1992," Box 74, PODER Records; Mike Ward, "Lead in Child Blamed on Old Radiator Shop," *Austin American-Statesman*, January 28, 1993, Folder "Newspaper Clips—Tank Farm 1992," Box 74, PODER Records; Mike Ward and Scott W. Wright, "Six More Wells Found Tainted in East Austin," *Austin American-Statesman*, May 2, 1992, Folder "Mobil Oil Press Clips, 1992," Box 74, PODER Records.

68. William John Young to Eliseo Garza, April 6, 1993, and "OENA Thanks City for Relocation of Garbage Trucks, Calls for Decisive Action to Move Rest of Hargrave Facility" (n.p., n.d.), both in Folder "Hargrave—Garbage City of Austin," Box 71, PODER Records.

69. "East Austin Environmental Initiative," Folder "East Austin Initiative—COA Chuck Lesniak," Box 28, PODER Records.

70. Smith, "One More Detour"; City of Austin Electric Utility Department, "Holly Power Plant Closure Options: RFPs for Replacement Resources," report, June 21, 1994, Folder "HPP Closure Options," Box 46, PODER Records.

71. Smith, "One More Detour."

72. Susana Almanza to Judge Bill Aleshire, January 11, 1994, Folder "PODER Correspondence HPP," Box 47, PODER Records.

73. Smith, "One More Detour"; City of Austin Electric Utility Department, "Plant Closure Options."

74. John Moore to Anthony C. Grigsby, March 8, 1994, Folder "PODER Correspondence HPP," Box 47, PODER Records.

75. Herrera and Almanza to Mayor and City Council Members," August 24, 1994, Folder "PODER Correspondence HPP," Box 47, PODER Records.

76. Herrera to Mayor Will Wynn, October 30, 2003, Folder "HPP Press Release—Health Survey 10-30-03, Austin City Council," Box 46, PODER Records;

"Magnetic Field Strengths near Holly Power Plant" (n.p., n.d.), Folder "HPP Press Briefing Material EUD 8/11/94," Box 47, PODER Records; assorted letters, Folder "City of Austin Memos 1994," Box 47, PODER Records.

77. John Moore to Mayor and Council Members, August 2, 1994, Folder "PODER Correspondence HPP," Box 47, PODER Records.

78. "Holly Street Noise Abatement Program," n.d., Folder "PODER Correspondence HPP," Box 47, PODER Records; "Holly Street Power Plant Monthly Report," (June, 1993), Folder "HPP Press Statement 3/31/99," Box 47, PODER Records.

79. "Segregation Dictated City's First Zoning Map," *Austin American-Statesman*, July 20, 1997, Folder "Zoning Articles '97," Box 29, PODER Records.

80. For a list of promised economic development initiatives in the 1970s and 1980s, see Eunice Moscoco and Tara Trower, "East Austin Taking Steps to Stop Problems before They Start," *Austin American-Statesman*, July 23, 1997.

81. "A Community Speaks," Naked City: Off the Desk, *Austin Chronicle*, December 6, 1996.

82. Ibid.

83. A.S., "Off the Desk: Recycle This," *Austin Chronicle*, August 9, 1996.

84. Ibid.

85. Kate Van Scoy, "Keeping Industry Out," *Austin Chronicle*, July 18, 1997.

86. Moscoco and Trower, "East Austin Taking Steps."

87. Laylan Copelin, "Five Former Leaders Say Holly Closure Is a Mistake," *Austin American-Statesman*, March 27, 1997, Folder "HPP Fire 6/14/99," Box 46, PODER Records.

88. "Close Holly Power Plant Now!" (pamphlet, 2003), Folder "Holly Plan Rezoning 3/26/03," Box 46, PODER Records.

89. "East Austin Land Use/Zoning Report," February 20, 1997, Folder "Land Use Zoning—Austin City Council 1997," Box 27, PODER Records.

90. "Zoning," n.d., Folder "East Austin Overlay 5/99," Box 26, PODER Records.

Epilogue

1. So-called because all seven members were dedicated to and had been part of Austin's mainstream environmental movement of the 1990s.

2. Greater Austin Chamber of Commerce, *Next Century Economy.*

3. Duany, Speck, with Lydon, *Smart Growth Manual*; Passell, *New Urbanism.*

4. Barna, "The Rise and Fall."

5. City of Austin, *Preliminary Agreement of Environmental Protection and Development.*

6. Mike Clark-Madison, "A City with Smarts: Austin Wising Up to Growth Plans," *Austin Chronicle*, April 17, 1998, accessed July 20, 2011, http://www.austinchronicle.com/news/1998-04-17/523318.

7. Grodach, "The Creative City." For more on the creative class, see Florida, *Rise of the Creative Class.*

8. Barna, "The Rise and Fall."

9. Florida, *Cities and the Creative Class.* Many studies have found that Austin has had the best local economy through the current economic malaise. Mason, "Austin, Texas Leads Nation in Job Growth"; MSNBC, Kotkin, "The Next Big Boom Towns in the U.S."; Levy, "Cities Where the Recession Is Easing"; Kiviat, "The Workforce."

10. City of Austin, "Balcones Canyonlands Preserve," accessed July 27, 2015, http://www.austintexas.gov/department/balcones-canyonlands-preserve-bcp; Swearingen, *Environmental City*, 193–196; City of Austin, *Imagine Austin Comprehensive Plan*, 58.

11. Swearingen, *Environmental City*, 196–199.

12. Mother Nature Network, "Top Ten Green U.S. Cities."

13. City of Austin, "Imagine Austin Wins."

14. Walsh, "Red State, Green State"; Moore, *Sustainable City*; Witt, "Austin's Green Ambitions."

15. Kate Van Scoy, "Natural Born Leader: Mayor Hits the Ground Running His First One Hundred Days," *Austin Chronicle*, October 3, 1997.

16. Emily Pyle, "Council Watch: Living in a Downtown World," Naked City, *Austin Chronicle*, December 22, 2000, accessed September 14, 2011, http://www.austinchronicle.com/news/2000-12-22/79923; Mike Clark-Madison, "The Best-Laid Plans," Naked City, *Austin Chronicle*, June 30, 2000, accessed September 14, 2011, http://www.austinchronicle.com/news/2000-06-30/77787; McCann, "Framing Space and Time."

17. For the relatively slow gentrification in the 1990s, see City of Austin Historic Gentrification Task Force, "Gentrification Task Force Data Sheet," n.d., Folder "COA Historic Gentrification Task Force Findings 12/4/02," Box 32, PODER Records, AHC. From 1990 to 2000, the central Eastside lost about twenty-three hundred African Americans and gained about eight hundred whites. The area of the study, bounded by Manor Road, IH-35, Town Lake, and U.S. 183, was 7.6 percent white in 2000.

18. Neil Carson, "Urban Gentry" (n.p., n.d.), Folder "Gentrification Housing Crisis 2004," Box 33, PODER Records; PODER, "SMART Growth, Historic Zoning and Gentrification of East Austin: Continued Relocation of Native People from Their Homeland" (2003), Folder "Gentrification Report 2003," Box 33, PODER Records. For property value increases on the Eastside, see Susan Smith, "City's Historic Homes Spark Tax Protest," *Austin American-Statesman*, May 15, 2002.

19. PODER, "Gentrification of East Austin"; "East Austin Housing Forum," 2006 , Folder "PODER Housing Forum 2006," Box 32, PODER Records. The University of Texas School of Architecture and Community and Regional Planning did the study of land values and property taxes. Wells Dunbar, "How Not to Gentrify: HRC Asks for Eastside Moratorium," *Austin Chronicle*, November 4, 2005.

20. Ryan Robinson, "Income and Neighborhood Planning Areas" (2006), accessed September 20, 2011, http://www.ci.austin.tx.us/demographics/downloads

/income_npas_collection.pdf. Website now defunct. Requests for the data can be made through the City of Austin demographer homepage at http://www.austintexas.gov/demographics.

21. Dunbar, "How Not to Gentrify."

22. Diana Welsh, "Protesters' Message: 'Stop Gentrifying the Eastside,'" Naked City, *Austin Chronicle*, April 8, 2005, accessed September 14, 2011, http://www.austinchronicle.com/news/2005-04-08/265699. For a more thorough description of the changes, see Busch, "Crossing Over."

23. Susana Almanza, "Speaking Up," Letter to the Editor, *Austin Chronicle*, December 2, 2005.

24. Ana Villalobos, "In Favor of PODER," Feedback, *Austin Chronicle*, November 28, 2005.

25. David Smith, "Grow Up Austin," Letter to the Editor, *Austin Chronicle*, October 22, 1999.

26. Rudolph Williams, "Now Is the Time for Justice!," Letter to the Editor, *Austin Chronicle*, June 12, 2007.

27. Externalities associated with sustainable planning in Austin have been the subject of much fruitful inquiry; research emphasizes how sustainable practices undermine social justice for vulnerable residents, often emphasizing displacement and community deterioration. Tretter, "Contesting Sustainability"; Long, "Constructing the Narrative"; McCann, "Inequality and Politics"; Mueller and Dooling, "Sustainability and Vulnerability"; Busch, "Crossing Over."

28. Mike Clark-Madison, "New Rules for Old Buildings: The Historic Task Force," *Austin Chronicle*, April 2, 2004; "Summary of New Provisions to the City's Historic Preservation Zoning Ordinances," n.d., Folder "Historic Zoning New Ordinances 1/2005," Box 32, PODER Records.

29. PODER, "Historic Tour 5/11/02," 2002, Folder "Historic Tour 5/11/02," Box 32, PODER Records.

30. Quoted in Kate Van Scoy, "Get on the Bus," *Austin Chronicle*, July 25, 1997.

31. Anita Quintanilla, "Cultural Center of Their Own," *Austin Chronicle*, December 2, 2005.

32. Brant Bingamon, "PODER vs. H-Zoning: Ready for Round Two?," *Austin Chronicle*, November 1, 2002; Bingamon, "Old Homes = New Yuppies?," *Austin Chronicle*, July 19, 2002.

33. Tretter makes this important point in "Contesting Sustainability" (308).

34. City of Austin, "District 1 Demographic Profile," n.d., accessed March 18, 2015, http://www.austintexas.gov/sites/default/files/files/Planning/Demographics/District_1_demographic_profile_2000_2010.pdf.

35. City of Austin, "Changing African American Landscape—Eastern Core," n.d., accessed March 18, 2015, http://www.austintexas.gov/sites/default/files/files/Planning/Demographics/afam_change00_10_eastern_core.pdf.

36. City of Austin, "Change in the White Percentage of Total Population, 2000 to 2010" (map, n.d.), accessed September 27, 2011, http://www.ci.austin.tx.us

/demographics/downloads/travis_t2000_change_white.pdf. n.d. Website now defunct. Requests for the data can be made through the City of Austin demographer homepage at http://www.austintexas.gov/demographics.

37. City of Austin, "Tract-Level Change, 2000 to 2010, Total Population, Race and Ethnicity," spread sheet, n.d., accessed March 18, 2015, http://www.austintexas.gov/page/demographic-data.

38. Ryan Robinson, "The Top Ten Big Demographic Trends in Austin, Texas" (unpublished paper, n.d.), available at the City of Austin website, accessed September 27, 2011, http://www.ci.austin.tx.us/demographics. Website now defunct. Requests for the data can be made through the City of Austin demographer homepage at http://www.austintexas.gov/demographics.

39. For suburban poverty, see Kneebone and Berube, *Confronting Suburban Poverty*.

40. City-Data.com, "Austin, Texas (TX) Poverty Rate Data."

41. Florida and Mellander, *Segregated City*.

42. See, e.g., Morgan Brennan, "America's Fastest Growing Cities," *Forbes*, January 23, 2013. *Forbes* named Austin its fastest-growing city in 2012. Greg Barr, "Austin's Economy Named No. 1 in the Country," *Austin Business Journal*, October 24, 2013.

43. Swearingen, *Environmental City*, 164–167.

44. Karvonen, *Politics of Urban Runoff*, 84–85.

Bibliography

Books, Dissertations, and Peer-Reviewed Essays

Abbott, Carl. *The Metropolitan Frontier: Cities in the Modern American West.* Tucson, University of Arizona Press, 1993.

——. *The New Urban America: Growth and Politics in Sunbelt Cities.* Chapel Hill: University of North Carolina Press, 1981.

Adams, John A. *Damming the Colorado: The Rise of the Lower Colorado River Authority.* College Station: Texas A&M University Press, 1990.

Allen, Ruth A. *Chapters in the History of Organized Labor in Texas.* Austin: University of Texas, 1941.

Bachin, Robin F. *Building the South Side: Urban Space and Civic Culture in Chicago, 1890–1919.* Chicago: University of Chicago Press, 2004.

Banks, James H., and John E. Babcock. *Corralling the Colorado: The First Fifty Years of the Colorado River Authority.* Austin: Eakin Press, 1988.

Barker, Paul, Rayner Banham, Peter Hall, and Cedric Price. "Non-plan: An Experiment in Freedom." *New Society* 13, no. 38 (March 1969): 435–443.

Barna, Joel Warren. "The Rise and Fall of Smart Growth in Austin." *Cite* 53 (Spring 2002): 22–25.

Baughn, William Hubert. *Changes in the Structure of Texas Commercial Banking, 1946–1956.* Austin: Bureau of Business Research, University of Texas, 1959.

Bayor, Ronald H. *Race and the Shaping of Twentieth-Century Atlanta.* Chapel Hill: University of North Carolina Press, 1996.

Beauregard, Robert A. *When America Became Suburban.* Minneapolis: University of Minnesota Press, 2006.

Bluestone, Barry, and Bennett Harrison. *The Deindustrialization of America: Plant Closings, Community Abandonment, and the Dismantling of Basic Industries.* New York: Basic Books, 1982.

Bolan, R. S. "Emerging Views of Planning." *Journal of the American Institute of Planners* 33, no. 4 (1967): 233–245.

Boyer, M. Christine. *Dreaming the Rational City: The Myth of American City Planning.* Cambridge, MA: MIT Press, 1983.

Boyer, Paul. *Urban Masses and Moral Order in America, 1820–1920.* Cambridge, MA: Harvard University Press, 1978.

Bridges, Amy. *Morning Glories: Municipal Reform in the Southwest.* Princeton, NJ: Princeton University Press, 1997.

Brilliant, Mark. *The Color of America Has Changed: How Racial Diversity Shaped Civil Rights Reform in California, 1941–1978*. New York: Oxford University Press, 2010.
Bullard, Robert D. *Dumping in Dixie: Race, Class, and Environmental Quality.* Boulder, CO: Westview Press, 1990.
———. Introduction to *Growing Smarter: Achieving Livable Communities, Environmental Justice, and Regional Equity*, edited by Robert D. Bullard, 1–19. Cambridge, MA: MIT Press, 2007.
———. *Unequal Protection: Environmental Justice and Communities of Color.* San Francisco: Sierra Books, 1994.
Burke, Edmund M. “Citizen Participation Strategies.” *Journal of the American Institute of Planners* 34 (September 1969): 287–294.
———. *A Participatory Approach to Urban Planning.* New York: Human Sciences Press, 1979.
Burnett, Jonathan. *Flash Floods in Texas.* College Station: Texas A&M University Press, 2008.
Busch, Andrew M. “Building ‘A City of Upper-Middle Class Citizens’: Labor Markets, Segregation, and Growth in Austin, Texas, 1950–1973.” *Journal of Urban History* 39, no. 5 (September 2013): 975–996.
———. “Crossing Over: Sustainability, New Urbanism, and Gentrification in Austin, Texas.” *Southern Spaces*, August 19, 2015. http://southernspaces.org/2015/crossing-over-sustainability-new-urbanism-and-gentrification-austin-texas.
———. “The Perils of Popular Planning: Space, Race, Environmentalism, and History in a ‘Austin Tomorrow.’ ” *Journal of Planning History* 15, no. 2 (May 2016): 87–107.
Carle, David. *Water and the California Dream: Choices for the New Millennium*. San Francisco: Sierra Club Books, 2000.
Catton, Theodore. *National Park, City Playground: Mount Rainier in the Twentieth Century.* Seattle: University of Washington Press, 2006.
Cobb, James C. *Industrialization and Southern Society, 1877–1984*. Lexington: University of Kentucky Press, 1984.
———. *The Selling of the South: The Southern Crusade for Industrial Development, 1936–1990*. Baton Rouge: Louisiana State University Press, 1982.
Cohen, Lizabeth. *A Consumer’s Republic: The Politics of Mass Consumption in Postwar America*. New York: Knopf, 2003.
———. “From Town Center to Shopping Center: The Reconfiguration of Community Marketplaces in Postwar America.” *American Historical Review* 100, no. 1 (October 1996): 1050–1081.
Cole, Luke W., and Sheila R. Foster. *From the Ground Up: Environmental Racism and the Rise of the Environmental Justice Movement.* New York: New York University Press, 2001.
Colton, Craig. *An Unnatural Metropolis: Wresting New Orleans from Nature.* Baton Rouge: Louisiana State University Press, 2005.

Connolly, N. D. B. “Sunbelt Civil Rights: Urban Renewal and the Follies of Desegregation in Greater Miami.” In *Sunbelt Rising: The Politics of Space, Place, and Region*, edited by Michelle Nickerson and Darren Dochuk, 164–187. Philadelphia: University of Pennsylvania Press, 2011.

Cowie, Jefferson, and Joseph Heathcott, eds. *Beyond the Ruins: The Meanings of Deindustrialization*. Ithaca, NY: ILR Press, 2002.

Cronon, William. *Nature’s Metropolis: Chicago and the Great West*. New York: W. W. Norton, 1991.

Culver, Lawrence. *The Frontier of Leisure: Southern California and the Shaping of Modern America*. New York: Oxford University Press, 2010.

Cunningham, Sean P. *Cowboy Conservatism: Texas and the Rise of the Modern Right*. Lexington: University of Kentucky Press, 2010.

Davidson, Chandler. *Race and Class in Texas Politics*. Princeton, NJ: Princeton University Press, 1990.

Davis, Charles Edwin. “United States v. Texas Education Agency, et al.: The Politics of Busing.” PhD diss., University of Texas at Austin, 1975.

Delaney, David. *Race, Place, and the Law, 1836–1948*. Austin: University of Texas Press, 1998.

Donaldson, Gary A. *Abundance and Anxiety: America, 1945–1960*. Westport, CT: Praeger Press, 1997.

Duany, Andres, Jeff Speck, with Mike Lydon. *The Smart Growth Manual*. New York: McGraw-Hill, 2010.

Duffy, John. *The Sanitarians: A History of American Public Health*. Urbana: University of Illinois Press, 1990.

Feagin, Joe R. *Free Enterprise City: Houston in Political-Economic Perspective*. New Brunswick, NJ: Rutgers University Press, 1988.

Feagin, Joe R., and Robena Jackson. “Delivery of Services to Black East Austin and Other Black Communities: A Socio-Historical Analysis.” Unpublished report, University of Texas at Austin, n.d., ca. 1978–1982.

Findlay, John M. *Magic Lands: Western Cityscapes and American Culture after 1940*. Berkeley: University of California Press, 1992.

Fishman, Robert. *Bourgeois Utopias: The Rise and Fall of Suburbia*. New York: Basic Books, 1989.

Flanagan, Maureen A. *America Reformed: Progressives and Progressivisms, 1890s–1920s*. New York: Oxford University Press, 2006.

Florida, Richard. *Cities and the Creative Class*. New York: Routledge, 2004.

———. *The Rise of the Creative Class: And How It’s Transforming Work, Leisure, Community, and Everyday Life*. New York: Basic Books, 2002.

Florida, Richard, and Charlotta Mellander. *Segregated City: The Geography of Economic Segregation in America’s Metros*. Toronto: Martin Prosperity Institute, 2015.

Fogelson, Robert M. *Bourgeois Nightmares: Suburbia, 1870–1930*. New Haven, CT: Yale University Press, 2005.

Foley, Neil. *The White Scourge: Blacks, Mexicans, and Poor Whites in Texas Cotton Culture*. Berkeley: University of California Press, 1997.
Fones-Wolf, Elizabeth. *Selling Free Enterprise: The Business Assault on Labor and Liberalism, 1945–60*. Urbana: University of Illinois Press, 1994.
Forman, Christopher H. *The Promise and Peril of Environmental Justice*. Washington, DC: Brookings Institution, 1998.
Formisano, Ronald P. *Boston against Busing: Race, Class, and Ethnicity in the 1960s and 1970s*. Chapel Hill: University of North Carolina Press, 1991.
Geiger, Roger L. *Research and Relevant Knowledge: American Research Universities since World War II*. New York: Oxford University Press, 1993.
Gibson, David V., and Everett M. Rogers. *R&D Collaboration on Trial: The Microelectronics and Computer Technology Corporation*. Boston: Harvard Business Review Press, 1994.
Gillette, Howard, Jr. *Civitas by Design: Building Better Communities, from the Garden City to New Urbanism*. Philadelphia: University of Pennsylvania Press, 2010.
Godschalk, David R. *Participation, Planning, and Exchange in Old and New Communities: A Collaborative Paradigm*. Chapel Hill: Center for Urban and Regional Studies, University of North Carolina, 1971.
Godschalk, David R., and William E. Mills. "A Collaborative Approach to Planning through Urban Activities." *Journal of the American Institute of Planners* 32, no. 3 (1966): 86–94.
Goldfield, David R. *Cotton Fields and Skyscrapers: Southern City and Region, 1607–1980*. Baton Rouge: Louisiana State University Press, 1982.
Gordon, Colin. *Mapping Decline: St. Louis and the Fate of the American City*. Philadelphia: University of Pennsylvania Press, 2008.
Gotham, Kevin F. *Race, Real Estate, and Uneven Development: The Kansas City Experience, 1900–2000*. Albany: State University of New York Press, 2002.
Gottdiener, Mark. *The Social Production of Urban Space*. Austin: University of Texas Press, 1985.
Gottdiener, Mark, and Alexandros Ph. Lagopoulos, eds. *The City and the Sign: An Introduction to Urban Semiotics*. New York: Columbia University Press, 1986.
Graff, Harvey. *The Dallas Myth: The Making and Unmaking of an American City*. Minneapolis: University of Minnesota Press, 2008.
Grodach, Carl. "Before and after the Creative City: The Politics of Urban Cultural Politics in Austin, Texas." *Journal of Urban Affairs* 34, no. 1 (2012): 81–97.
Grossman, James R. *Land of Hope: Chicago, Black Southerners, and the Great Migration*. Chicago: University of Chicago Press, 1989.
Haddon, Susan G. "The East Austin Tank Farm Controversy." In *Public Policy and Community: Activism and Governance in Texas*, edited by Robert H. Wilson, 69–94. Austin: University of Texas Press, 1997.
Hall, Peter. *Cities of Tomorrow: An Intellectual History of Urban Planning and Design in the Twentieth Century*. Oxford, UK: Blackwell, 1988.

———. "The Geography of the Fifth Kondratieff." In *Silicon Landscapes*, edited by Peter Hall and Ann Markusen, 1–19. Boston: Allen and Unwin, 1985.
Hall, Peter, and Ann Markusen, eds. *Silicon Landscapes*. Boston: Allen and Unwin, 1985.
Hardy, Dennis. "Utopian Ideas and the Planning of London." *Planning Perspectives* 20, no. 1 (January 2005): 35–49.
Harvey, David. "Flexible Accumulation through Urbanization: Reflections on 'Post-Modernism' in the American City." In *Post-Fordism: A Reader*, edited by Ash Amin, 361–386. Oxford, UK: Blackwell, 1995.
———. "The 'New' Imperialism: Accumulation by Dispossession." In *Socialist Register* 40 (2004): 63–87.
———. *Social Justice and the City*. Baltimore: Johns Hopkins University Press, 1973.
———. *The Urbanization of Capital: Studies in the History and Theory of Capitalist Urbanization*. Baltimore: Johns Hopkins University Press, 1985.
Hatch, Richard. *The Scope of Social Architecture*. New York: Van Nostrand Reinhold, 1984.
Hayden, Dolores. *Building Suburbia: Green Fields and Urban Growth, 1820–2000*. New York: Pantheon, 2003.
Hays, Samuel P. *Beauty, Health, and Permanence: Environmental Politics in the United States, 1955–1985*. New York: Cambridge University Press, 1989.
———. *Conservation and the Gospel of Efficiency: The Progressive Conservation Movement, 1890–1920*. Cambridge, MA: Harvard University Press, 1999.
Hernandez, Paul. "Defending the Barrio." In *No Apologies: Texas Radicals Celebrate the '60s*, edited by Daryl Janes, 122–130. Austin: Eakin Press, 1992.
Hirsch, Arnold. *Making the Second Ghetto: Race and Housing in Chicago, 1940–1960*. New York: Cambridge University Press, 1983.
Horan, Cynthia. "Community Development, Racial Empowerment, and Politics." *Annals of the American Academy of Political and Social Science* 594 (2004): 158–170.
HoSang, Daniel Martinez. *Racial Propositions: Ballot Initiatives and the Making of Postwar California*. Berkeley: University of California Press, 2010.
Howard, Christopher. *The Hidden Welfare State: Tax Expenditures and Social Policy in the United States*. Princeton, NJ: Princeton University Press, 1997.
Hoyt, Homer. *One Hundred Years of Land Values in Chicago: The Relationship of the Growth of Chicago to the Rise in Its Land Values, 1830–1933*. Chicago: University of Chicago Press, 1934.
Hughes, Thomas P. *Networks of Power*. Baltimore: Johns Hopkins University Press, 1983.
Hundley, Norris, Jr.. "Water and the West in Historical Imagination." *Western Historical Quarterly* 27, no. 1 (Spring 1996): 4–31.
Hurley, Andrew. *Environmental Inequalities: Class, Race, and Industrial Pollution in Gary, Indiana, 1945–1980*. Chapel Hill: University of North Carolina Press, 1995.

Hurt, Sonia. *Zoned in the USA: The Origins and Implications of American Land Use Regulation*. Ithaca, NY: Cornell University Press, 2014.
Jackson, Kenneth T. *Crabgrass Frontier: Suburbanization of the United States*. New York: Oxford University Press, 1985.
Jackson, Robena. "East Austin: A Socio-Historical View of a Segregated Community." Master's thesis, University of Texas at Austin, 1979.
Janes, Daryl, ed. *No Apologies: Texas Radicals Celebrate the '60s*. Austin: Eakin Press, 1992.
Jordan, Terry. *Texas: A Geography*. Boulder, CO: Westview Press, 1984.
Joss, Simon. *Sustainable Cities: Governing for Urban Innovation*. New York: Palgrave Macmillan, 2015.
Kahn, Terry, and Josh Farley. *The Impact of MCC: Economic, Population, and Land Use Trends*. Austin: Bureau of Business Research, 1985.
Kahrl, William L. *Water and Power: The Conflict over Los Angeles' Water Supply in the Owens Valley*. Berkeley: University of California Press, 1982.
Kantrowitz, Stephen. *Ben Tillman and the Reconstruction of White Supremacy*. Chapel Hill: University of North Carolina Press, 2000.
Karvonen, Andrew. *The Politics of Urban Runoff: Nature, Technology, and the Sustainable City*. Cambridge, MA: MIT Press, 2011.
Kelman, Ari. *A River and Its City: The Nature of Landscape in New Orleans*. Berkeley: University of California Press, 2003.
Kennedy, David M. *Over Here: The First World War and American Society*. New York: Oxford University Press, 1980.
Kerr, Jeffrey Stuart. *Seat of Empire: The Embattled Birth of Austin, Texas*. Lubbock: Texas Tech University Press, 2013.
Kevles, Daniel J. *The Physicists: The History of a Scientific Community in Modern America*. Cambridge, MA: Harvard University Press, 1995.
Kirby, Jack Temple. *Darkness at the Dawning: Race and Reform in the Progressive South*. Philadelphia: J. B. Lippincott, 1972.
Kirschner, Don S. *The Paradox of Professionalism: Reform and Public Service in Urban America, 1900–1940*. Westport, CT: Greenwood Press, 1986.
Klingle, Matthew. *Emerald City: An Environmental History of Seattle*. New Haven, CT: Yale University Press, 2007.
Kneebone, Elizabeth and Alan Berube. *Confronting Suburban Poverty in America*. Washington, D.C.: Brookings Institution Press, 2014.
Kruse, Kevin M. *White Flight: Atlanta and the Making of Modern Conservatism*. Princeton, NJ: Princeton University Press, 2005.
Kruse, Kevin M., and Thomas J. Sugrue, eds. *The New Suburban History*. Chicago: University of Chicago Press, 2006.
Kupel, Douglas E. *Fuel for Growth: Water and Arizona's Urban Environment*. Tucson: University of Arizona Press, 2003.
Lassiter, Matthew P. *The Silent Majority: Suburban Politics in the Sunbelt South*. Princeton, NJ: Princeton University Press, 2006.

Lécuyer, Christophe. *Making Silicon Valley: Innovation and the Growth of High Tech, 1930–1970*. Cambridge, MA: MIT Press, 2006.

Lefebvre, Henri. "The Right to the City." In *Writings on Cities*, trans. and edited by Eleonore Kofman and Elizabeth Lebas, 147–59. Oxford, UK: Blackwell, 1996.

Lerner, Steve. *Diamond: A Struggle for Environmental Justice in Louisiana's Chemical Corridor*. Cambridge, MA: MIT Press, 2005.

Leslie, Stuart W. *The Cold War and American Science: The Military-Industrial-Academic Complex at MIT and Stanford*. New York: Columbia University Press, 1993.

Lessoff, Alan. *Where Texas Meets the Sea: Corpus Christi and Its History*. Austin: University of Texas Press, 2015.

Limbacher and Godfrey Architects. *Barton Springs Pool Master Plan: Concepts for Preservation and Improvement*. Austin: City of Austin Parks and Recreation Department, 2008.

Lindsey, Brink. *The Age of Abundance: How Prosperity Transformed America's Politics and Culture*. New York: HarperCollins, 2007.

Link, William A. *The Paradox of Southern Progressivism, 1880–1930*. Chapel Hill: University of North Carolina Press, 1992.

Lipsitz, George. "The Possessive Investment in Whiteness: Racialized Social Democracy and the 'White' Problem in American Studies." *American Quarterly* 47, no. 3 (September 1995): 369–387.

Logan, John R., and Harvey L. Molotch. "Homes: Exchange and Sentiment in the Neighborhood." In *Urban Fortunes: The Political Economy of Place*, 99–146. Berkeley: University of California Press, 1987.

———. *Urban Fortunes: The Political Economy of Place*. Berkeley: University of California Press, 1987.

Long, Joshua. "Constructing the Narrative of the Sustainability Fix: Sustainability, Social Justice, and Representation in Austin, TX." *Urban Studies*, 2014. doi:10.1177/0042098014560501.

———. *Weird City: Sense of Place and Creative Resistance in Austin, Texas*. Austin: University of Texas Press, 2010.

Mann, Seymour Z. "Planning-Participation." *National Civic Review* 59, no. 4 (April 1970): 182–190.

Martin, Roscoe C. "The Municipal Electorate: A Case Study." *Southwestern Social Science Quarterly* 14, no. 3 (December 1933): 193–237.

Marx, Leo. *The Machine in the Garden: Technology and the Pastoral Ideal in America*. New York: Oxford University Press, 1964.

McCann, Eugene J. "Framing Space and Time in the City: Urban Policy and the Politics of Spatial and Temporal Scale." *Journal of Urban Affairs* 25, no. 2 (2003): 159–178.

———. "Inequality and Politics in the Creative City-Region: Questions of Livability and State Strategy." *International Journal of Urban and Regional Research* 31, no. 1 (March 2007): 188–196.

McDonald, Jason. *Racial Dynamics in Early Twentieth-Century Austin, Texas.* Lanham, MD: Lexington Books, 2012.
McGerr, Michael E. *A Fierce Discontent: The Rise and Fall of the Progressive Movement in America, 1870–1920.* New York: Free Press, 2003.
McHarg, Ian. *Design with Nature.* Garden City, NY: Natural History Press, 1969.
McKee, Guian A. *The Problem of Jobs: Liberalism, Race, and Deindustrialization in Philadelphia.* Chicago: University of Chicago Press, 2008.
McWhirter, Cameron. *Red Summer: The Summer of 1919 and the Awakening of Black America.* New York: Henry Holt, 2011.
Melosi, Martin V. *Garbage in the Cities: Refuse, Reform, and the Environment.* 1981. Reprint, Pittsburgh: University of Pittsburgh Press, 2005.
——. *The Sanitary City: Urban Infrastructure in America from Colonial Times to the Present.* Baltimore: Johns Hopkins University Press, 2000.
Miles, Steven, and Malcolm Miles. *Consuming Cities.* New York: Palgrave Macmillan, 2004.
Mohl, Raymond. "Making the Second Ghetto in Metropolitan Miami, 1940–1960." *Journal of Urban History* 21 (1995): 395–427.
——, ed. *Searching for the Sunbelt: Historical Perspectives on a Region.* Knoxville: University of Tennessee Press, 1990.
Molotch, Harvey L. "The Growth Machine: Towards a Political Economy of Place." *American Journal of Sociology* 82, no. 2 (1976): 309–332.
Monkkonen, Eric H. *America Becomes Urban: The Development of U.S. Cities and Towns, 1780–1980.* Berkeley: University of California Press, 1988.
Moore, Steven A. *Alternative Routes to the Sustainable City: Austin, Curitiba, and Frankfort.* Lanham, MD: Lexington Books, 2007.
Mueller, Elizabeth J., and Sarah Dooling. "Sustainability and Vulnerability: Integrating Equity into Plans for Central City Redevelopment." *Journal of Urbanism: International Research on Placemaking and Urban Sustainability* 4, no. 3 (2011): 201–222.
Myers, Dowell. *Quality of Life, Austin Trends1970–1990: A Research Report of the Spring 1984 Course, Measuring Local Quality of Life.* Austin: Community and Regional Planning Program, School of Architecture, University of Texas at Austin, 1984.
Nash, Gary B. *The Urban Crucible: Social Change, Political Consciousness, and the Origins of the American Revolution.* Cambridge, MA: Harvard University Press, 1979.
Nash, Gerald. *The Federal Landscape: An Economic History of the Twentieth-Century West.* Tucson: University of Arizona Press, 1999.
Nash, Linda Lorraine. *Inescapable Ecologies: A History of Environment, Disease, and Knowledge.* Berkeley: University of California Press, 2006.
Nash, Roderick. *Wilderness and the American Mind.* 1967. Repr., New Haven, CT: Yale University Press, 2001.

Nelson, Daniel. *Fredrick W. Taylor and the Rise of Scientific Management*. Madison: University of Wisconsin Press, 1980.

Newman, Robert J. *Growth in the American South: Changing Regional Employment and Wage Patterns in the 1960s and 1970s*. New York: New York University Press, 1984.

Nickerson, Michelle, and Darren Dochuk, eds. *Sunbelt Rising: The Politics of Space, Place, and Region*. Philadelphia: University of Pennsylvania Press, 2011.

Northcutt, Kaye. "Austin: The Perils of Popularity." *American Planning Association Journal* 50, no. 11 (November 1984): 4–10.

Nye, David E. *American Technological Sublime*. Cambridge, MA: MIT Press, 1994.

Oakey, Ray. "High Technology Industries and Agglomeration Economies." In *Silicon Landscapes*, ed. Peter Hall and Ann Markusen, 94–117. Boston: Allen and Unwin, 1985.

Olmsted, Frederick Law. *A Journey through Texas; or, A Saddle-Trip on the Southwestern Frontier, with Statistical Appendix*. New York: Dix, Edwards, 1857.

———. "Public Parks and the Enlargement of Towns." Address to the American Social Science Association, Lowell Institute, Boston, 1870. Reprinted in *The City Reader*, ed. Richard LeGates and Frederic Stout, 302–308. New York: Routledge, 2003.

O'Mara, Margaret Pugh. *Cities of Knowledge: Cold War Science and the Search for the Next Silicon Valley*. Princeton, NJ: Princeton University Press, 2005.

Orsi, Jared. *Hazardous Metropolis: Flooding and Urban Ecology in Los Angeles*. Berkeley: University of California Press, 2005.

Orum, Anthony R. *Power, Money, and the People: The Making of Modern Austin*. Austin: Texas Monthly Press, 1987.

Osofsky, Gilbert. *Harlem, the Making of a Ghetto: Negro New York, 1890–1930*. New York: Harper and Row, 1966.

Ovetz, Robert. "Entrepreneurialization, Resistance, and the Crisis of the Universities: A Case Study of the University of Texas at Austin." PhD diss., University of Texas at Austin, 1996.

Passell, Aaron. *Building the New Urbanism: Places, Professions, and Profits in the American Metropolitan Landscape*. New York: Routledge, 2013.

Pellow, David Naguib. *Garbage Wars: The Struggle for Environmental Justice in Chicago*. Cambridge, MA: MIT Press, 2002.

Piore, Michael J., and Charles F. Sabel. *The Second Industrial Divide: Possibilities for Prosperity*. New York: Basic Books, 1984.

Portney, Kent E. "Sustainability in American Cities: A Comprehensive Look at What Cities Are Doing and Why." In *Toward Sustainable Communities: Transition and Transformations in Environmental Policy*, edited by Daniel A. Mazmanian and Michael E. Kraft, 227–254. Cambridge, MA: MIT Press, 2009.

———. *Taking Sustainable Cities Seriously: Economic Development, the Environment, and Quality of Life in American Cities*. Cambridge, MA: MIT Press, 2003.

Preer, Robert W. *The Emergence of Technopolis: Knowledge-Intensive Industries and Regional Development*. New York: Praeger, 1992.
Pritchard, Sara B. *Confluence: The Nature of Technology and the Remaking of the Rhone*. Cambridge, MA: Harvard University Press, 2011.
——. "Joining Environmental History with Science and Technology Studies: Promises, Challenges, and Contributions." In *New Natures: Joining Environmental History with Science and Technology Studies*, ed. Dolly Jorgensen, Finn Arne Jorgensen, and Sara B. Pritchard, 1–17. Pittsburgh: University of Pittsburgh Press, 2013.
Pritchard, Sara B., and Thomas Zeller. "The Nature of Industrialization." In *The Illusory Boundary: Environment and Technology in History*, ed. Martin Reuss and Stephen H. Cutcliffe, 69–100. Charlottesville: University of Virginia Press, 2010.
Pulido, Laura. *Environmental and Economic Justice: Two Chicano Struggles in the Southwest*. Tucson: University of Arizona Press, 1996.
Rawson, Michael A. *Eden on the Charles: The Making of Boston*. Cambridge, MA: Harvard University Press, 2010.
Rice, Jenny. *Distant Publics: Development Rhetoric and the Subject of Crisis*. Pittsburgh: University of Pittsburgh Press, 2012.
Righter, Robert W. *The Battle over Hetch Hetchy: America's Most Controversial Dam and the Birth of Modern Environmentalism*. New York: Oxford University Press, 2005.
Riis, Jacob. *How the Other Half Lives: Studies among the Tenements of New York*. New York: Scribner's and Sons, 1890.
Roediger, David. *The Wages of Whiteness: Race and the Making of the American Working Class*. London: Verso, 1991.
Rome, Adam. *The Bulldozer in the Countryside: Suburban Sprawl and the Rise of American Environmentalism*. New York: Cambridge University Press, 2001.
——. "'Give Earth a Chance': The Environmental Movement and the Sixties." *Journal of American History* 90, no. 2 (September 2003): 525–554.
Rosenblum, Susan. "Mayor Offers Strategy for Leading New Economy." *Nation's Cities Weekly*, June 12, 2000.
Rossinow, Douglas. *The Politics of Authenticity: Liberalism, Christianity, and the New Left in America*. New York: Columbia University Press, 1998.
Rothman, Hal K. "Selling the Meaning of Place: Entrepreneurship, Tourism, and Community Transformation in the Twentieth-Century American West." In "Tourism and the American West," special issue, *Pacific Historical Review* 65, no. 4 (November 1996): 525–557.
——. "Stumbling towards the Millennium: Tourism, the Postindustrial World, and the Transformation of the American West." *California History* 77, no. 3 (Fall, 1998): 140–155.
Rugh, Susan Sessions. *Are We There Yet? The Golden Age of Family Vacations*. Lawrence: University Press of Kansas, 2009.

Sauer, Carl O. "The Morphology of the Landscape." 1925. In *Land and Life: A Selection from the Writings of Carl Ortwin Sauer*, ed. John Leighly, 315–350. Berkeley: University of California Press, 1963.
Saxenian, AnnaLee. *Regional Advantage: Culture and Competition in Silicon Valley and Route 128*. Cambridge, MA: Harvard University Press, 1994.
Schmidt, Peter H. *Back to Nature: The Arcadian Myth in Urban America*. Baltimore: Johns Hopkins University Press, 1969.
Schulman, Bruce J. *From Cotton Belt to Sunbelt: Federal Policy, Economic Development, and the Transformation of the South, 1938–1980*. Durham, NC: Duke University Press, 1994.
Schultz, Stanley K. *Constructing Urban Culture: American Cities and City Planning, 1800–1920*. Philadelphia: Temple University Press, 1989.
Schultz, Stanley K., and Clay McShane. "To Engineer the Metropolis: Sewers, Sanitation, and City Planning in the Late Nineteenth Century." *Journal of American History* 65 (September 1978): 389–411.
Scott, Allen J. "Capitalism and Urbanization in a New Key? The Cognitive-Cultural Dimension." *Social Forces* 85, no. 4 (June 2007): 1465–1482.
———. "Creative Cities: Conceptual Issues and Policy Questions." *Journal of Urban Affairs* 28, no. 1 (2006): 1–17.
Scribner, Christopher MacGregor. *Renewing Birmingham: Federal Funding and the Promise of Change, 1929–1979*. Athens: University of Georgia Press, 2002.
Self, Robert O. *American Babylon: Race and the Struggle for Postwar Oakland*. Princeton, NJ: Princeton University Press, 2003.
Sellers, Christopher C. *Crabgrass Crucible: Suburban Nature and the Rise of Environmentalism in Twentieth-Century America*. Chapel Hill: University of North Carolina Press, 2012.
Shaffer, Marguerite S. *See America First: Tourism and National Identity, 1880–1940*. Washington, DC: Smithsonian Institution Press, 2001.
Shermer, Elizabeth Tandy. *Sunbelt Capitalism: Phoenix and the Transformation of American Politics*. Philadelphia: University of Pennsylvania Press, 2013.
Sides, Josh. *L.A. City Limits: African American Los Angeles from the Great Depression to the Present*. Berkeley: University of California Press, 2003.
Silver, Christopher. *Twentieth-Century Richmond: Planning, Politics, and Race*. Knoxville: University of Tennessee Press, 1984.
Smith, Carl. *The Plan of Chicago: Daniel Burnham and the Remaking of an American City*. Chicago: University of Chicago Press, 2006.
Smith, Jason Scott. *Building New Deal Liberalism: The Political Economy of Public Works*. New York: Cambridge University Press, 2006.
Soja, Edward, Rebecca Morales, and Goetz Wolff. "Urban Restructuring: An Analysis of Social and Spatial Change in Los Angeles." *Economic Geography* 59, no. 2 (April 1983): 195–230.
Spear, Allan H. *Black Chicago: The Making of a Negro Ghetto, 1890–1920*. Chicago: University of Chicago Press, 1967.

Spirn, Anne Whiston. *The Granite Garden: Urban Nature and Human Design*. New York: Basic Books, 1984.

Steinberg, Ted. *Acts of God: The Unnatural History of Natural Disasters in America*. 2nd ed. New York: Oxford University Press, 2006.

Student Geology Society. *Guidebook to the Geology of Travis County*. Austin: University of Texas at Austin, 1977.

Sugrue, Thomas J. "Crabgrass-Roots Politics: Race, Rights, and the Reaction against Liberalism in the Urban North, 1940–1964." *Journal of American History* 82, no. 2 (September 1995): 551–578.

———. *Origins of the Urban Crisis: Race and Inequality in Postwar Detroit*. Princeton, NJ: Princeton University Press, 1996.

Swearingen, William Scott. *Environmental City: People, Place, Politics, and the Meaning of Modern Austin*. Austin: University of Texas Press, 2010.

Szasz, Andrew. *EcoPopulism: Toxic Waste and the Movement for Environmental Justice*. Minneapolis: University of Minnesota Press, 1994.

Sze, Julie. *Noxious New York: The Racial Politics of Urban Health and Environmental Justice*. Cambridge, MA: MIT Press, 2007.

Tang, Eric, and Chunhui Ren. *Outlier: The Case of Austin's Declining African American Population*. Austin: Institute for Urban Policy Research and Analysis, 2014.

Tarr, Joel A. "The City as an Artifact of Technology and the Environment." In *The Illusory Boundary: Environment and Technology in History*, ed. Martin Reuss and Stephen H. Cutcliffe, 145–170. Charlottesville: University of Virginia Press, 2010.

———. *The Search for the Ultimate Sink: Urban Pollution in Historical Perspective*. Akron, OH: University of Akron Press, 1996.

Teaford, Jon C. *The Unheralded Triumph: City Government in America, 1870–1900*. Baltimore: Johns Hopkins University Press, 1984.

Tretter, Eliot M. "Contesting Sustainability: 'SMART Growth' and the Redevelopment of Austin's Eastside." *International Journal of Urban and Regional Research* 37, no. 1 (January 2013): 297–310.

———. *Shadows of a Sunbelt City: The Environment, Racism, and the Knowledge Economy in Austin*. Athens: University of Georgia Press, 2016.

Tretter Eliot M., and M. Anwar Sounny-Slitine. *Austin Restricted: Progressivism, Zoning, Private Racial Covenants, and the Making of a Segregated City*. Austin: Institute for Urban Policy Research and Analysis, 2012.

Trotter, Joe R., ed. *The Great Migration in Historical Perspective: The New Dimensions of Race, Class, and Gender*. Bloomington: Indiana University Press, 1991.

Turner, Fred. *From Counterculture to Cyberculture: Stewart Brand, the Whole Earth Network, and the Rise of Digital Utopianism*. Chicago: University of Chicago Press, 2006.

Tuttle, William, Jr. *Race Riot: Chicago in the Red Summer of 1919*. New York: Atheneum, 1970.

Urry, John. *The Tourist Gaze: Leisure and Travel in Contemporary Societies*. Thousand Oaks, CA: Sage, 1990.

Vance, Rupert K. *All These People: The Nation's Human Resources in the South.* Chapel Hill: University of North Carolina Press, 1945.

Vogel, Eve. "Defining One Pacific Northwest among Many Possibilities: The Political Construction of a Region and Its River during the New Deal." *Western Historical Quarterly* 42, no. 1 (Spring 2011): 28–53.

Vose, Clement. *Caucasians Only: The Supreme Court, the NAACP, and the Restrictive Covenant Cases.* Berkeley: University of California Press, 1959.

Walker, Richard. *The Country in the City: The Greening of San Francisco.* Seattle: University of Washington Press, 2008.

Walmsley, John Charles. "City Planning, the Press, and the Government: Çitizen Participation in the Austin Tomorrow Program in Austin, Texas." Master's thesis, University of Texas at Austin, 1975.

Wang, Jessica. *American Science in an Age of Anxiety: Scientists, Anticommunism, and the Cold War.* Chapel Hill: University of North Carolina Press, 1993.

Waring, George E. *Modern Methods of Sewage Disposal: For Towns, Public Institutions, and Isolated Houses.* New York: D. Van Nostrand, 1903.

———. *Street Cleaning and the Disposal of a City's Wastes: Methods and Results and the Effects upon Public Health, Public Morals, and Municipal Prosperity.* New York: Doubleday and McClure, 1898.

Weiss, Marc A. *The Rise of the Community Builders: The American Real Estate Industry and Land Planning.* New York: Columbia University Press, 1987.

Wheeler, Steven M. *Planning for Sustainability.* New York: Routledge, 2004.

White, Richard. "'Are You an Environmentalist or Do You Work for a Living?' Work and Nature." In *Uncommon Ground: Rethinking the Human Place in Nature,* ed. William Cronon, 171–185. New York: W. W. Norton, 1996.

———. *The Organic Machine: The Remaking of the Columbia River.* New York: Hill and Wang, 1996.

Whyte, H. William. *The Last Landscape.* Garden City, NY: Doubleday, 1968.

Wilson, William Julius. *When Work Disappears: The World of the New Urban Poor.* New York: Knopf, 1996.

Winkler, Earnest C. "The Permanent Location of the Seat of Government." *Texas Historical Quarterly* 10 (January 1907): 217–218.

Wright, Gavin. *Old South, New South: Revolutions in the Southern Economy since the Civil War.* New York: Basic Books, 1986.

Wrobel, David M. *Promised Lands: Promotion, Memory, and the Creation of the American West.* Lawrence: University Press of Kansas, 2002.

Published Primary Sources

Newspapers and Periodicals

Austin American

Austin American-Statesman

Austin and Industry

Austin Business Executive

Austin Chronicle

Austin Citizen

Austin Herald
Austin Home Builder
Austin in Action
Austin Progress
Austin Semi-Weekly Statesman
Austin Statesman
Business Week
Change
Charities and the Commons: A Weekly Journal of Philanthropy and Social Advance
Christian Science Monitor
Daily Texas
Dallas Morning News
Economist
Engineering News
Engineering-Science News
Forbes
Fort Worth Star-Telegram
Grist
Harper's Weekly
Harvard Business Review
Holiday Inn Magazine
Journal of Business Venturing
LCRA News
Monthly Business Review
Municipal Perspective
Nation's Cities Weekly
On Campus
Phoenix Gazette
Prevention
San Antonio Express
Science
Scientific American
Seattle Times
Southwest Real Estate News
Sunday Morning News
Texas Business and Industry
Texas Business Review
Texas Highways
Texas Monthly
Texas Observer
Texas Parade
Texas Public Employee
Third Coast
Time
Top Spot for Fun

Books, Reports, and Planning Documents

Addams, Jane. *Democracy and Social Ethics.* New York: Macmillan, 1902.
American National Bank. "A Citizen of No Mean City." Pamphlet, 1922. Courtesy Austin History Center.
Arbingast, Stanley A., Otis D. Horton, and Robert H. Ryan. "Austin, Texas R&D Nucleus." In *Texas Business Review.* Austin: Bureau of Business Research, 1968.
Austin Area Economic Development Foundation. *Annual Report, 1949.* Austin: The Foundation, 1949.
Austin and Travis County, Texas: Charms of the Capital City and Its Environs; Attractiveness and Resources of Texas and Peculiar Organic Code of the Great Commonwealth. Austin: Democratic Statesman and Book and Job Stream Print, 1877. Courtesy Dolph Briscoe Center for American History (BCAH), University of Texas at Austin.
Austin Chamber of Commerce. *The Austin Area Lakes.* Austin: Austin Chamber of Commerce, 1938.
———. *Austin in a Nutshell.* Austin: Austin Chamber of Commerce, 1942.
———. *Austin in a Nutshell, the Friendly City.* Austin: Austin Chamber of Commerce, 1949.

———. *Austin Invites You to Share Texas' Scientific, Educational, and Recreational Center.* Austin: Austin Chamber of Commerce, 1960.
———. *Austin, the Friendly City.* Austin: Von Boeckmann-Jones, 1925.
———. *Lakes in the Austin Area.* Austin: Industrial Bureau, 1937.
———. *Progressive Austin.* Austin: Austin Chamber of Commerce, 1915.
———. *Progressive Austin.* Austin: Austin Chamber of Commerce, 1916.
Austin Commercial Club. *Austin, Texas: The Future Great Manufacturing Center of the South; the Healthiest City in the South; Facts for Consideration of Tourists, Home-Seekers, Investors, Manufacturers, and Merchants.* Austin: Ben C. Jones, 1891.
Austin Housing Authority. "From Slums through Public Housing to Home Ownership." Annual report. Austin: Austin Housing Authority, 1948.
———. *Report of the Housing Authority of the City of Austin for the Years 1938–1939.* Austin: Austin Housing Authority, 1939.
Austin Human Relations Commission. *A Housing Patterns Study of Austin, Texas.* Austin: Austin Human Relations Commission, 1979.
Austin Planning Commission. *The Austin Master Plan.* Austin: Austin Planning Commission, 1958.
Austin, Texas, Illustrated: Famous Capital City of the Lone Star State. Houston: Southwest, 1909.
Austin, the Capital of Texas, and Travis Country. Austin: Dupre and Peacock, 1876. Courtesy BCAH.
Barker, Eugene C. "Description of Texas by Stephen F. Austin." *Southwestern Historical Quarterly* 29 (October 1924): 98–121.
Bureau of Research in the Social Sciences. *Population Mobility in Austin, Texas, 1929–1931.* Austin: University of Texas, 1941.
Carson, Rachel. *Silent Spring.* New York: Houghton Mifflin, 1962.
Carson, William J., ed. *The Coming of Industry to the South.* Philadelphia: American Academy of Political and Social Science, 1931.
City of Austin. *Imagine Austin Comprehensive Plan.* Austin: City of Austin, 2013.
City of Austin. *1977 Third Year Housing and Community Block Grant Application.* Austin: City of Austin, 1977.
City of Austin. *Preliminary Agreement of Environmental Protection and Development.* Austin: City of Austin, 1999.
City of Austin, Department of Planning. *Austin Tomorrow Comprehensive Plan.* Austin: City of Austin, 1980.
Dumble, E. T. *Second Annual Report of the Geological Survey of Texas.* Austin: State Printing Office, 1890.
East Austin Chicano Economic Development Corporation. *Barrio Unido Neighborhood Plan.* Austin: East Austin Chicano Economic Development Corporation, 1984.
Freeman, Martha Doty. *East Austin: An Architectural Survey.* Austin: Freeman and Doty Associates, 1980.

Galbraith, John Kenneth. *The Affluent Society*. Boston: Houghton Mifflin, 1958.
Greater Austin Chamber of Commerce. *Next Century Economy: Sustaining the Austin Region's Economic Advantage in the Twenty-First Century*. Austin: Greater Austin Chamber of Commerce, 1997.
Hahn, L. Albert. *Deficit Spending and Free Enterprise*. N.p.: Chamber of Commerce USA, 1944.
Hamilton, William B. *A Social Survey of Austin, Texas*. Austin: University of Texas, 1913.
——. *Social Survey of Austin #2*. Austin: University of Texas, 1917.
Harold F. Wise Associates. *The Austin Plan*. Austin: City Planning Commission, 1958.
Haw John W., and F. E. Schmitt. "Report on Federal Reclamation to the Secretary of the Interior." Washington, DC, 1934.
Howard, Ebenezer. *Garden Cities of To-morrow*. London: S. Sonnenschein, 1902.
Ickes, Harold L. *Back to Work: The Story of the PWA*. New York: Macmillan, 1935.
Johnson, Lyndon B. *Tarnish on the Violet Crown, Extension of Remarks of Hon. Morris Sheppard of Texas in the Senate of the United States, February 3, 1938, Appendix to the Congressional Record*. Washington, DC: Government Printing Office, 1938.
Koch and Fowler. *A City Plan for Austin, Texas*. Austin: Department of Planning, 1928.
Konecci, Eugene B., et al. *Commercializing Technology Resources for Competitive Advantage*. Austin: IC2 Institute, 1986.
Kozmetsky, George, Frederick Williams, and Victoria Williams. *New Wealth: Commercialization of Science and Technology for Business and Economic Development*. Westport, CT: Praeger, 2004.
Long, Walter E. *Flood to Faucet*. Austin: Steck, 1956.
——. *Something Made Austin Grow*. Austin: Austin Chamber of Commerce, 1948.
McDonald, John. *The Great Dam and Water and Light System at Austin, Texas*. Austin: Eugene von Boeckmann, 1893.
McKinstry, William C. *The Colorado Navigator, Containing a Full Description of the Beds and Banks of the Colorado River from the City of Austin to Its Mouth*. Matagorda, TX: Office of the Colorado Gazette, 1840.
Mead, Daniel W. "Report on the Dam and Water Power Development at Austin, Texas." Daniel W. Mead and Charles V. Seastone, consulting engineers, 1917.
Moore, G. S. "Planned or Unplanned Growth." Unpublished paper, 1943. Courtesy BCAH.
NAACP (National Association for the Advancement of Colored People). *Mobbing of John R. Shillady, Secretary of National Association for the Advancement of Colored People, at Austin, Texas, Aug. 22, 1919*. New York: NAACP, 1919.
National Parks Service, Lower Colorado River Authority (LCRA), and Texas State Parks Board. *The Highland Lakes of Texas*. Washington, DC: Government Printing Office, 1941.

Office of Civil and Defense Mobilization. "Ten Steps to Industrial Dispersal." April 1960.

Packard, Vance. *The Waste Makers*. 1960. Reprint, Brooklyn, NY: Ig Publishing, 1988.

Park, Robert E., Ernest W. Burgess, and Roderick D. McKenzie. *The City*. Chicago: University of Chicago Press, 1925.

Pinchot, Gifford. *The Fight for Conservation*. New York: Doubleday, 1910.

Potter, David M. *People of Plenty: Abundance and the American Character*. Chicago: University of Chicago Press, 1954.

Sinclair, Upton. *The Jungle*. New York: Doubleday, Page, 1906.

Smilor, Raymond J., George Kozmetsky, and David V. Gibson. "Creating and Sustaining the Technopolis: High Technology Development in Austin, Texas." *Journal of Business Venturing* 4 (1988): 49–67.

——, eds. *Creating the Technopolis: Linking Technology, Commercialization, and Economic Development*. Cambridge, MA: Ballinger, 1988.

Taylor, Thomas U. *The Austin Dam*. Washington, DC: Government Printing Office, 1900.

Thompson, J. Neils. "Integrating Sponsored Research into the University Research Program." Unpublished paper, 1949. Courtesy BCAH.

Wood, Marshall K. "Industry *Must* Prepare for Atomic Attack." *Harvard Business Review*, May–June (1955): 115–128.

Wood, Richardson. "Outline for a Plan for the Future Economic Development of Austin, Texas." Unpublished paper, 1948. Courtesy BCAH.

Index

FSC
www.fsc.org
MIX
Paper from
responsible sources
FSC® C013483